Genetically Yours

Bioinforming

Biopharming

Biofarming

Genetically Yours

Bioinforming

Biopharming

Biofarming

Hwa A. Lim

D'Trends, Inc., Silicon Valley
California, USA

World Scientific
New Jersey • London • Singapore • Hong Kong

Published by

World Scientific Publishing Co. Pte. Ltd.

P O Box 128, Farrer Road, Singapore 912805

USA office: Suite 1B, 1060 Main Street, River Edge, NJ 07661

UK office: 57 Shelton Street, Covent Garden, London WC2H 9HE

British Library Cataloguing-in-Publication Data
A catalogue record for this book is available from the British Library.

The cover illustration is a fusion of art and science. It depicts a spiraling DNA with high-tech features representing nanotechnology. The 0/1 globe indicates computation: the upper right globe contains chromosomes to show the importance of genomes in biotechnology, while the lower right globe is *Copy Cat* — the first cloned cat born on December 22, 2001, at Texas A&M University.

The lower left image on the back cover connotes stem cell research and cloning.

GENETICALLY YOURS
Bioinforming, Biopharming, and Biofarming

ISBN 981-02-4938-1
ISBN 981-02-4939-X (pbk)

Printed in Singapore.

ABOUT THE AUTHOR

Dr. Hwa A. Lim has a Ph.D. (Science) and an M.A. (Science), and an MBA (strategy and business laws) from the United States, a B.Sc. (Honours) and ARCS from the United Kingdom. In 1997, he founded D'Trends, Inc., a professional consulting company serving the biotech, pharmaceutical and healthcare sectors. Prior to this new venture, he was Vice President of Sc. & Tech., DoubleTwist (1996–1997), Director of Bioinformatics, Hyseq, Inc. (1995–1996), a tenured faculty member and Program Director, Supercomputer Computations Research Institute, Florida State University (1987–1995). He has authored/edited eleven books, and is a well sought after keynote speaker at international meetings. He is credited with coining the word "bioinformatics" and for initiating the very first bioinformatics conference series (Bioinformatics & Genome Research), which is still ongoing. In addition to his scientific background, he has extensive experience in market research and developing strategic alliances internationally. Currently he also serves as a bioinformatics expert for the United Nations, a review panelist for the US National Science Foundation and the US National Cancer Institute. He has been a consultant for McKinsey, Prudential, VAXA, Eli Lilly and Company, Monsanto and Company, Taipei government, institutions in China, institutions in India and other organizations. Besides traveling for business, he also travels extensively to mingle with locals to experience and learn firsthand.

Dr. Hwa A. Lim is Chairman of Board of two Silicon Valley-based companies — AbMetrix, Inc. and D'Trends, Inc., and a Michigan-based company — GeneGo, LLC.

Also by Hwa A. Lim

1. Charles R. Cantor and Hwa A. Lim (eds.), *Electrophoresis, Supercomputing and the Human Genome*, (World Scientific Pub. Co., New Jersey, 1991), 325 pages.
2. Hwa A. Lim, James W. Fickett, Charles R. Cantor and Robert J. Robbins (eds.), *Bioinformatics, Supercomputing and Complex Genome Analysis*, (World Scientific Pub. Co., New Jersey, 1993), 648 pages.
3. Clement L. Markert, John G. Scandalios, Hwa A. Lim and Oleg L. Serov (eds.), *ISOZYMES: Organization and Roles in Evolution, Genetics and Physiology*, (World Scientific Pub. Co., New Jersey, 1994), 309 pages.
4. Nikolay A. Kolchanov and Hwa A. Lim, *Computer Analysis of Genetic Macromolecules: Structure, Function and Evolution*, (World Scientific Pub. Co., New Jersey, 1994), 556 pages.
5. Hwa A. Lim, Charles R. Cantor, (eds.), *Bioinformatics and Genome Research*, (World Scientific Pub. Co., 1995), 529 pages.
6. Roger S. Holmes, Hwa A. Lim, (eds.), *GENE FAMILIES: Structure, Function, Genetics and Evolution*, (World Scientific Pub. Co., New Jersey, 1996), 220 pages.
7. Ralf Hofestädt, and Hwa A. Lim (Hrsg.), *Molecular Bioinformatics- Sequence Analysis*, (Shaker Verlag, Aachen, Germany, 1997), 60 pages.
8. John R. McCarrey, John L. VandeBerg, and Hwa A. Lim (eds.), *Genes, Gene Families, and Isozymes*, The Journal of Experimental Zoology, Vol. 282, No. 1/2, (Wiley-Liss, New York, 1998), 283 pages.
9. Guoxiong Xue, Yongbiao Xue, Zhihong Xu, Roger Holmes, Graeme Hammond, Hwa A. Lim (eds.), *GENE FAMILIES: Studies of DNA, RNA, Enzymes and Proteins*, (World Scientific Pub. Co., New Jersey, 2001), 305 pages.
10. Hwa A. Lim (Guest editor), *Pathways of Bioinformatics: From data to diseases*, Briefings in Bioinformatics, Vol. 3(1), (Henry-Stewart Publishers, London, 2002).

Dedicated to:

The Argentinean Lady
Who wrote me in 1992

My Father
Who became a statistic of a botched hospital operation in 1993

My Mother
A courageous fighter and survivor of breast cancer

My Family
Beneficiaries of modern biotechnology and healthcare system

PREFACE

The scene was cast in 1992 while I was still a tenured faculty member at a university and program director at a supercomputer institute. I received a letter from a lady in Argentina. The lady had just lost her only beloved son and was looking for someone to help clone her son. Still young and early in my career, I made a number of phone calls to try to get some leads. I called experts, including a personal physician of a former president. Of the experts I called, one was less polite. He snarled skeptically that "someone must have scribbled your name on a toilet wall". In spite of the few days of futile attempts, I had a colleague who could speak the lady's native tongue to call and tell her it was not doable. At that time, the closest we came to cloning was the very popular flick by Steven Spielberg — The Jurassic Park.

Then in 1993, my father became a statistic in a botched routine hospital operation while I was on a United Nations mission to help set up a biotech and bioinformatics science park along the Pacific Rim. My mother was more fortunate. She is a courageous fighter and survivor of breast cancer. My family members, like most people, are fortunate beneficiaries of the wonders of modern biotechnology and healthcare system.

As I penned the first word of this book, I thought of the Argentinean lady. If the same real-life scene could be recast in the current prevailing climate, the whole episode would have unfolded very differently. She would in all likelihood not have written me. In all probability, she would have contacted Claude Vorlihon of the Rael Group, or Panayiotis Zavos in Kentucky, or Severino Antinori in Rome. As for the skeptic expert who snarled that "someone must have scribbled your name on a toilet wall", he probably could not foresee what he thought was impossible less than a decade ago is now within grasp.

I also thought of my father. Had the healthcare system in Malaysia been a little more advanced, my father would likely have survived the operation to see this book. So I decided that I would title the book in a letter format, in response to the Argentinean lady's letter, and in a form that I can convey to my father.

By 1992, the human genome project had already begun. The neologism "bioinformatics" had been coined for almost five years. Though bioinformatics was still not popular, international conference on the subject had been convened at least twice. In certain sense, I was then a pioneer in the field. The field exploded after 1996. Like most pioneers of the field, I soon had a hard time keeping up with the progress. Soon, I find myself falling into being a student. As the eclectic field grows larger, becomes more encompassing and ubiquitous, I can hardly catch a breath just to keep abreast. Thanks to all the wonderful publications: BioCentury, Genetic Engineering News, Genome Technology, Nature, Science, Trends in Biotechnology, CNN News, New York Times, San Francisco Chronicle,

Washington Post, and many other publications and books too numerous to mention each by name, I manage to keep myself afloat in the ocean of bioinformation.

The ideas of the book come from many sources, including the above-mentioned publications, only some of which I have been able to acknowledge specifically in the text as footnotes. Fellow peers, experts, colleagues, and friends, whom I meet at conferences, conventions and workshops around the world, and many other cyber friends whom I have never met have all provided insightful and enlightening perspectives. I am very grateful to many friends who regularly bring to my attention or forward electronically to me interesting articles. I appreciate the many penetrating and often times difficult to answer questions — particularly when I am on the air live — from reporters of TV and radio stations.

This book, in part, builds on my previous writings. They all — genome projects, agbiotech, aquaculture, cloning, stem cell research, animal bioreactor, genomic and proteomic technologies, patentability, bioinformatics, and modern warfare — I now realize, hang together as different facets or parcels of life. To separate them would be to miss the meanings and messages of the forest in a minute examination of its trees.

The topics of discussion are marching ahead at an unprecedented pace. The torrents of news of new breakthroughs, of new discoveries such as there are only 30,000–40,000 human genes instead of the long-held 100,000 human genes necessitate a revision. There is always something to update and add. Without the encouragement of colleagues and a self-imposed deadline, the last chapter would never have been completed.

I applaud the generosity of D'Trends for the time while I am on trips, in flights, between flights, late evenings, and on numerous occasions during work when inspiration strikes. I like to thank all my ballroom dancer friends who help me unwind every Friday and Saturday evening after a week of hectic chores. Their company is the ballast that keeps my sanity in balance. I also admit the child in me when I take a breather and watch Tom and Jerry cartoons in the wee hours.

This book is addressed primarily to experts who are very busy and would like to have a different and comprehensive perspective, and to laypeople who are interested in learning more about bioinformatics, the biotechnology, pharmaceutical and healthcare industries, and the roles of biotechnology and information technology in modern warfare. Writing this book is my way of learning a very vast and fast progressing subject. I hope that reading it will help others do the same.

February 2002
D'Trends, Inc.
Silicon Valley, California, USA

Hwa A. Lim
hal@d-trends.com
http://www.d-trends.com

CONTENTS

Chapter 2

Post-Genomics Matters: Send out A Thousand Ships, Not Catalogue a Thousand Genes   **33**

Chapter 3

AgBiotechnology: Genetically Modified   **55**

Chapter 4

Aqua Biotechnology   **79**

Chapter 8

Genomic And Proteomic Technologies And Beyond   **199**

Chapter 9

Submicron Technology And Nanotechnology   235

Chapter 10

Patentability in Biotechnology   253

Chapter 11

**Biological and Biology-Related Information as a Commodity and the
Rise of Bioinformatics**   **273**

Chapter 13

A New Kind Of War: Biowarfare And Info Warfare   **331**

Chapter 14

Two Decades of Biotechnology and a Decade of Bioinformatics   371

Chapter 1

THE GENOME PROJECT: ITS IMPACTS AND IMPLICATIONS

"We wish to suggest a structure for the salt of deoxyribonucleic acid (DNA). This structure has novel features which are of considerable biological interest."
J.D. Watson and F.H.C. Crick, *Nature*, 25 April 1953.

1 Looking Back In The Future

The new millennium just started. It may sound utterly ridiculous at this juncture to ask the question by the year 3000 when historians look back on the 21st century and try to record important events that occurred during the century, what they will have to jot down in their history books. Though only a few months old, the new millennium is already very eventful — the human genome project (June 2000), Bush-Gore presidential tug-of-war (November 2000), collapse of the World Trade Center towers (September 11, 2001) — not to mention all the excitements and social upheavals about the anticipated implications, impacts and repercussions.

In historical records there is a plentiful account of lives of societies, royalty, empire builders, but not much is known about the average people. History books remember ancient civilizations like the Mayans, Incas, Egyptians, Chinese and Indians; conquerors like the Greeks and Romans; maritime powers like the Portuguese, Spanish, Dutch and English; and space explorers like Yuri Gagarin, Neil Armstrong. History also remembers individual empire builders: Alexander the Great, Caesar, Qin Huang Ti, Akhbar the Great, Genghis Khan, Peter the Great, and Napoleon. History books are also populated with people with great ideas: Aristotle, Archimedes, Confucius, Shakespeare, Newton, and Einstein. Noteworthy in history books are explorers: Columbus, Magellan, Marco Polo, Admiral Cheng Ho and others. In historical footnotes are many inventions and presidents.

Events worth recording have come far in between in the past. In modern times, historical records will be very different. First, with rapid advances in technology, historical records will be recorded more often. Second, modern recording mechanisms are much more efficient, and the recording mediums are much more sophisticated, reliable and accurate. Third, royal pedigrees are a disappearing breed, and more room will be allocated for empire builders of another kind. As we understand genetics better, we also disprove the existence of royal blood.

Today building geographical empires has become irrelevant. With land and natural resources much less important to building a wealth pyramid since geographical empires do not create the wealth they used to. It costs more to build them than gaining from having them. This is why in the past half a century, the French, the British, the Russians have all given up their empires. But it is equally certain that as knowledge rises in importance to the construction of a modern wealth pyramid, those that create big technological breakthroughs will be remembered as the empire builders of our current knowledge-based economy. They redefine humanity's future.

2   Lewis-Clark Dé Jàvu

On June 26, 2000, at a photo-opportunity ceremony at the White House attended by Craig Venter, President of Maryland-based Celera Genomics,[1] Francis Collins, director of the National Human Genome Research Institute, and the U.S. President William Jefferson Clinton, a historical moment was defined. As the doors to the ornate East Room of the White House slid open, President Clinton strode in flanked by Collins and Venter. "Today the world is joining us here in the East Room to behold a map of even greater importance," Clinton said, "Without a doubt, this is the most important, most wondrous map ever produced by humankind."

Figure 1. President Bill Clinton, flanked by Celera Genomics head Craig Venter (to his right) and Francis Collins, head of the Human Genome Project of the National Institutes of Health, meets reporters in the East Room of the White House. (Photo: Courtesy of Associated Press).

For people familiar with history and the White House, two centuries ago in 1806, President Thomas Jefferson had stood in the same room to view "a magnificent map" of North America produced by the Lewis-Clark expedition. The map Bill Clinton was referring to is the first draft of the human genome. Note the words used by Clinton, "of even greater significance", is there a deeper implication in the message? Was President Clinton trying to put his footprints in the sands of history by drawing the parallel between the human genome project and the Lewis-Clark expedition?[2] The Lewis-Clark expedition would later open up national trade in the U.S. The genetic map marks an epic and is expected to put the U.S. in a dominant position in the global trade of a different kind in a very different knowledge-based economy.

---

[1] http://www.celera.com
[2] Irving W. Anderson, "A history of Lewis and Clark Expedition", http://www.lewisandclark.org/pages/story0.htm

Besides the uncannily identical "Jefferson" in the presidents' names, the ceremony leaves no doubt that it was set up to be a photo opportunity and the president was bathing in the glory of the accomplishment. It is ironic that the same technique used to sequence the human genome was also used in a 1999 DNA test to prove that President Thomas Jefferson could have fathered a child with a slave Sally Hemmings. And here we recall the outcome of the 1999 DNA fingerprinting of a stain on the famous Monica Lewinsky's blue dress, an evidence in the very politically motivated allegation of Bill Clinton's White House infidelity.

We believe the White House ceremony will be a footnote to this momentous event. Though President Clinton will be remembered for many other great achievements during his tenure, his name will likely be forgotten in this context. The same can be said of his counterpart, Prime Minister Tony Blair of United Kingdom, who held in tandem a ceremony in London to underscore the international nature of the genome project. But the wind of time will never erase the line etched into the sand of history by the completion of the first draft of the human genome.

3 The Human Genome Tour

The human body has 10 trillion cells. With the exception of red blood cells, each cell contains the entire human genome, all the genetic information needed to "build" a human being.

The entire human genome has 22 pairs of chromosomes (autosomes), and the X and the Y chromosome (sex chromosomes). A male has 22 pairs plus an X and a Y. A female has 22 pairs plus two X's. The 23 pairs of chromosomes that make up the human genome contain the complete information for human development. The chromosomes themselves are largely made up of very long DNA molecule, which has been elaborately wound up. The DNA, if stretched out, will be about 5 feet long and 50 trillionths of an inch wide. Somewhere hidden in the DNA coils and strung out in definite order are genes. The genes are simply sections of a chemical message which runs along the spine of the DNA molecule. Four chemical bases or nucleotides: Adenine (A), Guanine (G), Cytosine (C) and Thymine (T), run the full length of the DNA molecule. The order of these chemicals constitutes a code, instructing cells to make different proteins. The total information stored in the genome, therefore, can be most simply thought of as a very large encyclopedia containing the recipe for life. Despite the complexity of a human being, the recipe is written with an alphabet of only the four letters, A, C, G, and T. Using this analogy, *Encyclopedia Humanica* is three billion letters long from end to end. Somewhere in the encyclopedia are the clues to genetic diseases and phenotypes.

Pushing the encyclopedia analogy further, *Encyclopedia Humanica* contains all of the instructions needed to go from a fertilized egg to a complete human being. In this analogy, *Encyclopedia Humanica* has 22 volumes, and volume X and volume Y.

Much of what is found in *Encyclopedia Humanica* is very much similar to the encyclopedia of the mouse, for example. Many sentences that appear in a human's DNA also appear in a mouse's DNA. This reflects the fact that the two texts were the same until some point in the past when variations were introduced and the two species diverged from their common ancestry. In fact, the best way to read *Encyclopedia Humanica* is to read it alongside other encyclopedias, such as those of the mouse, fruit fly or nematode. Fruit flies are good model organisms for they reproduce in large quantities, and their life cycle is short. Besides ethical issues associated with experimenting with humans, short life cycle is a good attribute for experimentation with effects that can be passed on (inherited), and reproducing in large number is an advantage for statistical analysis. For experimentation purposes, humans reproduce too slowly and too few in numbers.

Genetic variations in genomes, various possible combinations of different possible genes, like different spellings of the same word, create the infinite variety that we see among individuals of a species.[3] For example, between the genomes of

❑ Bart and Lisa Simpson, two blood siblings, there is a difference in 2 million base pairs (bp)

❑ Tarzan and Jane, two individuals of the same species, there is a difference of 6 million bp in their genomes.

❑ Tarzan and chimpanzee, two organisms of closely related species, there is a difference of 50 million bp in their genomes.

❑ Popeye and spinach, two organisms of distant species, there is a difference of 2 billion bp in their genomes.

Besides these genotype differences, there are also other variations. Genetic variations lead to diseases:

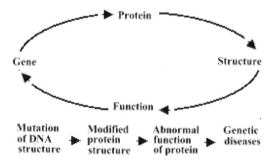

Figure 2. Genes code for proteins. Proteins have particular structures leading to different functions. A mutated DNA can lead to a modified protein structure, thus abnormal protein function. This is manifested as abnormalities or diseases.

---

[3] Hwa A. Lim, "Lecture Note at Science & Computer Teachers Summer Training", Supercomputer Computations Research Institute, Tallahassee, Summer 1992.

4 Historical Perspective — The Timeline Of The Human Genome Project

The beginning of the international human genome project (HGP) is a point of contention. There are numerous accounts.

4.1 Pre-genomic Sequencing Cottage Industry

Prior to the 1970s, scientists knew of no way of finding which part of the DNA molecule caused diseases such as Huntington's disease. The only detective techniques they had available were those associated with sex, or isolated cases with telltale signs. For example, one in 15 men is color-blind while almost all women are not. This fact indicates that this predominantly male disease must be caused by a defective gene somewhere on the single X chromosome. Women would have to have both X chromosomes defective to get the disease. Men, having one X and one Y chromosome, do not have the luxury of a spare X chromosome, and if the only X chromosome is defective, that is it. Other diseases, such as Down's syndrome, have telltale signs. Children with Down's syndrome have three copies of chromosome 21 (trisomy 21). This narrows down the location of the Down's syndrome gene to chromosome 21. This is a special case of aneuploidy, a defect that arises because of too many or too few chromosomes.

However, diseases with telltale signs are exceptions. Genetic diseases like Huntington's disease, a degenerative condition of the brain, leave no such clues for genetic detective Sherlock Holmes. In 1977, biologists made a breakthrough. They hit on a brilliant simple way of doing detective job to orient themselves in the DNA molecule using genetic markers. Any sort genetic differences will do, but by far the easiest kinds are tiny spelling differences in the DNA text. Much like milestones and road signs on freeways help highway patrols locate accident sites, these genetic markers help biologists navigate along the DNA and locate disease genes.

Despite the successes in diagnosing genetic diseases, biologists were novice at reading the genetic script as late as the 1980s. DNA sequencing — the job of deciphering short stretches of the four chemical letters of DNA — was a time-consuming, tedious, and error-prone task. Before the DNA could be sequenced, it had to be extracted from cells and grown up in bacteria colonies. Then it was chemically sliced into pieces of varying lengths. To sort the pieces into different sizes, they were dragged electrically through a gelatinous medium in a process called electrophoresis. Then after radioactively labeling the pieces, they were visualized on an X-ray plate. The resulting pattern of bars could be read off against a template, letter by letter, as the sequence. A tiny stretch of 10,000 letters could take a laboratory a year.

4.2   Scaling up a la Big Science — A Biological Moonshot, A Biological Particle Smasher

Biologists then were working in a cottage industry manner. They worked in small groups in independent laboratories. The human genome involves some 3 billion letters. At the rate of 10,000 per year per laboratory, it would take a laboratory working alone about 300,000 years to complete the task of sequencing the entire human genome! One way to tackle such a massive project was for biology to scale up. Physicists took on enormous international projects. CERN (European Organization for Nuclear Research) in Switzerland is one good example. SSC (Superconducting Super Collider) was a Department of Energy atom smasher. The construction cost was estimated at US$12 billion and on top of that, it would have cost US$500 million annually for operations. It went defunct when the U.S. House of Representatives decided in 1993 to halt the project after 14 miles of tunneling were completed and $2 billion dollars had been spent.

Fifty years ago, finding a single gene was an interminable process. Genes are small sections of DNA that tell cells how to function. Each gene contains instructions on producing or controlling a specific protein. The proteins, in turn, carry out the cell's work in the body. In 1985, a biologist and chancellor at the University of California at Santa Cruz, Robert Sinsheimer had been searching for a large project to take biology to another level and to put Santa Cruz on the scientific map. He began to wonder if there were any large-scale biological projects that were being missed.[4] Sinsheimer invited a dozen of the world's experts in molecular genetics to Santa Cruz to fathom their reactions to the idea of mapping and sequencing the entire human genome. At that point in time, many of those present thought it was a proposal way ahead of its time. On the face of it, Sinsheimer's idea seemed absurd.

Biologists in several quarters also began to campaign for a systematic gene discovery program. The proposal proved quite elusive because of divergent agendas and ideas. From a more cultural perspective, biology had always been a cottage industry science, populated by independent researchers and financed by modest grants. The culture of biology was hostile to the central planning implicit in a big project like unraveling the human genome.

Walter Gilbert, a physicist-turned-biologist and a Nobel Laureate for inventing a method of sequencing DNA, argued that the project had to be done, even if it meant biologists have to change the way they work. At a heated meeting at Cold Spring Harbor, Gilbert wrote what he thought would be the cost of sequencing the entire human genome on a blackboard, $3 billion. Many feared the size and cost would undermine ordinary cottage biology. At that time, the fear, especially among younger scientists, the ones who were really the key to make the genome project work, became palpable. They wonder what would happen to them because they

---

[4] "Decoding the Book of Life", NOVA, WGBH Educational Foundation, October 31, 1989.

would find a situation in which not just the funding would be tight and the best ideas got funded, but also they would be required to join the sequencing effort or leave the field for good.

On a more technical level, many scientists thought it was a waste of time to study the entire genome because an estimated 95% of it is believed to be junk DNA, stretches of chemical letters with no apparent purpose. In the words of Sydney Brenner, then director of Molecular Genetics Institute at Cambridge, "More than 95% of the DNA is junk. But let me point out that it's not garbage because the difference between junk and garbage is exactly the same difference you make. Garbage you throw away and junk you keep because you think you might want to do something useful with it, and of course you never do. So, 95%, or more than 95% is junk, and I think that is a valid argument to say against the idea of sequencing the entire genome, because we'd spend a lot of time doing this. Against this, people said, well, you don't know until you've done it whether it is or isn't junk."[5]

In this regard, the human genome may be thought of as a software program evolved over 4 billion years. It is a language not too different from any human conversation. There is a lot of redundancy, a lot of hemming, hawing and stuttering. The minor difference is that the English alphabet has 26 letters, while the human genome alphabet has only 4 letters. Unlike most texts, the genome is a historical text in the sense that it has been passed on from generation to generation, much like in the Middle Ages, monks would make copies of a text one at a time and passed on to other monks. Despite their best efforts, over the course of the copying iterations, they introduced unintentional mistakes into the text. In an analogous way, the genome is a historical text recopied at each generation. Some of the variations that occurred during the reproduction process have turned out to be harmless (no visible effects), some have turned out to be advantageous (for example, adaptation), and some have manifested adverse effects (for example, deformities and diseases).

4.3   Enters the Black Knight

Improbable it may sound, the Human Genome Project (HGP) was saved from oblivion by the U.S. Department of Energy (DOE). DOE administered a dozen of national laboratories, including Los Alamos (LANL), Livermore (LLNL), and Lawrence Berkeley National Lab (LBNL). DOE was used to doing big science, had vast technical resources and unrivalled computing facilities. Its interest in biology stemmed from monitoring damage and inherited damage caused by radiation or other elements in the environment, especially after Hiroshima and Nagasaki. Despite all criticisms, DOE secured funding to start work on three chromosomes and set up three human genome centers at Los Alamos, Berkeley and Livermore. Soon, the initiative attracted the interest of the U.S. Congress for two possible reasons:

---

[5] "Decoding the Book of Life", NOVA, WGBH Educational Foundation, October 31, 1989.

1. The intuitive feel that the initiative was going to make fundamental advances in human medicine, and
2. The belief that the initiative would serve as a stimulus to spin off a new industry for which the U.S. would have supremacy for decades.

At a 1988 National Research Council (NRC) committee meeting, headed by Bruce Alberts, settled the scientific debate. The committee recommended a 15-year, $3-billion program to decipher the human genome. In the Summer of 1988, the National Institutes of Health (NIH), in a political and legerdemain move to regain its role in the center stage of biological science, appointed James Dewey Watson to head up the NIH new genome office. Watson is a Nobel laureate of 1962 for co-discoveries concerning the molecular structure of nucleic acids and its significance for information transfer in living material with Francis Crick and Maurice Wilkins. The U.S. Congress adopted the NRC strategy in 1990.

4.4   Divide-and-Conquer Strategy

As mentioned above, back in the 1980s, gene discovery was a labor-intensive and time-consuming exercise. Scientists used radioactive tags to reveal the sequence and it could take researchers a decade to decipher a single gene using the technique developed by the British scientist Frederick Sanger, who shared the 1980 Nobel Prize with Walter Gilbert for their contributions concerning the determination of base sequences in nucleic acids.

In 1986, Applied Biosystems (the new name is PE Biosystems) of Foster City developed a machine that improved upon Sanger's method by using fluorescent tags instead of radioactive ones. Fluorescent tags can be detected by lasers, which feed the data to computers, thus automating the sequencing process. Craig Venter, then at NIH, was quick to capitalize on the new machine. He used the machine to randomly hunt for gene fragments. At the time, the accepted approach was to look for specific genes whose functions were known, such as the insulin gene. Detractors scoffed that it was aimless to know the sequences of genes for which the functions were unknown. By the early 1990s, Venter had already found as many genes or gene fragments as all other scientists combined. In 1992, NIH caused a stir when it tried to patent the thousands of genes Venter had discovered.

Concurrently, a team of government scientists led by James Watson had already embarked on a different strategy to understand the genome. In contrast to Venter's approach to sequence genes at random, Watson's human genome project proposed to map the genome first. The original strategy for genome sequencing is a divide-and-conquer approach. In this approach, the genome is first dissected into smaller manageable pieces. During the dissection process, the order of the pieces is determined. An analogy will be taking apart a huge jigsaw puzzle in a sensible way so that the information of connectivity is not completely lost in the disassembly process. If the human genome is regarded as *Encyclopedia Humanica*, then the dissection is equivalent to dividing the encyclopedia into sections of a volume (chromosomal fragments). Genetic mapping will be equivalent to dividing each book into chapters, a chapter for brown eye for example. Physical mapping will be like dividing the chapters into pages. Sequencing will then be reading text off each of the pages. In other words, mapping involved isolating genes, finding out something about them, and locating them on one of the 24 chromosomes.

The two polar strategies would soon lead Watson and Venter into irreconcilable disagreement. In April 1992 Watson left NIH, to be replaced by Acting Director Michael Gottesman, and Venter founded The Institute for Genomic Research (TIGR), where he continued pursuing his machine-driven, rapid-fire hunt for random genes.

4.5 Shotgun Strategy

Since its inception, the premise of the government project had been to find key genes and map their location on the 24 chromosomes. In other words, the more genes scientists could find and map, the smaller it would be the unknown stretches in between and the easier the eventual task of reading the entire genome.

Venter's vision was quite different. He envisioned the genome as a huge instruction book that was far too long for his machine to read. In his approach, the book is shredded into smaller readable fragments. By shredding several copies of the book and taking care to completely randomize the process, a solution is obtained. Each shredded book would result in a pile of fragments. The randomization process guarantees that some fragments in each puzzle will overlap. This is the so-called shotgun strategy.

The shotgun strategy involves the following steps. Genome center researchers first take a whole DNA from an anonymous donor. The DNA is chopped up into large strands of 150,000 bp using restriction enzymes or DNA text-cutters, which was discovered in 1968. Some strands constitute whole chromosome and others are parts of large chromosomes. If necessary, like in the international human genome effort which involves various institutions, these large strands are distributed to the five main and 11 contributing centers worldwide. These large unknown strands are cut into smaller pieces of roughly 2,000 bp. The smaller pieces are then cloned thousand of times. Technicians add in As, Ts, Gs, and Cs to match up with the strand under study. Done enough number of times, a fragment of every possible length will be created. The end of each fragment is tagged with a fluorescent dye. These fragments are put into a machine to sort them by size using a process called electrophoresis. In the electrophoretic process, smallest strand reaches a laser detector first, followed by one that is one bp longer, and so forth. The laser detects whether the fluorescent marker is an A, a C, a G, or a T. Eventually, it reads the whole fragment. The process is repeated 4–10 times for each 2,000 bp fragment. The result is a huge puzzle. Computers are used to detect patterns and assemble the fragments in a way that reveals the sequence of the larger unknown strand. By computationally comparing overlapping segments, scientists can reassemble the complete book.

This shotgun process, first used in the late 1970s to read the genome of a virus with a DNA just a few thousand letters long, with the blessing of higher throughput of modern sequencing machines, proves to be more cost-effective and less time-consuming than the original divide-and-conquer strategy.

4.6   The Traveling Salesman Is a Chinese Postman

All sequencing projects involve breaking up a genome into manageable sizes and putting it back together again. For example, Celera sequenced the human genome by breaking it up at random and piecing together the resulting 27,271,853 sequences. Although the public Human Genome Project took a more structured approach, both groups faced similar problems when reassembling their sequences.

Chief among the problems is that large genomes such as the human genome are very repetitive, like a jigsaw with many identically shaped pieces. Sequencing errors compound the problem, making it difficult to tell if one is looking at different stretches of DNA or not.

Sequence assembly is analogous to finding the shortest route through many cities that passes through each only once, like a traveling salesman. The path traversed is officially known as a Hamiltonian path. Mathematicians call problems like this NP-complete: the only way to solve them is to try every possible route. This requires massive computer power as the number of possibilities rises exponentially with the number of towns — or DNA pieces.

By breaking the chunks of DNA into smaller fragments of equal size, Pavel Pevzner of University of California, San Diego, and his colleagues have transformed the Hamiltonian path of genome assembly into a Eulerian path.[6]

In an Eulerian path, instead of visiting every city once only, the traveler must travel down every road once only — passing through each junction as often as needed. Finding the shortest route through this network is called the Chinese postman problem.

Chinese postman problems are mathematically much more tractable than traveling salesman problems. The traveling salesman problem is yet intractable except resorting to computers by forcing assemblers to make arbitrary decisions, which lead unavoidably to potential further errors. The Eulerian path is almost the same in formulation, but there is a dramatic difference in complexity.

In a play-off against other genome assemblers including PHRAP, used by the Human Genome Project, Pevzner's program, christened EULER, was the only one to make no errors piecing together fragments of the *Neisseria meningitidis* genome,

---

[6] P.A. Pevzner, H. Tang, and M. Waterman, "An Eulerian path approach to DNA fragment assembly", *Proceedings of the National Academy of Sciences USA*, 98, 2001, pp. 9748–9753.

the bacterium that causes meningitis. Bacterial genomes are relatively nonrepetitive, so the researchers are in the process of giving EULER stiffer challenges using data from higher organisms.[7]

4.7 The Race to the Finishing Line — Bigger is Better

The turning point came in 1999 when Venter, now at Celera Genomics formed in 1988 by PE Biosystems, used shotgun sequencing to sequence *hemophilus influenza*, a microbe that caused ear infection and meningitis. The sequencing of *hemophilus* convinced the government camp to switch from mapping to shotgun sequencing. *Hemophilus* is very small in size compared to the human genome. In the book analogy, if *hemophilus* were a book page, then the human genome would be War and Peace. Scientists soon realized that size did not matter. To do shotgun sequencing of the human genome, what the community needs is stronger machinery.

When PE Biosystems came up with a new generation of faster sequencers, Francis Collins, who had assumed directorship of National Human Genome Research Institute (NHGRI) in 1993 from Michael Gottesman, decided it was time to halt mapping and go full force on sequencing. Collins managed to get $250 million to buy hundreds of the new sequencing machines to turn the five main labs into production line factories running 24 hours a day, seven days a week, nonstop. These five main labs are:

1. Baylor College of Medicine
2. Washington University at St. Louis, Missouri
3. Whitehead Institute at the Massachusetts Institute of Technology.
4. The Sanger Institute, United Kingdom
5. The Joint Genome Institute (JGI) of the U.S. Department of Energy.

Francis Collins calls these main labs the G5 Alliance.

Together, the five labs were to produce 85% of the public program genome sequence. The remainder would be handled by international collaborators over time. Eventually, the genome project involves sixteen institutions from six countries that form the Human Genome Sequencing Consortium (HGSC):

1. Baylor College of Medicine, Houston, Texas, USA
2. Beijing Human Genome Center, Institute of Genetics, Chinese Academy of Sciences, Beijing, China
3. Gesellschaft fur Biotechnologische Forschung mbH, Braunschweig, Germany
4. Genoscope, Evry, France
5. Genome Therapeutics Corporation, Waltham, MA, USA
6. Institute for Molecular Biotechnology, Jena, Germany

---

[7] P.A. Pevzner, "A new approach to fragment assembly in DNA sequencing", Tenth Annual Bioinformatics and Genome Research, San Francisco, June 17–19, 2001, Chair: Hwa A. Lim.

7. Joint Genome Institute, U.S. Department of Energy, Walnut Creek, CA, USA
8. Keio University, Tokyo, Japan
9. Max Planck Institute for Molecular Genetics, Berlin, Germany
10. RIKEN Genomic Sciences Center, Saitama, Japan
11. The Sanger Centre, Hinxton, U.K.
12. Stanford DNA Sequencing and Technology Development Center, Palo Alto, CA, USA
13. University of Washington Genome Center, Seattle, WA, USA
14. University of Washington Multimegabase Sequencing Center, Seattle, WA, USA
15. Whitehead Institute for Biomedical Research, MIT, Cambridge, MA, USA
16. Washington University Genome Sequencing Center, St. Louis, MO, USA

The previous mapping effort did not completely go to waste. The public's mapping program had already broken the 3-billion-letter genome into fragments corresponding to the 24 chromosomes. This enabled them to distribute the work among the sixteen public sequencing centers, and adopted a hierarchical shotgun strategy, a complementary strategy to the whole-genome shotgun of Celera Genomics.

4.8   Complementary Strategies — Ally or Foe

The public and the private sectors use similar automation and sequencing technology, but different strategies to sequence the human genome. The public project uses a "hierarchical shotgun" strategy in which individual large DNA fragments of known position are subjected to shotgun sequencing. Celera uses a "whole-genome shotgun" strategy in which the entire genome is shredded into small fragments that are sequenced and put back together on the basis of sequence overlaps.

Pros and cons of either approach have been a subject of debate.[8] The hierarchical shotgun approach has the advantage that the global location of each individual sequence is known with certainty, but it requires constructing a map of large fragments covering the genome. The whole-genome shotgun approach does not require this step but has challenges in the assembly of fragments. Both approaches align the sequences along the human chromosomes by using markers contained in the physical map. Indeed, the two approaches are complementary and in the future, a hybrid of the two approaches may prove to be even more effective.

---

[8] Hwa A. Lim, and T.V. (Venky) Venkatesh, "Bioinformatics in the pre- and post-genomic eras", *Trends in Biotechnology*, April 2000, Vol. 18 No. 4(195), pp. 133–135.

4.9   From Genome Marathon to Sprint to the Finishing Line

Officially, the genome project started in 1990 when the U.S. Congress decided to fund the initiative. The first ten years were a long marathon and progress was slow because the infrastructure and ancillary technologies were not in place. In 1999, partly because the development phase of the project had matured, and partly because of fierce competition from the private sector, the pace picked up. The healthy competition between the public and the private sectors fueled the sprint to the finishing line.

By November 1999, the public Human Genome Project had read one third of the genome. In March 2000, they reached the two-thirds milestone, on track to complete a first working draft in June 2000. On April 6, 2000, Celera announced that it had completed sequencing the human genome, but would need another three to six weeks to assemble the fragments into one book, thus completing a first working draft of the human genome. The working draft of the human genome will have thousands of blank spots or gaps. These genome gaps occur where the structure of DNA is hard for machines to read. It will take scientists another two years or so to go back and fill in the blanks.

By April 2000, the public effort had used $1.9 billion. But one has to be careful in interpreting the expenses. Much of the money was spent on developing better tools, techniques and infrastructure in the years 1990–1999. Thus a fairer comparison will be the $250 million Collins acquired to buy machines to pull even with Celera, which obtained $300 million from PE Biosystems.

4.10   The Gutenberg Press of the Human Genome Project

Much as in the 15th century, Europe's awakened desire for learning created a huge demand for books, the genetic revolution has created a rapidly growing market for genomic research instrument and chemicals. According to a study by William Blair & Co., the market for genomic instrument and chemicals is about $2.7 billion worldwide in 2000, and it is expected to grow at a rate of 22% per annum for the next 3–5 years.

The DNA sequencing machine, a gray box about the size of a small refrigerator, has positioned its manufacturer Perkin-Elmer Applied Biosystems of Foster City, California, in solid monopoly in the galvanic field of genomics. Central to the success is the premise that in most industries except biomedical instrumentation, standardized equipment is used to unify the process and churn out conforming data.

In the field long hindered by expensive and labor-intensive procedures, the machine has cut sequencing time by 60% and labor costs by 80% (year 2000 statistics). Its round the clock automation of once tedious tasks has embraced not just giant laboratories involved in sequencing genomes, but also average scientists who undertake more focused projects that were once off limit because of cost and time. Besides being the machine that scientists used to complete for the first time the first draft of the human genome, it has also been used to identify the gene responsible for baldness and the gene responsible for aging. It also helped identify the subspecies of chimpanzee that is likely the source of AIDS epidemic. This machine is so ubiquitous it is hard at work in three-quarters of the international human genome laboratories. Analysts estimate that 40,000 laboratories in 100 countries rely on PE Biosystems instrumentation in 2000. In the USA, nine out of 10 university-based genetics laboratories use the instrument.

The DNA sequencing machine is usually paralleled with the 15th-century Gutenberg press.[9] As far as necessity is the mother of invention is concerned, the analogy between the Gutenberg press and sequencing machine is good. However, we argue that the analogy is off in terms of reproduction. The subtle difference between the Gutenberg press and the sequencing machine is that in the former, the press mass prints books written by humans for other humans to read. In the latter, the machine reads the books of life of humans and other organisms for the benefit of humankind. In this sense, the press propagates, whereas the sequencer deciphers.

The real reproduction process in biology is the polymerase chain reaction (PCR), invented by Kary Mullis while at Chiron Corporation in the late 1980s. Mullis was awarded the 1993 Chemistry Nobel Prize for this invention. PCR revolutionizes biotechnology by allowing a genetic fragment to be reproduced indefinitely.

Table 1. Timeline of the genome project — from the beginning to the U.S. White House and U.K. Whitehall announcement.

1985:
> The concept of the Human Genome Project (HGP) materializes. Robert Sinsheimer holds a meeting on human genome sequencing at University of California, Santa Cruz.
> Charles DeLisi and David Smith commission the first Santa Fe conference to assess the feasibility of a Human Genome Initiative.

1986:
> Following the Santa Fe conference, U.S. Department of Energy (DOE) announces the Human Genome Initiative with $5.3M to develop critical resources and technologies at DOE national labs.

1987:
> Senator Pete Domenici of New Mexico introduced legislation for about $28 million to fund the Human Genome Project.[*]
> Congressionally chartered DOE advisory committee recommends a 15-year, multidisciplinary, scientific and technological undertaking to map and sequence the human genome.
> The National Institutes of Health (NIH) begins funding genome projects.

---

[9] Lisa M. Krieger, "DNA sequencing machine puts PE Biosystems at forefront", *San Jose Mercury News*, June 10, 2000.
[*] Senator Pete Domenici calls the human genome project "the greatest wellness project".

Table 1. (Continued)

Hwa A. Lim of Supercomputer Computations Research Institute (SCRI), a DOE supercomputer institute, coins bioinformatics.

1988:

National Research Council (NRC) recommends a concerted genome program.
DOE and NIH outline plans for cooperation.
Human Genome Organization (HUGO) founded by scientists to coordinate.
First Cold Spring Harbor meeting on genome mapping and sequencing starts.

1989:

DOE and NIH establish Joint ELSI Working Group. ELSI: Ethical, Legal, and Social Issues.

1980s:

Maynard Olson invents YAC. Leroy Hood invents four-color sequencing.

1990:

DOE and NIH present a joint 5-year plan to Congress.
The 15-year project formally begins in the U.S.
James Watson is appointed director of National Human Genome Project, NIH.
Hwa Lim convenes the very first "Bioinformatics & Genome Research" conference series, funded by DOE and NSF (National Science Foundation), Technology Research & Development Authority (TRDA) of Florida, and computer companies.

1991:

Genome Database (GDB) is established.

1992:

James Watson, Leroy Hood, Maynard Olson, Robert Waterston, Bill Gates, Paul Allen met in Seattle, USA to discuss HGP. Olson and Hood establish University of Washington Human Genome Center.
Low-resolution genetic linkage map of the entire human genome is published.
DOE and NIH establish guidelines for data release and resource sharing.
Watson and Craig Venter ran into disagreement on EST. Watson resigns from directorship. Venter leaves NIH to found the Institute for Genomic Research (TIGR).
The Sanger Centre is established in United Kingdom, funded by Wellcome Trust and the British Government. John Sulston and Mike Morgan are co-directors.

1993:

Genome centers in the U.S. are established: Washington University (WU), MIT, Baylor College, University of Washington (UW).
DOE and NIH revise 5-year goals of the genome project.
French Généthon provides mega-YACs to the genome community.

1994:

Genetic mapping 5-year goal achieved 1 year ahead of schedule.
Genetic Privacy Act is proposed to regulate collection, analysis, storage, and use of genetic information.
Sequencing by hybridization technology funded by DOE commercialized through Hyseq. Hwa Lim is appointed director of bioinformatics at Hyseq.

Table 1. (Continued)

1995:

Los Alamos and Livermore National Labs announce high-resolution physical maps of chromosomes 16 and 19, respectively.

Moderate-resolution maps of chromosomes 3, 11, 12, and 22 are published.

Sequence of the smallest bacterium *Mycoplasma genitalium* completed and provides a model of the minimum number of genes needed for independent existence.

First nonviral whole genome is sequenced, bacterium *Haemophilus influenzae*.

1996:

Saccharomyces cerevisiae (yeast) genome sequence is completed by international consortium.

Methanococcus jannaschii genome is sequenced and it confirms existence of third major branch of life on Earth.

Healthcare Portability and Accountability Act prohibits use of genetic information in healthcare insurance eligibility decisions.

PE 377 sequencing machines (48 lanes and 64 lanes) first came on the market.

Venter used a shotgun strategy to sequence bacteria.

1997:

Escherichia coli genome is sequenced.

High-resolution physical maps of chromosomes X and 7 are complete.

DOE forms Joint Genome Institute (JGI) consisting of DOE genome centers.

National Center for Human Genome Research (NCHGR) becomes National Human Genome Research Institute (NHGRI).

UNESCO adopts Universal Declaration on the Human Genome and Human Rights.

1998:

Caenorhabditis elegans genome sequence is complete.

Mycobacterium tuberculosis sequence is complete.

DOE and NIH revise 5-year plan through 2003.

Human Genome Project passes mid point.

Hospital for Sick Children, Toronto, assumes GDB data collection and curation.

Incyte Pharmaceuticals announces to sequence human genome in two years.

1998, May:

PE invests $300M and 300 automated 96-lanes PE 377 to form Celera with Venter. The plan is to use shotgun sequencing to finish the human genome in 3 years, most of which will be made public, others will be patented.

1998, Aug:

Pharmacia 96-lane MegaBACE capillary sequencing machine debuts on market.

1998, Oct:

The U.S. and U.K. agree to speed up HGP, planned to have a first draft complete by Spring of 2000, and a detail sequence by 2003.

1999, Jun:

With 300 PE 377 and high performance computers, Celera announces completion of rice genome in 6 weeks, and completing of the human genome in 2–3 years.

Table 1. (Continued)

Dec 1, 1999
  Chromosome 22 is the first human chromosome to be completely sequenced. Biotechnology stocks skyrocketed.
Jan 10, 2000:
  A day before the International Human Genome Conference, Celera announces using shotgun sequencing to cover 1.8×80% of the human genome, more cost effective and less time-consuming than HGP can do the work.
Jan 11, 2000
  International Human Genome Meeting in San Francisco. Participants talk a lot about Celera. NIH and Celera enter into confidential negotiations to share data.
Mar 6, 2000:
  NIH-Celera negotiation breaks down.
Mar 12, 2000:
  U.S. President Bill Clinton and U.K. Prime Minister Tony Blair jointly announce that genetic data should be freely available. News media misconstrue their message. Biotechnology stocks crash.
Mar 13, 2000:
  The U.S. Patent Office announces that Bill Clinton speech is misconstrued. It does not forbid patent applications.
April 6, 2000:
  Venter and Waterston appear at a U.S. Congress Hearing. Celera announces completion of 90% of first human genome draft.
May 8, 2000:
  To avoid the dilemma of January Celera announcement, HGP announce entering the second phase of sequencing a day before the Cold Spring Harbor Meeting.
June 26, 2000:
  The U.S. and U.K. jointly announce completion of the first draft of the human genome.
Feb 12, 2001
  Science publishes online the data from the private sector, Nature[10] publishes online data from the public sector.

5   A Small Step For Genome, A Giant Step For Humankind

Deciphering the human genome has been an epic task. Scientists have used the most powerful computers and developed a new generation of automated machines to read the 3 billion chemical letters. The task has been likened to putting a man on the moon. But is it?

---

[10] http://www.nature.com/genomics

While this metaphor is useful in political and funding purposes, the comparison does not convey the true significance of HGP. The moon shot affected a negligible number of people substantially and had only negligible effect on a substantial number of people. The impact of HGP will be substantial and it will affect a substantial number of people.

Table 2. Comparison and contrast of moon shot and the human genome project.

Moon Shot	The Human Genome Project
• An interdisciplinary project involving scientists, engineers, legal experts, and social scientists.	• Ditto. In this case, mostly wet bench scientists and computer scientists.
• The project could not have begun without concurrent advances in other fields. In this case, in engineering such as rocketry and material science.	• Ditto. In this case, high throughput instrumentation and computer technology.
• A huge project that requires involvement of a nation or an international effort. This is a U.S. effort.	• Ditto. In this case, it is an international effort, involving 5 main and 11 contributing centers worldwide. Celera Genomics is a private enterprise.
➤ A project that will amount to nothing if the project is half-complete. It will serve no one any good to shoot a rocket half way up the sky and have it come through Earth's atmosphere in a flurry. Though not a moon shot, the Superconducting Super Collider project is a case in point.	➤ A project that something will be learned even if the project should have to be abandoned. We will still obtain DNA sequences from which we can find genes.
➤ It affects the lives of very few, mostly those who go up to the moon.	➤ It will affect everyone, especially those who are genetically predisposed to diseases.
➤ This is a Cold War competition between the U.S. and the USSR, mainly politically motivated. A classified project.	➤ This is a peacetime effort. It is motivated by the goal to improve quality and productivity of life. A non-classified project.
➤ It opened up the stratosphere commons. To date, U.S. has six American flags on the moon to mark U.S. "territories". This is an exploratory project.	➤ It opens up the biosphere commons, and a floodgate of patentability of biological commons. This is a deciphering project.
➤ It was funded primarily by the U.S. government. Direct benefits from the project will not be gained for a long time to come.	➤ It is funded both by the public and the private sectors. The private sector is interested in the information, an asset in the current prevailing knowledge-based economy.

Table 2. (Continued)

➢ Private sector involvement was from the high tech sector to manufacture parts for the project, mostly from aerospace companies. Only the most advanced developed countries can spearhead in this project.	➢ Private sector involvement is from high-tech, medium-tech, and low-tech sectors. Almost any underdeveloped, developing, and developed countries can be a part.
➢ Highly capital intensive. Only a few qualify to be truly a part of the mission.	➢ Mostly labor intensive. It involves long hours, repetitive and laborious production-line effort.
➢ Being exploratory in nature, the end result is unknown.	➢ The end result is known, a huge sequence of about 3 billion letters.

5.1 Impacts and Implications — Deciphering the Rosetta Tablet

Knowledge of the human genetic sequence will help scientists understand more about the causes of human ailments and fashion powerful new treatments. But some fear the knowledge will be misused because the human genome can give scientists potent tools to manipulate human traits and behavior before society has sorted out the implications. Elbert Branscomb, Director of the Joint Genome Initiative, Walnut Creek, said, "For the first time we have rolled back the big stone and peered into the sepulcher with our tiny flashlights, reading the sacred script off the tablet." Branscomb was apparently referring to the famous Rosetta Tablet found in 1799, which has consumed vast amount of energy of many a scholar to interpret the hieroglyphic texts.

5.1.1 Genetic Engineering

Genetic engineering will allow us to correct misspelled individual words in *Encyclopedia Humanica*. In this regard, it is important to make a distinction between two very different kinds of genetic engineering:

1. Somatic genetic engineering — when we correct misspellings in *Encyclopedia Humanica*, the corrections will be made in the cells carrying out the particular function. The alterations will have been made in only the individuals and will not be passed into future generations.

2. Germline genetic engineering — when we correct misspellings in germ cells that create new individuals, we are rewriting *Encyclopedia Humanica* for the new traits will be passed on to future generations and the books will have been altered permanently.

Though genetic engineering has been a subject of debate because of its ethical implications, the full utility of genetic engineering will probably not be felt for a few decades.

5.1.2   Humpty Dumpty

As the genomes of yeast, nematode, fruit fly and many other organisms have been deciphered, and the first draft of the human genome has been completed, there is going to be a tidal wave of diagnostic tests that will tell patients their predisposition, but offer no therapeutic cure. The gap between diagnostics and therapeutics is no secret to biotechnology insiders. This is what the community calls the Humpty Dumpty dilemma.[11] This makes anxiety the most likely first byproduct of the genomics revolution.

Table 3. Human chromosomes and the associated diseases.[12]

Chromosome Number	Diseases
Chromosome 1	Alzheimer's disease, Glaucoma, Prostate cancer,
Chromosome 2	Colon cancer, Memory, Muscular dystrophy
Chromosome 3	Colon cancer, Dementia, Lung cancer
Chromosome 4	Ellis-van Creveld syndrome, Huntington's disease, Parkinson's disease
Chromosome 5	Colorectal disease, Diastrophic dysplasia, Spinal muscular atrophy
Chromosome 6	Dyslexia, Epilepsy, Schizophrenia
Chromosome 7	Cystic fibrosis, Diabetes, Obesity
Chromosome 8	Burkitt lymphoma, Hemolytic anemia, Werner syndrome
Chromosome 9	Chronic myeloid leukemia, Malignant melanoma, Tangier disease
Chromosome 10	Cowden disease, Hermansky-Pudlak syndrome, Jackson-Weiss syndrome,
Chromosome 11	Albinism, Diabetes, Sickle-cell anemia
Chromosome 12	Darier disease, Inflammatory bowel disease, Rickets
Chromosome 13	Breast cancer, Pancreatic cancer, Retinoblastoma
Chromosome 14	Alzheimer's disease, Goiter, Leukemia/T-cell lymphoma
Chromosome 15	Angelman syndrome, Juvenile epilepsy, Marfan's syndrome
Chromosome 16	Chron's disease, Familial Mediterranean fever, Polycystic kidney disease
Chromosome 17	Breast cancer, Cataract, Muscular dystrophy
Chromosome 18	Niemann-Pick disease, Pancreatic cancer, Tourette's syndrome
Chromosome 19	Arteriosclerosis, Diamond-Blackfan anemia, Myotonic dystrophy
Chromosome 20	Bubble-boy disease, Creutzfeldt-Jakob's disease, no immunity to viruses

---

[11] This is attributed to Eric Shulse, director of the molecular diagnostics unit at PE Biosystems, Foster City, who said "You know Humpty Dumpty is crashed, that's your diagnostic, but it's a lot harder to put Humpty Dumpty back together again."
[12] Genome by Matt Ridley, National Institutes of Health, Department of Energy Joint Genome Institute (JGI).

Table 3. (Continued)

Chromosome 21	Amytrophic lateral sclerosis, Down syndrome, Usher syndrome.
Chromosome 22	Chronic myeloid leukemia, DiGeorge syndrome, Ewing's sarcoma
X Chromosome	Colorblindness, hemophilia, Gout
Y Chromosome	Gonadanal dysgenesis

5.1.3 Single Nucleotide Polymorphism

The first draft of the human genome will show genetic sequences that are considered normal. The next step, already underway in public and private laboratories, is to look for the many slight differences in the genetic codes of any two humans. This is called single nucleotide polymorphisms or SNPs. SNPs will help explain why some people, and not others, are susceptible to complex diseases likely caused by the interplay between environmental factors and genetic factors. Drug giants are already talking about using SNPs to tailor remedies to individuals, i.e., personalized medicine, and thereby lessen the side effects coming from one-size-fit-all medicine.

5.1.4 Patentability of Biocommons

The patent office interpreted the Chakrabarty decision[13] as permitting gene patents. The logic is that if living organisms can be patented, small pieces of chemicals isolated and purified from them should be patentable. Isolated and purified are the key words. After the Chakrabarty decision, biotechnology firms raced to isolate, purify, and patent genes. For example, genes that produce insulin, growth hormones and blood production factors have been granted patents. Today, hundred of genes have been patented, both by private firms and universities. As John Doll, chief of the biotechnology division at the U.S. Patent and Trademark Office in Washington, DC explains, "The 100,000 or so genes (we now know that this estimate of the number of human genes is way too high) in the human body are clumped inside the nucleus of every cell. Envision a piece of steel wool, made up of filaments wound round and round. The entire steel wool would be the human genome. A gene would be a single tiny snippet. It is not just floating around. Man goes in, clips it out of the genome, isolates it, gives it to us and says this is what it does."[14]

The rush to decipher the human genome has triggered a money-making free-for-all akin to the claim-jumping disputes of the Gold Rush era. The only difference is that instead of racing to the land office, gene hunters stake their claims at the U.S.

---

[13] Sidney A. Diamond, Commissioner of Patents and Trademarks, petitioner, v. Ananda M. Chakrabarty et al., *65L ed 2d 144*, June 16, 1980, 144–47.

[14] Tom Abate, "Gene-patent opponents not licked yet, debate about human genome far from over", *San Francisco Chronicle*, March 20, 2000, pp. B1.

Patent & Trademark Office (PTO). A patent will provide them a 20-year monopoly on any drug or therapy that comes from the gene.

Table 4. Names of institutions holding patents on genes as of Y2000.[15]

Institutions	Number of Patents
U.S. Government	388
Incyte Genomics	356
University of California	265
SmithKline Beecham	197
Genentech	175
Eli Lilly	145
Novo Nordisk	142
Chiron	129
American Home Products	117
Isis Pharmaceuticals	108
Massachusetts General Hospital	108
Human Genome Sciences	104
University of Texas	103
Institut Pasteur	101

The PTO now faces a backlog of 30,000 biotechnology patent applications. Many of the gene patent applications involve genes which were discovered using machines and their functions are not known. Some observers argue that if monopoly is given over a machine-discovered gene of unidentified function, the patent holder can demand royalties from researchers who put the gene to work. This might discourage their innovation. After years of bickering from various sectors, the PTO is now proposing to change the rule. To win a patent, applicants will have to describe a "substantial, specific and credible" use for their gene. Using this guideline, the huge backlog of 30,000 patent applications can essentially be categorized into three classes:[16]

1. The first class includes genes isolated in the laboratory (i.e., wet biology) whose purpose is likely to be known. Wet biology has been used to patent genes for more than twenty years. There is no controversy over this class.
2. At the other extreme is a class which includes "naked" DNA sequences. These are machine-generated gene discoveries. The functions of the genes are not known. The new guideline rules them out.

As the controversy over "nakedness" festers, biotechnology companies get smarter and overcome the legal rhetoric by employing software to analyze the structure to deduce functions of machine-discovered genes. These "naked" sequences now

---

[15] Source: PricewaterhouseCoopers.

[16] Tom Abate, "Call it the gene rush — Patent stakes run high", *San Francisco Chronicle*, April 20, 2000, pp. A8.

become *"in silico"* sequences because their functions are deduced via computational means. A third class of patent applications begin to arrive on the scene:

3. In the middle ground between wet-biology discovered gene patent applications and machine-generated gene patent applications is the large number of applications that are machine-discovered genes but they are not "naked" DNA sequences because their functions have been computationally derived.

The crux of the matter is that the recent huge increase in the number of applications is due largely in part to speed of *in silico* analyses. Though the exact number of applications falling into the third category is unknown, it is easy to infer that the majority of the 30,000 applications do fall into this murky area.

5.1.5 Ethical, Social, and Legal Issues

Just as genetic knowledge can be used to better medicine, the project is fraught with potential perils. The knowledge can also be misused. On top of the list is genetic privacy. What is to stop employers, insurers, or anyone else from using genetic information in decision-making, leading to discrimination or genetocracy. This incidentally has become a main issue in the U.S. Y2000 Presidential Campaign.

Then there is the issue of perfect babies. The concept of what is perfect and what is abnormal are often merely reflections of the cultural stereotypes and prejudices of ourselves and our society.[17] The whole definition of normal could well be changed — the issue becoming not the ability of the child to be happy, but rather our ability to be happy with the child.[18] That is, our ability to be happy with the child's expected height, eye color, intelligence level or weight, rather than the child's expected welfare if it is genetically predisposed to diseases.

6 What Is In Store?

Most common people misunderstand that once the first draft of the human genome is complete, the work is done. Quite the contrary, the completion is just a milestone marking the beginning of efforts to unravel the secrets of the book of life.

6.1 Gaps Filling

The human genome project aims at determining the sequence of the euchromatic portion of the human genome. This portion excludes certain regions consisting of long stretches of highly repetitive DNA that encode little genetic information.

---

[17] Andrew Kimbrell, *The Human Body Shop: The engineering and marketing of life*, (Harper Collins, San Francisco, 1993).
[18] Geoffrey Cowley, "Made to order babies", *Newsweek*, Winter/Spring 1990, pp. 98.

Repetitive regions account for about 10% of the genome and are said to be heterochromatic. Heterochromatic regions include the end (tolemeres) and middle (centromeres) of each of the chromosomes. They are impervious to current sequencing technologies.

Thus there are still many gaps in the human genome draft. As an example, there are 11 gaps in chromosome 22. The public team has a June 2003 target for a complete, correct and end-to-end copy of the human genome with an overall accuracy of 99.99%. Although the working draft is already useful for most biomedical research, a highly accurate sequence is critical for obtaining all the information. In shotgun sequencing, a single pass will lead to 64% accuracy. Four passes will lead to 90% accuracy, while 10 passes will yield an accuracy of 99%. For an accuracy of 99.99%, 12.5 passes will be required. This is tautology to saying that to get an accuracy of 99.99%, statistically, a genome has to be sequenced repeatedly 12.5 times.

6.2   Gene Finding

To the naked eye, the genetic sequence seems an inscrutable text containing a jumble of As, Ts, Gs, and Cs. Within this jumble of text, 95% of them are junk, with no apparent purpose. The consensus is that nature did not lard the genome with junk without a good reason and that junk will not have been preserved through the evolution process. One goal of the genome effort is to determine the purpose of these junks. For now, the focus is on finding genes, which are of more immediate interest because genes direct the body to make proteins and enzymes.

6.3   Other Genomes

On March 24, 2000, Gerald M. Rubin (University of California, Berkeley) and Craig Venter (Celera Genomics) announced decoding the cryptic genetic sequence of the fruit fly, *Drosophila melanogaster*. The fruit fly provides a dictionary for learning much about human genes because 60% of the fruit fly's 13,600 genes are identical to those in humans. By studying the genes of fruit flies during decades of experiments, scientists have already acquired powerful insight into genetic disorders that afflict millions of people.

The fruit fly is the third organism whose genetic code has been completely deciphered. Scientists have deciphered the genomes of a single microscopic yeast strain *Saccharomyces* as well as the common roundworm *C. elegans*. By comparing human genetic code with that of yeast, the roundworm and the fruit fly, scientists hope to better understand the function and origin of human genes and gain a deeper understanding of how human bodies are designed.

The similarity is even more pronounced between the human and the mouse genomes. The mouse genome is one of the next targets of the genome project. Many more genomes are on the list.

Table 5. Sizes of various known genomes.[19]

Genomes	Length ($\times 10^6$bp)
Human	3,000
Mouse	3,000
Fugu fish	400
Rice	400
Drosophila	120
Nematode	100
Arabidopsis	100
Yeast	15
E. Coli	5
Mycobacteria	3–4
Mycoplasma	1
T_4 phage	0.16
λ phage	0.05
$\phi\chi$ phage	0.0005

As of June 17, 2000, the following bacterial genomes have been completed.

Table 6. Bacterial genomes completely sequenced as of June 27, 2000.[20]

Aeropyrum pernix K1	*Methanobacterium thermoautotrophicum*
Aquifex aeolicus	*Methanococcus jannaschii*
Archaeoglobus fulgidus	*Mycobacterium tuberculosis*
Bacillus subtilis	*Mycoplasma genitalium*
Borrelia burgdorferi	*Mycoplasma pneumoniae*
Campylobacter jejuni	*Neisseria meningitidis MC58*
Chlamydia pneumoniae CWL029	*Neisseria meningitidis Z2491*
Chlamydia pneumoniae AR39	*Pyrococcus abyssi*
Chlamydia trachomatis	*Pyrococcus horikoshii*
Chlamydia muridarum	*Rickettsia prowazekki*
Deinococcus radiodurans R1	*Synechocystis PCC6803*
Escherichia coli	*Treponema pallidum*
Haemophilus influenzae	*Thermotoga maritima*
Helicobacter pylori 26695	*Ureaplasma urealyticum*
Helicobacter pylori J99	

---

[19] Hwa A. Lim, lecture note at the 9th International Conference on Mathematical and Computer Modeling, University of California at Berkeley, July 26–29, 1993.
[20] Source: http://www.ncbi.nlm.nih.gov/PMGifs/Genomes/micr.html.

6.4   Gene Function

Knowing the sequence of a gene is just the beginning. Without knowing the function, besides not being able to acquire a patent for the gene, the gene is rather useless in fashioning a medical remedy. Gene hunters have computer software that can provide hints on a new gene's function by comparing its sequence to that of genes with known functions. Such comparisons may shed light that the gene may be involved in making a certain type of protein. To confirm these hunches, fashion drugs and therapies, researchers may have to spend months or years in laboratories working with comparable genes in a model organism (mouse for example). If the experiment with model animals pans out, researchers may move on to do clinical trials on humans.

6.5   Single Nucleotide Polymorphisms

At the genetic level, humans differ from one another at the rate of one letter in every thousand in the book of life. These differences are called single nucleotide polymorphisms or SNPs. Most SNPs are inconsequential because they occur in the junk DNA. If the difference falls on a critical spot of a gene, this may cause the gene to produce the wrong protein or run amok. In 1999, fourteen drug companies have formed a consortium to begin building a public SNP database. These SNPs are obtained from comparisons of genomes of both sexes and of different ethnicities.

SNP data feeds into the budding field of pharmacogenomics or personalized medicine. Some drugs may help one patient but cause side effects in others. Researchers believe that some of these outcomes are attributable to genetic variations in patients.

6.6   The Eclectic Bioinformatics

In the computer and telecommunication sectors, digitization is everything. In the biotechnology sector, genetization is everything. Once genetized, the gene data can be digitized and hence the marriage of computer and biotechnology sectors. This is where bioinformatics comes in. Bioinformatics is the study of information content and information flow in biological systems.[21]

International efforts have played a critical role in the human genome project's success. Currently there are more than 18 countries supporting programs for analyzing the genomes of a variety of organisms ranging from microbes to economically important plants and animals to humans.[22]

---

[21] Hwa A. Lim and Tauseef Butt, "Bioinformatics Takes Charge", *Trends in Biotechnology*, 16 No. 3 (170), 104–107 (1998).
[22] T.V. Venkatesh, B. Bowen, and Hwa A. Lim, "Bioinformatics, Pharma and Farmers", *Trends in Biotechnology*, 17 No. 3 (182), 85–88 (1999).

6.7   An Array of Possibilities — New Genomic Technologies

Biochip, DNA chip, DNA microarray, gene array, or gene chip closely resembles computer chips. They are packed with DNA and are designed to read the reams of genetic information in the genomes.

Highly parallel quantitative measurements of gene expression patterns are possible using different methods to produce DNA microarrays. An array is an orderly arrangement of samples. Base-pairing or hybridization is the underlining principle of DNA microarray. DNA microarray provides a means for matching known and unknown DNA samples based on base-pairing rules and automating the process of identifying the unknowns. Applications areas include:

❑   Functional genomics — Functions of genes can be deduced from their expression patterns in different tissues.

❑   Gene discovery — This focus uses the chip to discover the expression pattern of unknown genes.

❑   Disease diagnosis — This is not the traditional chip for gene expression analysis. It is usually designed to study unique disease genes.

❑   Drug discovery — Gene chips can be used to study correlations between therapeutic responses to drugs and the genetic profiles of patients.

❑   Toxicological research — The goal of toxicogenomics is to find correlations between toxic responses and the gene expression profile.

❑   Environmental research — The goal is to identify presence of certain organisms.

7   The Next Fifty Years

Francis Collins and the genome community make the following predictions for the upcoming five decades:[23]

2010

• Genetic testing will be available for 25 common conditions such as colon cancer.

• Interventions will be available to decrease a person's risk of most of these genetic diseases.

• Gene therapy will prove successful for several conditions.

• Most doctors will begin practicing genetic medicine.

• Preimplantation diagnosis will be widely available.

• The limitations of genetic testing will be fiercely debated. Questions will include what applications are appropriate?

---

[23] http://www.msnbc.com/news/415085.asp

- Effective legislative solutions to medical record privacy will be in place in the United States.
- Access to genetic screening and therapies remains inequitable, especially in the developing world.

2020
- Gene-based designer drugs will be available for common conditions such as diabetes and high blood pressure.
- Cancer therapy will be targeted to the molecular fingerprint of the tumor.
- Pharmacogenetic applications will be standard practice for the diagnosis and treatment of many diseases.
- Genetic diagnosis and treatment of mental illness will be available.
- Geneticists will learn how to perform germline gene therapy without affecting other genes and hence human germline therapy will be declared safe and ethical.

2030
- Genes involved in aging will be fully catalogued.
- Clinical trials will be underway to extend maximum human life expectancy.
- Use of a full computer model of human cells will replace laboratory experiments.
- The complete genomic sequencing of an individual will be routine, cost less than $1,000.
- Major anti-technology movements will be active in the United States and elsewhere.

"Here's my sequence..."

Figure 3. Customer to pharmacist, "Here is my sequence..." (Reproduced with permission from Don Halbert of Abbott Laboratories).

2040

- Complete genome-based health care will be the norm.
- Individualized preventive medicine will be available and largely effective.
- Illness will be detected earlier, before symptoms develop, by molecular surveillance.
- Gene therapy and gene-based drug therapy will be available for most diseases.
- The average life expectancy will reach 90 years.
- The debate about the role of humans in taking charge of their own evolution grows louder. Who is to decide what is a good characteristic or trait?

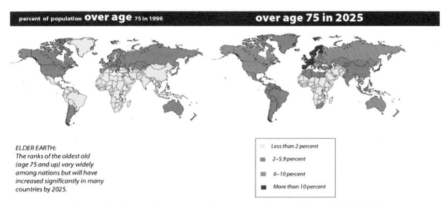

Figure 4. The world growing older. Source: Nature Online, Feb 12, 2001

8   The Test Of Time

Returning to the test of time. Today, with inventions and breakthroughs appearing regularly, and superstars being born almost daily, the common people forget all of these events soon afterwards. Inventors of new gadgets and gizmos are forgotten almost as soon as a better substitute or improvement is developed. Superstars (in sports and Hollywood for example) are purged from memory almost the day after they retire. Great scientists come infrequently and far in between. Their theories are normally so abstract that they will not affect human lives for years to come.

The watershed completion of the first draft of the human genome stands out in these respects. It will be remembered for a long time since it affects directly our very own existence. Reporters liken the human genome project (HGP) with the moonshot. While this metaphor is useful in political and funding purposes, the comparison does not convey the true significance of HGP. The moonshot affected a negligible number of people substantially and had only negligible effect on a substantial number of people. The impact of HGP will be substantial and it will affect a substantial number of people.

In President Clinton's photo-op analogy, the Lewis-Clark expedition, summoned by President Thomas Jefferson, started from the mouth of Wood River on the Illinois side of the Mississippi, opposite the entrance to the Missouri River on May 14, 1804. The expedition reached the westernmost point of Fort Clatsop at the mouth of Columbia, Oregon in mid November, 1805. By then Clark recorded that 554 days had elapsed and 4,132 miles had been traversed since they left Wood River.

Figure 5. The route of Lewis-Clark expedition, St. Louis to Fort Clatsop.

The human genome expedition, if typed in 10-pitch font, would start in San Francisco (HAL's office) to Chicago, on to Baltimore, Houston, and end in Los Angeles. The sequencing journey officially commenced in 1990, planned to take 15 years, and would end in 2005. In the past few years, due mainly to concurrent technology breakthroughs in computer technology and ultra-high throughput sequencing technology, the sequencing journey reached its destination five years ahead of the original schedule.

Figure 6. The Lewis-Clark expedition of the 21st Century, reproduced from Cincinnati Enquirer, December 16, 1999. The author thanks Don Halbert of Abbot Laboratories.

So will the group of international contributors be remembered as empire builders, conquerors, or explorers? These people are not empire builders for when the human genome sequence is digitized or genetized, it lives in cyberspace which recognizes no geo-political borders. These people are not conquerors, for before them there have been many groups and individuals in other disciplines who have slaved through to improve on instrumentation, computer and ancillary technologies. These people are not explorer either. For exploration, the final frontier is unknown. In this case, we know what the end result would be, the sequence of the human genome. These people are not royal family either. Quite the contrary, they help disprove the existence of royal blood. These people are not president. Quite the contrary, presidents love to bathe in the glory of their accomplishment.

But these people certainly have opened a terrain for empire builders for they have opened up the biosphere commons and started the biotechnology century. They have conquered long hours performing the laborious production line repetitive task of sequencing. They have pushed the envelope of our legal system. In the words of Bill Clinton, they have produced "a map of greater significance". If the whole genome were to be written on an average 8½" by 11" paper (U.S. size paper) of average thickness and the sheets of paper were stacked one on top of the other, the entire genome would stack as high as the Washington Monument.

The expedition is just beginning. And bioinformatics will be a big part in the next phase of the expedition. To highlight the significance of bioinformatics, we can perhaps end this chapter by quoting a paragraph from Yoneji Masuda, a leading figure in the Japanese plan to become the first fully developed information society,[24]

Δ

Unlike
material goods,
information does not
disappear by being consumed,
and even more important, the value of
information can be amplified indefinitely by
constant additions of new information to the existing
information. People will thus continue to utilize information
which they and others have created, even after it has been used.
$¢£¥MF$¢£¥MF$¢£¥MF$¢£¥MF$¢£¥MF$¢£¥MF$¢£¥MF$¢£¥MF$¢£¥MF$

This is the major difference between the industrial economy and knowledge-based economy. We have intentionally arranged Matsuda's statement into a pyramid, a wealth pyramid of the information era. This also distinguishes the Human Genome Project from the Lewis-Clark expedition. The race to the finishing line of the map of the human genome between the private and the public sectors,

---

[24] Yoneji Masuda, "Managing in the Information Society: Releasing Synergy Japanese Style", *World Future Society*, Washington, DC, 1980.

particularly the Human Genome Project led by Francis Collins and Celera Genomics led by Craig Venter, was a carefully prearranged tie. Though there was no clear winner between the competitors, the competition has certainly galvanized the public project to finish five years ahead of the original schedule. The biggest winner of all from this race is humankind, the one who is going to enjoy a better quality and more productive life, provided we do not abuse the data.

A good thing about digital revolution is it is status-blind, that electronic mail strikes the highest executive and the lowliest supply clerk alike. After all, that was the promise from the very beginning: e-mail was going to be the great democratizing agent of corporate America, allowing an instant exchange of ideas unimpeded by gatekeepers and protocols. You were either in or you were out in the new corporation and, thanks to e-mail, you could be in anywhere — all distances, whether geographic or hierarchical, would vanish into electronic ether.

Hopefully, in a similar way, once the genomes of organisms, humans inclusive, have been digitized and genetized, all organisms will be democratized. In the language of SNPs, each *Homo sapiens* is not too different from the other, only in 0.1% of the genomes. Similarly, in the language of genes, *Homo sapiens* are not too different from other organisms. The *Home sapiens* has only one and a half times as many genes as a *Caenorhabditis elegans*. The catch phrase of condescension is "The *Homo sapiens* are just more complex".

Chapter 2

POST-GENOMICS MATTERS: SEND OUT A THOUSAND SHIPS, NOT CATALOGUE A THOUSAND GENES

> *"The idea is to send a thousand ships, not catalogue a thousand genes."*
> Eric Lander, Director, Whitehead Institute, MIT, 2000.

1 Genome Armada

When laypeople think of the post-genomic era, they think of a closed chapter in human genome sequencing. In reality, the post-genomic era is a state of excitements and opportunities. Amidst all the initial excitements are states of confusion as new findings run counter to long-held biology textbook dogmas. In this state, the best solution is to spur further efforts to resolve all apparent inconsistencies. As the Director of Whitehead Institute, Eric Lander rightly said after the publication in *Science*[1] and *Nature*[2] of the findings of the private and public genome efforts, "The idea is to send a thousand ships, not catalogue a thousand genes," borrowing the analogies from "explorers" or "conquerors".

2 Perplexity And Enlightenment

There is no doubt that the human genome is one of the most precious bodies, if not the most precious body, of information. It affects all of us directly. The February 15, 2001 joint report by the private sector representative, Celera Genomics and the public sector, the Human Genome Sequencing Consortium (HGSC) of the analyses of the human genome sequence have found both perplexity as well as enlightenment.[3] Among the more important findings are

❑ The human genome, with its 3 billion letters of DNA code, is 25 times larger than any other single genome studied so far. But the number of protein-coding genes, those that determine the roles of cells and organs, are far fewer than scientists expected — only about 30,000 or 40,000, rather than earlier estimates of about 100,000 genes.[*] A fruit fly, by comparison, has about 13,000 such genes.

---

[1] *Science*, 291, 5507(16) February 16, 2001.
[2] *Nature*, 409, February 15, 2001.
[3] Nicholas Wade, "Genome's riddle: few genes, much complexity", *New York Times*, February 13, 2001.
[*] The International Human Genome Sequencing Consortium found evidence for 29,691 human transcripts, while the commercial genome project of Celera Genomics found 39,114 genes. In October 2001, Bo Yuan, Director, Bioinformatics, Ohio State University suggests humans possess between 65,000 and 75,000 genes.

33

❑   Human beings are 99.9% identical in the overall make-up of their DNA. But, so far, 1.4 million genetic differences — called single nucleotide polymorphisms or SNPs — have been identified in the remaining 0.1% of human DNA.

❑   The evolution of humans from smaller and outside organisms can be traced by looking at our DNA. For example, more than 200 genes in the modern human genome have been inherited from bacteria.

2.1   *The New Genetic Math — 1.5 × Caenorhabditis elegans = Homo sapiens*

The overestimate of 100,000 genes for the human genome is a number derived from the long-held textbook dogma of "one gene one protein".   From the relative complexity of the *Homo sapiens*, 100,000 was still a good guesstimate even after the genomes of the first two model organisms, the worm and the fruitfly, were deciphered.   The worm genome was completely sequenced in December 1998[4] and found to have 19,098 genes.   The fruit fly was completely sequenced in March 2000 and found to have 13,601 genes.[5]

Figure 1. *Caenorhabditis elegans* has been found to have 19,098 genes.   The human has about one and a half times as many genes. (Courtesy: *Nature Online*, Feb 12, 2001).

Figure 2. *Drosophila melanogaster* has been found to have 13,601 genes. (Courtesy: *Nature Online*, Feb 12, 2001).

---

[4] *Nature* 396, 620–621, 1998.
[5] *Nature* 403, 817, 2000.

Craig Venter and colleagues at Celera Genomics reported in *Science*[6] that they had identified 26,588 human genes, with another 12,731 candidate genes. There was a panic when they first screened the gene families likely to have new members of interest to pharmaceutical companies but could not find the genes. The publicly funded consortium of academic centers, HGSC, has arrived at a similar conclusion. Its report in *Nature*[7] pegs the probable number of human genes at 30,000 to 40,000. Because the current gene-finding methods tend to over predict, each side prefers the lower estimate of 30,000 as the likely number of human genes.

As the modest number of human genes became apparent, biologists in both camps were forced to think how to account for the greater complexity of the human, given that they seem to possess only one and a half times as many genes as the roundworm. It is not foolish pride to suppose there is something more to *Homo sapiens* than *Caenorhabditis elegans*. The roundworm is a little tube of a creature with a body of 959 cells, of which 302 are neurons in what passes for its brain. Humans have 10 trillion cells in their body, including 100 billion brain cells.

Though the estimate of 30,000–40,000 human genes has been in circulation since June 2000, the public joint announcement and publications in authoritative journals like *Science* and *Nature* carry tremendous weight. The puzzle comes from two angles:

❑   First, the "one-gene one-protein" dogma is in doubt.
❑   Second, the ego of the *Homo sapiens* as the most intelligent organism gets a shot.

We are not as "sophisticated" as we once thought ourselves to be. We have only 1.5 times as many genes as the lowly roundworm, or 3 times as many genes as the lowly fruitfly. In other words, we are far closer in genetic patrimony to these two tiny invertebrates than almost anyone had expected.

Figure 3. J. Craig Venter, left, and Dr. Francis Collins, right, appeared at a press conference on Feb 12, 2001 to announce the publication of their findings. (Courtesy: *New York Times*).

---

[6] *Science*, 291 5507(16), February 16, 2001.
[7] *Nature*, 409, February 15, 2001.

Several plausible explanations are emerging to explain the extra complexity of the human other than by having more genes. One is the concept of combinatorial complexity — with just a few extra proteins one can make a much larger number of combinations between them. Jean-Michel Claverie, of the French National Research Center, notes that with a simple combinatorial scheme, a 30,000-gene organism like the human can in principle be made highly complicated. But Claverie suspects humans are not that much more elaborate than some of their creations. In fact, with 30,000 genes, each directly interacting with four or five others on average, the human genome is not significantly more complex than a modern jet airplane, which contains more than 200,000 unique parts, each of them interacting with three or four others on average.[8]

Indeed, in general, for a N-gene organism with each gene interacting with other n genes, the total number of ways this can be achieved is

$$^{N}C_{n}$$

or N combination n. In general, an organism with a lower number of genes will have to have more genes interacting with each other to be as complex as an organism with a higher number of genes. With this in mind, we can now perform a numerical experiment and use the formula to enumerate the total number of ways for a 30,000-gene organism and a 100,000-gene organism, with each gene interacting with say (n+1) and n other genes, respectively.

Table 1. A table to show that for a 30,000-gene organism with each gene interacting with 7–8 other genes has about the same complexity as a 100,000-gene organism with each gene interacting with 6–7 other genes.

n	$^{30,000}C_{n+1}$	$^{100,000}C_{n}$
1	4.49985×10^{8}	10.0000×10^{4}
2	4.49955×10^{12}	4.99995×10^{9}
3	3.37432×10^{16}	1.66662×10^{14}
4	2.02432×10^{20}	4.16642×10^{18}
5	1.01199×10^{24}	8.33249×10^{22}
6	4.33624×10^{27}	1.38868×10^{27}
7	1.62571×10^{31}	1.98371×10^{31}
8	5.41759×10^{34}	2.47946×10^{35}
9	1.62478×10^{38}	2.75474×10^{39}

Note that $^{30,000}C_{n+1} > {}^{100,000}C_{n}$ until n attains a value somewhere between 6 or 7 when they are equal. Beyond that number, $^{100,000}C_{n} > {}^{30,000}C_{n+1}$. Note if we have used (n+2) instead of (n+1), the point of equality will be larger than 6 or 7.

This numerical experiment shows by example that an organism with a lower number of genes can be as complex as an organism with a higher number of genes as

---

[8] Nicholas Wade, "Genome's riddle: few genes, much complexity", *New York Times*, February 13, 2001.

long as the number of interacting genes in the former is higher than the number of interacting genes in the latter.

2.2 *Human Genome Archaeology*

The human genome itself is a huge archaeological artifact. Findings in this aspect are also full of surprises. Most of the repetitive DNA sequences in the 75% of the human genome that is essentially junk ceased to accumulate millions of years ago, but a few of the sequences are still active and may be of some biological significance. Large blocks of genes seem to have been extensively copied from one human chromosome to another, daring genetic archaeologists to figure out the order in which the copying occurred and thus to reconstruct the history of the genome.

First scanning of the genome by the two camps (Celera and HGSC) suggests two specific ways in which humans have become more complex than worms. One comes from protein domains and the other gene segments. Proteins, the working parts of the cell, are often multipurpose tools, with each role being performed by a different section or domain of the protein. Many protein domains are very ancient. Only 7 percent of the protein domains found in the human are absent in the worm and fly, indicating that few new protein domains have been invented in the vertebrate lineage. But these domains have been mixed and matched in the vertebrate line to create more complex proteins. In the words of Francis Collins, "The main invention seems to have been cobbling things together to make a multitasked protein. Maybe evolution designed most of the basic folds that proteins could use a long time ago, and the major advances in the last 400 million years have been to figure out how to shuffle those in interesting ways."

Evolution has devised another ingenious way of increasing complexity, which is to divide a gene into several different segments and use them in different combinations to make different proteins. The protein-coding segments of a gene are known as exons and the DNA in between as introns. The initial transcript of a gene is processed by a delicate piece of cellular machinery known as a spliceosome, which strips out all the introns and joins the exons together. Sometimes, perhaps because of signals from the introns that have yet to be identified, certain exons are skipped, and a different protein is made. The ability to make different proteins from the same gene is known as alternative splicing.[9] Alternative splicing is more common in human cells than in the fly or worm and that the full set of human proteins can be five times as large as the worm's. Another possible source of extra complexity is that human proteins have sugars and other chemical groups attached to them after synthesis.

Another interesting discovery is segmental duplications, or the copying of whole blocks of genes from one chromosome to the other. These block transfers are

---

[9] Victor B. Strelets and Hwa A. Lim, "Ancient splice junction shadows with relation to blocks in protein structure", *BioSystems*, 36, 37–41, (1995).

so extensive that they seem to have been a major evolutionary factor in the genome's present size and architecture. They may arise because of a protective mechanism in which the cell reinserts broken-off fragments of DNA back into the chromosomes. Chromosome 19 seems the biggest borrower, or maybe lender, with blocks of genes shared with 16 other chromosomes. Much the same set of large-scale block transfers seems to have occurred in the mouse genome, suggesting that the duplications appear to predate the divergence of the mouse and the human about 100 million years ago. Segmental duplication is an important source of innovation because the copied block of genes is free to develop new functions.

An idea enshrined in many textbooks is that the whole genome of early animals has twice been duplicated to form the vertebrate lineage. There are several cases in which one gene is found in the roundworm or fly and four very similar genes in vertebrates. The quadruplicated genes that failed to find a useful role would have been shed from the genome. However, to date no evidence for the alleged quadruplication of genes has been found. If this venerable theory is incorrect, the four-gene families may all arise from segmental duplication.

2.3   The Genome Topography

The map or landscape of the entire genome resembles a population map, with urban areas of dense habitation, and vast rural tracts occupied by few people. Not only are genes distributed unevenly across the chromosomes, but also are the types of noncoding repeat sequences that make up the bulk of the genome. Repetitive sequences with an excess of the nucleotides C (cytosine) and G (guanine), or GC-rich, tend to be found in the neighborhood of genes, while repeat sequences heavy on A (adenine) and T (thymine), AT-rich generally dominate throughout the non-gene "deserts".[10]

The nucleotide distribution on the chromosome offers a definitive explanation why we see the distinctive dark and light bands on chromosomes familiar to anybody who has seen a chromosome karyotyped. As it turns out, the light bands represent GC-rich regions, and so are areas of comparatively high gene concentration; while the dark bands signal neighborhoods of AT-rich and sparse in genes.

Gene distribution is also uneven across the chromosomes. For reasons that remain unknown, some chromosomes have very high gene concentration. Chromosome 19, for example, is among the smallest of the 23 chromosomes, yet it is the most densely packed with genes, as well as with noncoding sequences that are CG-rich. As a result, Chromosome 19 in a karyotype displays very few dark bands.

The ruggedness of the human genome distinguishes itself from that of other species whose genomes have been sequenced, including the roundworm, the fruit

---

[10] Natalie Angier, "For the microscopic genome, it's a big moment in biology", *New York Times*, February 13, 2001.

fly, yeast and several species of bacteria. Other genomes are like plains and tend to be flat. The human genome is more undulating. The architecture of the human genome is no accident, but has been shaped over millions of years by evolution. Otherwise, by simple laws of statistics, there would be deterioration toward the mean. Since the genome has maintained a broad spectrum of mountainous landscapes, this means that the landscapes must be good for us.

Like the real world, the exaggerated topography of the genome reveals nothing so much as an evolutionary stage for opportunism. It is a genuine jungle of wild ecosystem of competing "species" of DNA freeloaders, DNA symbionts, endangered DNA sequences and rotting DNA fossils. The genomes of organisms are spectacular records of the genetic history of life on Earth. A number of our sequences date back 700 million years or more, when we lived a tidy, unicellular life, while other chemical phrases bespeak our patrimony to the fraternity of apes, and may explain how we came to be human. Some stretches of base pairs are ancient remnants left behind by infecting viruses and bacteria. The human immune system, which is capable of cutting and pasting together different genetic segments to spin off a staggering variety of warrior immune cells and antibodies, may well have adopted the editing trick from a virus that was cutting and pasting DNA into its host many millions of years ago. The gene that encodes monoamine oxidase, an important degradative enzyme for the central nervous system, was bequeathed to an ancestor's cell by bacteria, as were about 230 other genes in our genomic toolbox.

The fact that there has been lateral transfer of genetic information from bacteria to vertebrates suggests that the architecture of the human genome is not sacrosanct or static, but extraordinarily fluid and dynamic, noted Leroy Hood of the Institute for Systems Biology in Seattle. For better or worse, the genome passes around reams of information intramurally as well. The great bulk of the noncoding sequences in human DNA are not foreign-born, but represent the offspring of bits of genetic material that long ago broke away from a chromosome or part of the cell's RNA and protein synthesis machinery and decided to go freelance. Called jumping genes or transposons, the little entrepreneurs figured out how to reproduce themselves and reinsert their copies — their progeny — back into the mother genome. Researchers examining the landscape of the human genome have identified four classes of transposons. One class, called DNA transposons, appears to be dead, mere fossils in the DNA that have lost the signals they need for effective replication, are incapable of ever jumping anywhere again and are gradually decaying out of existence. Another class, called the LTR transposons, is on the critically endangered list" and may soon go extinct.

Only two transposon families in the genome are considered active, replicating themselves tidily in the course of transmission from parent to child. One is a genuine parasite called a LINE[11], for Long Interspersed Element, which encodes

---

[11] Nikolay A. Kolchanov and Hwa A. Lim, *Computer Analysis of Genetic Macromolecules: Structure, Function and Evolution*, (World Scientific Pub. Co., New Jersey, 1994), 556 pages.

instructions for everything it needs, from copying its DNA into the intermediary form, RNA, and copying the RNA back again into DNA, and then hopping back onto its little chromosomal niche. It is the ultimate selfish element that evolved at the beginning of eukaryotes. It has been wildly successful as a perfect parasite. Significantly, most of the LINEs in the human genome are located in AT-rich regions of the chromosomes, far from genes, and therefore far from the DNA quality control machinery that attends to the genes' well-being and might try to sniff them out and eliminate them.

But as the food chain in the macroscopic world, the LINE parasite itself has a parasite — the infamous Alu element. With more than a million copies scattered across the chromosomes, Alu elements, which run about 300 bases in length, are the most abundant sequences in the human genome. Molecular biologists have long despised them for getting in the way of their efforts to clone genuine genes. Alus cannot replicate on their own, but instead "borrow" the machinery of the larger LINE elements to reproduce. At first glance, they seem to be even more perfect in their parasitism, and like a real parasite, they can harm their host. On occasion, in the course of the creation of an egg or sperm cell, a replicating Alu sequence will be inserted in the midst of a critical gene, resulting in a child with a genetic disease.

A scan of the human genome suggests that genes do benefit from having Alus in the neighborhood. Whereas most species of transposons are located in the AT deserts of the chromosomes, Alus are preferentially clustered in the GC regions, among genes. It looks like they are being selectively held onto as genetic companions because they are useful to human biology. That is quite surprising and the natural conclusion is this part of our junk DNA is not junk at all.

It may be just possible that Alu sequences helped make us human, argued Wanda F. Reynolds, a molecular biologist at the Sidney Kimmel Cancer Center in San Diego. Alu elements are found only in higher primates, and that the core of the Alu sequence is responsive to a large family of receptor proteins, the so-called nuclear receptor superfamily. These receptors are the cell's way of recognizing potent hormones like estrogen, retinoic acid and thyroid hormone. Hence the presence of an Alu sequence in or around a gene may result in the gene being pitched a little higher or lower, turned up or down, should the appropriate hormone come calling. This gives the genome a great deal of plasticity, and a lot of choices. Our genes are almost identical to chimpanzees. Something had to happen that made us different from chimpanzees. Reynolds believes that a change in the expression of genes, in the volume and timing of their activation, could have a more revolutionary effect on development than would a mutation in the genetic sequence proper.

The sequencing of the genome lends strong support to a theory proposed a couple of years ago by Carl Schmid of the University of California at Davis and his colleagues — that the Alus may have begun life long ago as rank parasites, but they have since been co-opted by the human genome to do useful work. Schmid proposed that the genome uses the Alu elements to help modulate the body's response to stress, from, say, excessive exposure to heat, or too much alcohol. Through experiments in which mice were dunked in hot tubs or made to drink lots of alcohol, Schmid and his co-workers observed that the rodent equivalent of Alu sequences were activated by such stressful events.

3 How The Statistics Stack Up?

At the time of the publication of the *Science* and *Nature* articles, Incyte Genomics advertised access to 120,000 human genes, including 60,000 not available from any other source. Human Genome Sciences said it had identified 100,000 human genes, and DoubleTwist 65,000 to 105,000 genes. Affymetrix was selling DNA analysis chips containing 60,000 genes. But now it turns out there might be only around 30,000–40,000 genes. If that is the case, what exactly have these companies been selling?[12]

The new gene tally — 30,000 to 40,000 — is among the most significant findings in rival papers by Celera Genomics and HGSC. For science and the genomics business, the papers are landmarks, containing both land mines and beacons of opportunities.

3.1 On the Pessimistic Side

In particular, the consensus of the two rival camps, Celera and HGSC, that humans have far fewer genes than anticipated could raise questions about the credibility of the genomics approaches used until now.

Investors are also having doubts about companies that sell tools, information or services to drug companies, rather than develop drugs themselves. The tools approach has been considered a quicker way to profitability, avoiding the long clinical trials, huge investments and risk of failure inherent in drug development. But the sentiment has shifted. Now drugs are seen as offering the biggest potential payoff. They can be sold for years, while gene analysis techniques can become obsolete quickly. And, some analysts say, there is too much competition in tools. There are at least 10 technologies, for instance, for detecting genetic variations among people. So virtually every genomics company is now tripping over itself to become a drug company.

In fact, a new study by Lehman Brothers and McKinsey & Company concludes that genomics could actually double the pharmaceutical industry's research and development costs per drug, at least over the next few years. New genes and proteins are being discovered so rapidly, the report says, that drug companies are being overwhelmed with potential paths to pursue. But since the roles of these

---

[12] Andrew Pollack, "Double helix with a twist", *New York Times*, February 13, 2001.

genes and proteins in the body are not well understood, there is a greater chance that drugs will fail after costly clinical trials.

3.2   On the Mixed Side

The lower estimate can also mean that developing drugs based on gene studies will be quicker than anticipated — but also present a smaller business opportunity. The lower number of genes also suggests that gene hunters have already received or applied for patents on a greater proportion of total genes than anticipated — leaving less room for newcomers. But if genes are not the whole story, it also means those patents could be worth less.

Investors in the year preceding the *Science* and *Nature* articles were enamored by the science, but there is a need to shift away from scientific discoveries to medical discoveries, as noted by William Blair & Company. There could be long-term implications because pharmaceutical companies can get to drugs and profits faster than if the human genome has fewer genes. This can also mean more limited prospects for genomics companies and less of a cornucopia for drug companies.

Jean-Michel Claverie, who heads a genetic information laboratory run jointly by the French government and the drug manufacturer Aventis, argues that only 10 percent of genes can be expected to provide good targets for drugs. If there are only 35,000 genes, which means only 3,500 targets or a number the drug industry can work its way through in a few years.

But other experts dismiss this argument. First, they say, even 3,500 targets would be a huge increase. All the drugs that exist today are aimed at a total of only 500 different protein targets in the body. Moreover, the new genome papers show that the complexity of the human body results not so much from having more genes than simpler creatures, but from having many more proteins. Genes are of interest to drug companies primarily because they are the recipes for making proteins. But it is the proteins that actually carry out bodily functions, and drugs are developed to bind to particular proteins. It was once thought that knowing the gene would be enough to know the protein. But in humans, more so than in simpler creatures, this is turning out not to be the case. Genes are made of pieces that can be spliced together in different combinations. So one gene can make more than one protein.

3.3   On the Positive and Opportunistic Side

Genomics alone, therefore, cannot answer everything drug companies need to know. This is giving new impetus to the emerging field of proteomics, which seeks to identify all proteins and how they relate to one another. But proteomics is far more daunting than genomics, because proteins are more complex than genes and also more plentiful, probably numbering in the hundreds of thousands.

Just before the *Science* and *Nature* articles, Large Scale Biology, based in Vacaville, California, said it had compiled a database of more than 115,000 human proteins. Hybrigenics, a French company, published a map showing about half the interactions of the proteins in the bacterium linked to ulcers and stomach cancer. And the Cytogen Corporation of Princeton, N.J., said it had mapped the interactions of one of the roughly 70 families of human proteins.

3.4 Comparatively Speaking

For some companies, namely those attempting to sell tools and databases claiming to have 60,000 to 120,000 human genes, the news that the genome may contain far fewer genes than previously expected may have sounded a bit harsh to the ears. But for others, the findings were cause for celebration. [13]

Companies selling comparative genomics services say that having fewer genes in the human genome can make the task of deciphering gene function more manageable, and the high degree of similarity across animal species demonstrates the value of studies using comparative techniques.

Exelixis, a South San Francisco-based seller of comparative genomics data using nematode and drosophila models, claims that fewer uniquely human genes, a finding supported by Celera's data showing that humans have only about 300 genes not found in mice, should allow comparative studies using other, simpler organisms to identify the function of human genes more accurately. Other companies in comparative genomics voiced similar optimism. Lexicon believes the lower number of genes means the function of each gene has more significant value. With 30,000-40,000 genes, suddenly the project [of determining gene function] is doable. Lion Bioscience adds the low number of genes implies that comparative genomics will be a good way to understand gene function and which gene functions are conserved.

But other comparative geneticists, namely those in academia, warn that the situation may not be so simple. The significance of the fewer-than-expected number of genes, some scientists said, lies in the complexity of gene networks and the post-translational modifications unique to human biology. Furthermore, the low number of uniquely human genes may be remarkable, but not proof that the genes we have in common with other animals act exactly the same way.

Lisa Stubbs of Lawrence Livermore National laboratory argues that it is not the number of genes that matters at all; but rather what the genes encode, how they are regulated, and what kinds of changes they have accumulated over eons of evolution. Although humans may have 300 or so genes not found in mice, the reverse is also true. The mouse has plenty of genes that are not represented in humans. What accounts for the aspects of human physiology that make the human distinctly so human is not the 300 unique genes, but minor variations in gene networks. William

---

[13] John S. MacNeil, "Will fewer genes help comparative genomics efforts?" *GenomeWeb News*, Feb 16, 2001.

Loomis, a developmental biologist at the University of California, San Diego believes that probably no new genes came into play as the human developed its neural faculties. It probably occurred just as a modulation of existing networks.

Nevertheless, the importance of comparative techniques in determining the function of genes involved in common biochemical pathways hinges on the amazing conservation of gene sequences and genome organizations among mammals, vertebrates, and other animals. According to John Postlethwait, a zebrafish genomics researcher at the University of Oregon in Eugene, it is these conserved features that allow information from all animals to provide strong suggestions about what human genes — known only by sequence and position — really do.

Public companies like Lion is seizing on the opportunity the completed human genome provides. Currently, Lion's GenomeScout software product performs similarity searches between microbial genomes, but the company is developing the ability to make human-mouse and human-microbial comparisons. Start-ups like Michigan-based GeneGo have already capitalized on the human genome project. GeneGo's proprietary database (GGDB) and supporting software are already operational. By mid Y2001, there are 1829 enzymes and 2861 pathways in GGDB, and more than 1800 probable human-specific pathways have been identified.

4   Genestimate Discrepancies

The fact that one gene can make more than one protein also partly explains the wide variation in estimates of gene numbers. Celera and the Human Genome Consortium independently estimated the number of genes by taking the entire human genome of about three billion letters and performing various computer analyses to try to determine which small parts of that sequence contain the code for proteins.

DoubleTwist, based in Oakland, California, also did a computer analysis yielding a much higher estimate, a sign that such computer models are subject to wide variations.

Incyte and Human Genome Sciences find genes by catching them in the act of making proteins. They search human cells for the messages sent by the genes to the cell's protein-making machinery. But since one gene can make different proteins, and therefore send out different messages, the assumption that each message comes from a different gene can lead to an overcount. Also the technique actually detects not whole messages but fragments of them. So different fragments of the same message might be incorrectly assumed to represent different messages. Incyte, which advertises that it has 120,000 genes, now says what it really means is 120,000 messages, which would translate into 40,000 genes if each gene were assumed to make three proteins. It emphasizes that the message information is more valuable than genes because it is more indicative of what proteins are being made.

Human Genome Sciences (HGS) remains unshaken in its estimate of 100,000 to 120,000 genes. HGS claimed to have captured and sequenced 90,000 full-length genes, from which all alternative splice forms and other usual sources of confusion have been removed. They have made and tested the proteins from 10,000 of these genes. They hold the opinion that the gene finding methods used by the two camps — Celera and Human Genome Consortium — depend in part on looking for genes like those already known, a procedure that may miss radically different types of genes, citing that 5 of the 10 genes in the AIDS virus were missed at first, and thus the methods are imperfect. As if confirming that, AlphaGene, a genomics company in Woburn, Mass., used the message technique to find 264 genes on chromosomes 21 and 22, the first chromosomes fully sequenced, that had been missed by the HGSC scientists.

Affymetrix, the leading manufacturer of DNA chips, has always made clear that its chips contain 60,000 genes or messages. Such chips are used to measure which genes are active, or "expressed" in a cell. Measuring which genes are turned on in a tumor cell but not in a healthy cell, for instance, could provide clues to the causes of cancer. The company would now begin producing chips using the completed genome sequence. Such chips will provide far more information than chips made using the messages. Rosetta Inpharmatics of Kirkland, Washington, concurs the usefulness of a similar technique.

No one could expect a text as vast and enigmatic as the human genome to yield all its secrets at first glance, and indeed it has not done so. But to Craig Venter of Celera, all this confusion just confirms his contention that the information produced until now is of limited value. The whole gene expression field is going to start over from scratch. He added that the principal purpose of the *Science* articles is to describe the sequence, and conferences of experts should be convened to help further interpret it. Eric Lander echoed the HGSC's analysis too is just preliminary and the consortium wrote the *Nature* paper that was not the last word on the genome but sketched all the directions researchers can go in. The goal was to launch a thousand ships, not to catalogue a thousand genes.

4.1 Humbled by Our Own Ego Genes

The two rivaling camps, Celera and HGSC, jointly released the authoritative *Science* and *Nature* reports of the human genome on February 15, 2001, the birthday of Charles Darwin, who jump-started our biological understanding of life's nature and evolution in "The Origin of Species" in 1859.

The fruit fly *D. Melanogaster*, the model organism of laboratory genetics, possesses 13,601 genes. The roundworm *C. elegans*, the staple of laboratory studies in development, contains only 959 cells of which 302 are neurons in what passes for its brain, looks like a tiny formless squib with virtually no complex anatomy beyond its genitalia, and possesses 19,098 genes. The general estimate for the number of genes of *Homo sapiens* — with 10 trillion cells in the body, including 100 billion brain cells, sufficiently large to account for the vastly greater complexity of humans under conventional views — had stood at well over 100,000, with a more precise figure of 142,634 widely advertised and considered well within the range of reasonable expectation. The *Science* and *Nature* reports indicate *Homo sapiens* possesses only between 30,000 and 40,000 genes, or that the humans have about one and half times as many genes as the worm. In other words, human bodies develop under the directing influence of those extra half as many genes as the tiny roundworm needs to manufacture its utter, if elegant, outward simplicity.

Knowing the number of genes of the worm, and its utter simplicity, it is difficult to conceive the human complexity generated by 30,000–40,000 genes under the old view of life embodied in what geneticists literally called their "central dogma" — a grotesquely oversimplified one direction of causal flow of "DNA makes RNA makes protein".[14] In other words, from code to message to assembly of substance. Each item of code — the gene, ultimately makes one item of substance — the protein, and the congeries of proteins make a body.

Those 142,634 messages no doubt exist, as they must to build our bodies' complexity if we accept the central dogma. Faced with the new facts of the *Science* and *Nature* reports, our previous error is now exposed as the assumption that each message came from a distinct gene, that is, the central dogma must fall.

To pole vault over this apparent discrepancy, we may envision several kinds of solutions for generating many times more messages (and therefore proteins) than genes. In the most plausible and widely discussed mechanism, a single gene can make several messages because genes of multicellular organisms are not discrete strings, but composed of coding segments (exons) separated by noncoding regions (introns). The resulting signal that eventually assembles the protein consists only of exons spliced together after elimination of introns. If in the process, some exons are omitted, or the order of splicing is permuted, then a distinct message can be generated by each permutation or alternative splicing.

The implications of the finding of a lower number of human genes cascade across several realms.[15]

The commercial effects are immediately obvious. Much of biotechnology - the rush to patent genes, diagnostic products, and others — has assumed the old view that "fixing" an aberrant gene would cure a specific human ailment.

Socially, we may finally liberate ourselves from the simplistic and harmful belief, which is false for many other reasons as well, that each aspect of our being, be it physical or behavioral, may be ascribed to the action of a particular gene for the trait in question.

But the deepest ramifications will be scientific or philosophical. Since the late 1600s, science has strongly privileged the reductionism mode of thought that breaks overt complexity into constituent parts. The totality is then explained by the

---

[14] F.H.C. Crick, "On protein synthesis", In: *Symposia of the Society for Experimental Biology*, (Cambridge University Press, Cambridge, 1958), pp. 138–153.
[15] Stephen J. Gould, "Humbled by the genome's mysteries", *New York Times*, February 19, 2001.

properties of these parts and simple interactions among them. This reductionism method works triumphantly for simple systems — two-body collision, planetary motions, but not the histories of their complex surfaces.

The fall of the doctrine of one direction of casual flow of "one gene for one protein", from basic codes to elaborate totality, marks the failure of reductionism for the complex systems, including biology. There are two major reasons.

❑ First, the key to human complexity is not more genes, but more combinations and interactions, and many of these interactions or emergent properties must be explained at the level of their appearance. So organisms must be explained as organisms, and not predicted from its constituent parts.

❑ Second, the unique contingencies of history, not the laws of physics, set many properties of complex biological systems. Our 30,000–40,000 genes make up only 1 percent or so of our total genome. The rest, including bacterial immigrants and transposons that can replicate and jump, originate more as accidents of history than as predictable necessities of physical laws.

Moreover, the noncoding regions, disrespectfully called "junk DNA", also build a pool of potential for future use that, more than any other factor, may establish any lineage's capacity for further evolutionary increase in complexity.

5   Genome — A Map To Profitability?

The prediction in the middle of Y2000 was that the 21st century is going to be the century of telecommunications and biotechnology. With the drastic downslide of technology stocks in the second half of 2000, and all the news splashes of biotechnological breakthroughs, investors have been turning in throngs to biotechnology, but with extreme cautions, especially after the dotcom hysteria and the subsequent demise.com.

Caught up in the ongoing technology stock wreck, biotechnology stocks have sold off despite the relative strength of other healthcare sectors and in the face of year 2000 enthusiasm.[16]   Although some market leaders have fared better, a broad index of biotechnology companies is down about 25% since peaking in September of 2000.   Other than investor sentiment, nothing much has changed since then. Biotechnology breakthroughs continue to grab headlines, and even though earnings are scarce for some of these companies, earnings warnings are even harder to come by.

In periods of uncertainty, people are less willing to pay for future earnings potential. The mindset that requires earnings sooner rather than later works against biotechnology stocks, which are mainly about the future. Drug and gene discoveries tend to capture headlines, but they are just the beginning of the road for biotechnology firms, which must then validate, perfect and market their products —

---

[16] Gayle Ronan, "Biotech for beginners", *CNBC*, March 5, 2001.

all with FDA regulatory approval. A new drug can take more than 10 years and $500 million to create.[17] That's rough even on an investor with a longer-term outlook, and it is definitely a disincentive for a momentum investor.

Further muting enthusiasm, most pharmaceutical products, about 80%, fail to ever exit the pipeline — a statistic that explains why so many biotechnology companies fold and why typical stock valuation methods often fail in this sector.

Most analysts in the sector believe this is the "Golden Era of Biology", and think investing in biotechnology is definitely worth the bother. Y2000 excitement over the human genome project is just a milepost in a multi-year cycle of discovery and the advancement of biology over the next 20 years is expected to be as dramatic as the last 20 years of advancements in electronics. In other words, the industry's fundamentals are improving, with the number of actual products coming to market accelerating along with profitability over the next several years. Microarray technology is a case in point. With concurrent advances in other supporting technologies, microarray products are moving into market.

Contrary to prior boom-or-bust environments, the number of products in late-stage development, new product approvals and companies reporting profitability, licensing and partnering deals are all historically high. That adds up to high potential rewards for patient investors.

But news releases and basic investing metrics, such as price-to-earnings ratio, are not particularly useful. The former may be a paid medium, and the reliability of the latter is questionable.

Table 2. Financial ratios of biotechnology companies on January 14, 2001.

Company	Ticker	Share price	Market cap (mm)	Revs 2000	Gain/ Loss	Mrk.cap/ Rev. ratio
Compugen	CPGN	6 1/4	158	3.6	-11.9	44
Curagen	CRGN	26 9/16	1,163	15.6	-27.7	74
DeCode Genetics	DCGN	9 1/2	424	14.0	-7.9	30
Exelixis	EXEL	9	418	17.7	-30.9	24
Gemini Genomics	GMNI	8 1/4	259	0.1	-9.7	2590
GeneLogic	GLGC	22 3/16	575	18.8	-8.0	31
Informax	INMX	9 15/32	184	11.4	-8.2	16
Myriad Genetics	MYGN	56	1,274	10.8	-2.1	118
Paradigm Genetics	PDGM	7 7/8	203	2.2	-10.6	92
Rosetta Inpharmatics	RSTA	13 7/16	427	4.9	-25.9	87

Assuming investing in biotechnology stocks is worth the bother, but news releases and financial metrics are not particularly useful, how does one evaluate biotechnology stocks?

---

[17] Hwa A. Lim, "Bioinformatics and cheminformatics in the drug discovery cycle", In: *Lecture Notes in Computer Science, Bioinformatics*, R. Hofestaedt, T. Lengauer, M. Loffler and D. Schomburg (eds.), (Springer, Heidelberg, 1997), pp. 30–43.

A safe bet is to look at larger companies first since these companies have measurable profits and revenues. It also helps to understand what a company's products in the pipeline are and what diseases they will treat. This will enable an investor to decide if the company makes sense both in terms of science and market opportunity. The next step is to assess whether the market opportunity will be big enough to justify the market capitalization of the company.

In evaluating pharmaceutical companies, it is important to know where in the pipeline the promising new drugs are. If they are still in or about to enter Phase I testing, the investor will likely face a 15-year wait and a high probability (80%) that the drug will never make a penny. Compare this to Phase III drugs, which are roughly 12 to 18 months away from becoming products and whose success is more probable.

Evidence of good management is another way of separating real earnings potential from easily deflated hype. Brilliant scientists are a requirement for a successful biotechnology company, but brilliant science does not always translate to brilliant business acumen. Look for executives, or announcements of executives joining the company, who were formerly with either more established, profitable biotechnology firms or big pharmaceutical companies. An A team hires A team players.

Another tip for evaluating biotechnology companies is to ride the coattails of those in the know. Look at which top ten biotechnology companies and other heavy hitters are licensing from or partnering with. Their involvement indicates a third party validation that the science is likely valid and the market potential is worthwhile. The risk of failure remains, but the odds of success are greatly enhanced.

Despite long-term enthusiasm for this sector, no one is forecasting less market volatility, which has been the case all along for the biotechnology sector.

5.1   Biotech Bay, Biotech Beach, Biotech Beltway, Biotech Corridor

As average people have become more conversant with the latest biotechnology efforts, more and more cities and regions are injecting more energy, if not more monies, into biotechnology.[18]

The troubling economy of Y2001 is not dampening investor enthusiasm. According to a latest MoneyTree survey, despite a significant drop in venture capitalists' investments overall in the United States, venture investments into life sciences and healthcare industries rebounded in the second quarter of Y2001. Biopharmaceuticals led U.S. venture growth by attracting $500 million in equity investment in the second quarter of Y2001 alone. Biotechnology investors are not typical high-risk rollers like wildcatters or high-tech investors. The industry's

---

[18] "Biotechnology Dallas needs to inject more energy into effort", Dallas Morning Post, September 2, 2001.

investors must have not only the ability to absorb losses, but also a sophisticated understanding of potential commercial applications as well as saintly patience. Commercialization of biotechnology research can take years, as many as ten years for pharmaceuticals. For every 10 efforts to capitalize on research, 8 efforts are unsuccessful, one breaks even, and if lucky, one is a clear winner.

Because private returns are apt to be much more certain if one is looking for an extension of existing knowledge (cf drilling a developmental well) than if one is looking for a major breakthrough (cf drilling a wildcat well to find a new big oil field), private firms tend to concentrate their money on the developmental end of R&D process. Time lags are also shorter, and in the business sector, speed is everything.

For the risk, investors as well as the region profit. The biotechnology industry offers host cities more high-skill, high-wage jobs, increased diversification of the local economy into an industry set for long-term growth, and a higher tax base from a relatively clean industry. Leading biotechnology regions include Silicon Valley (Biotech Bay), San Diego (Biotech Beach), Boston Area or Route 128, North Carolina's Research Triangle, Seattle, and Washington DC area.

Those thinking about investing in research and development have an incentive to wait to see what they can get for free and skip the risky phases of investment. Because of this proclivity, they prefer to dive straight into the clear path of development. Capturing this free knowledge is one of the reasons that concentations of high-tech companies exist in places like Silicon Valley and Route 128.[19] One learns sooner if one is in fact a neighbor. Knowledge, like fluids, finds its own level and eventually equalizes geographically. But eventually is not instantly. In fast-moving fields, the advantages of being inside the relevant learning communities are enormous. Companies establish windows, offices and subsidiaries in Silicon Valley to have listening posts. Other locations have much lower operating costs, but the same information is not as readily accessible.

5.2   Biotechnology Tools: Friends or Foes?

The biotechnology revolution has been touted as the transformation that would speed sluggish drug development times and boost efficiency in the pharmaceutical industry. A study finds so far it has actually slowed the process down. The finding comes from a new study at the Tufts Center for the Study of Drug Development.[20]

The report, based on data of the clinical drug development process during the last two decades, finds that the time spent on clinical trials has surged upward by nearly 80 percent since the mid-1980s, from an average of 33 months to an average

---

[19] Adam B. Jaffe, Patent citations and the dynamic of technological change", *Reporter*, Summer 1998, National Bureau of Economic Research.

[20] Kathleen McGowen, "Biotech tools are slowing down drug development process, study shows", *GenomeWeb*, November 14, 2001.

of 68 months, while the average time for regulatory approval by the U.S. Food and Drug Administration has shortened by about 40 percent, from 24 to 15 months.

The report author Janice Reichert, a senior research fellow at the Tufts Center explains that in the future development times may come down as a consequence of the use of these techniques, though at the moment, the techniques have contributed to lengthening the development times.

There are various causes, including the focus on more complicated diseases that have previously defied drug treatment, the demand for safer and more efficient drugs, and the scientific exploration of new drug mechanisms.

Other plausible causes include quirks of the industry. Some biotechnology drugs may bounce from company to company as they track through clinical trials, lengthening the overall development time. In other cases, a small company might quickly put a drug through phase I clinical trials as proof of concept, but spend years rounding up the cash to get the drug into expensive phase III trials.

The barrage of new techniques and new targets are also to blame. It may take a while to weed through the new techniques and targets. At this moment, people are in the learning phase. People are trying to figure out what things mean, such as what it means to have a number of mutations in one gene. What is reasonable and sufficient to design a diagnostic, and so on.

So much information is now flooding into the discovery stage that genomics may actually push development times up for a while. Then, as these technologies become more robust and deliver their promise — more intelligently designed diagnostics and the identification of better drug targets — the process will speed up again. Reichert predicts with current drug development times ranging around five years, acceleration of the process may not materialize for another eight to ten years.

Genomics will not likely to have a major impact on the most time-consuming part of the process: grinding through clinical trials. The genome is not going to revolutionize the process. It may revolutionize the information going into it, but not the process. It still takes the same amount of time to get through a phase I study. If a phase I requires a six-month study with a one-year follow up, it still takes six months and one year to do it.

6   The Other Sequencing Centers

Sequencing is not only for the United States and United Kingdom. Some of the biggest new customers of Amersham Pharmacia Biotech's MegaBACE 1000 DNA analyzer in Y2000 are in Asia and South America.[21]

The Organization for Nucleotide Sequencing and Analysis (ONSA), Brazil is a virtual sequencing operation that involves instruments dispersed across some 61 laboratories scattered over 13 cities in the state of Sãn Paulo. The enterprise is a

---

[21] Adrienne Burke, "Monster sequencing show", *Genome Technology*, April 2001, pp. 33.

U.S. Department of Agriculture contract to sequence and annotate a strain of *Xyllela fastidiosa* pathogen that plagues grape crops in California.[22]

The Beijing Genomics Institute (BGI), China, made its name sequencing a part of chromosome 3 as one of the 16 members of the International Human Genome Consortium. In October 2000, the Institute opened a branch in Hangzhou. The two locations, Beijing and Hangzhou are churning out 10 megabases a day. Aside from finishing chromosome 3, BGI has embarked on three new projects: finding single nucleotide polymorphisms (SNPs) of the Chinese population, shotgun sequencing the pig with a Danish consortium, and sequencing the *Indica* strain of rice.[23]

decode Genetics, Reykjavik, Iceland, is backed by Roche, a Swiss pharmaceutical company and staffed by scientists from 25 countries. It currently spreads over 5 locations. Six of the 56 ABI machines are dedicated to the good old-fashioned sequencing to search for mutations in the Icelandic population. The bulk of the machines are employed in 24/7/365 conducting genome-wide scans for microsatellite markers.[24]

Almost 16 years after Akiyoshi Wada had publicly promoted the idea of sequencing supercenters,[25] in a classic case of *gaiatsu* or outside pressure, the Japanese decided to back the project. RIKEN Genomics Science Center started in October 1998. GSC is about 30 km south of Tokyo in the city of Yokohama, made its name sequencing chromosome 21 as one of the 16 members of the International Human Genome Consortium. There are six groups working on mouse cDNA, chromosomes 11 and 18, and protein folds.[26]

Table 3. Large sequencing centers outside of the United States and United Kingdom. (Adapted from *Genome Technology*).

	Facility	Staff	Capacity	Projects	Remarks
ONSA, Brazil	11 MegaBACE, 8 ABI 3700, 41 ABI 377	350		Pathogens, bacterial genomes	61 labs in 13 cities
BGI, China	70 MegaBACE, 11 ABI 377	400	10 Megabase or 50,000 reactions per days	Rice, pig, human chromosome 3, SNP of Chinese population, proteomics of traditional Chinese medicines	2 locations

[22] Meredith Salisbury, "Genome carnival", *Genome Technology*, April, 2001, pp. 34–35.
[23] Aaron J. Sender, "Great spiral forward", *Genome Technology*, April 2001, pp. 36–37.
[24] Adrienne Burke, "Genotyping lagoon", *Genome Technology*, April 2001, pp. 38–39.
[25] Akiyoshi Wada, "The practicability of and necessity for developing a large-scale DNA-base sequencing system: Toward the establishment of international super DNA-sequencing centers", In: *Biotechnology and the Human Genome: Innovations and Impact*, Avril D. Woodhead, and Benjamin J. Barnhart (eds.), Plenum Press, New York, 1987), pp. 119–130.
[26] Sara Harris, "Razor sharp RIKEN", *Genome Technology*, April 2001, pp. 40–41.

Table 3. (Continued)

DeCode, Iceland	56 ABI 3700 of which 6 are for sequencing, 50 are for microsatellite detection	450	500,000 genotypes per day	Genotypes of the 300,000 Icelanders	Study 40 common diseases
RIKEN, Japan	31 RSA, 6 ABI 377, 13 ABI 3700, 10 LiCor, 18 MegaBACE	90	42.6 Megabase per 20-hour day	Mouse cDNA, human chromosome 11 and 18, polygenic adult diseases	

7   From The Book Of Life To Inventory Of Life

Now the first draft of the human genome is available and several industries have spun off the project, a natural next big biological science project will be the Human Proteome Project — a project to catalogue and characterize all the proteins in the human body.

Other audacious ideas are also being considered. Of particular note is the All-Species Project (ASP) — a project to catalogue all life on Earth within 25 years, the equivalent of a human generation.

The task is daunting. The scope of the project may be seen in the fact that in past centuries, scientists have identified and named only 1.8 million of the estimated 10 to 100 million extant species.

A comparison of All-Species Project (ASP) with the Human Genome Project (HGP) can put the projects in perspective. Fifteen years ago, the notion of sequencing the entire human genome was seen as essentially impossible. But ASP goals dwarf even that of HGP. HGP was a more focused challenge with an endpoint. ASP is very different because there are many challenges. For example, over the past 300 years, biologists have described only 2 percent of the world's biodiversity.

Then why ASP? An advisor to All-Species and director of the School of Biological Sciences at the University of Texas, Austin, David Hillis argues that asking why knowing all the species makes a difference is similar to asking why knowing all the chemical elements is important. Without knowledge of all the chemical elements, the predictive aspect of chemistry would be limited. Without knowledge of all the species, the predictive aspect of biology is limited.[27]

---

[27] Ricki Lewis, "Inventory of life", *The Scientist* 15(15), July 23, 2001.

Chapter 3

AGBIOTECHNOLOGY: GENETICALLY MODIFIED

"Twenty-five years ago, Herb Boyer and Bob Swanson had the incredible insight, bold genius, and unwavering commitment to start a different kind of company – one based on the belief that recombinant DNA technology would produce commercially viable, breakthrough medicines within a relatively short period of time. On April 7, 1976, the two founded Genentech – and in doing so launched the biotech industry. The adventure began."

Extracted from the website of Genentech celebrating its twenty-fifth anniversary.

1   The Biotech Revolution

In certain sense, biotechnology has been contributing to human well-being for thousands of years. It began with the use of microorganisms to ferment foods during the dawn of civilization. In the present era, biotechnology is being used to produce wonder drugs such as antibiotics, anthelmintics, and cholesterol-lowering agents. This is the "old biotechnology".

The technology has been instrumental in the development and implementation of processes for the manufacture of antibiotics and other pharmaceuticals, industrial sugars, alcohols, amino acids and other organic acids, foods, and specialty products through the application of microbiology, fermentation, enzymes, and separation technology.

Engineers, working with life scientists, often achieve scale-up to industrial production in remarkably short periods. A relatively small number help to catalyze, over a period of fifty years, the growth of the pharmaceutical, food, agricultural processing, and specialty-product sectors of the U.S. economy to the point where annual sales exceeded $100 billion in the early 1990s.[1]

The "new biotechnology" of the early 1970s enabled direct manipulation of the cell's genetic machinery through recombinant DNA technique and cell fusion. Its application on an industrial scale since 1976 has fundamentally expanded the utility of biological systems. This has also positioned a number of industries for explosive global growth. The principal impact of the new biotechnology has been in the pharmaceuticals arena.

Scientists and engineers can now change the genetic make-up of microbial, plant, and animal cells to confer new characteristics. Biological molecules, for which there is no other means of industrial production, can now be biologically manufactured or biofactured. Existing industrial organisms can be systematically

---

[1] *Putting Biotechnology to Work*, National Research Council, (National Academy of Sciences, Washington, DC, 1992).

altered or genetically engineered to enhance their function and to produce useful products in new ways. The new biotechnology, combine with the existing industrial, government, and university infrastructure in biotechnology and the pervasive influence of biological substances in everyday life, has set the stage for unprecedented growth in products, markets, and expectations.

In its broadest sense, biotechnology — as defined in a 1991 Office of Technology Assessment (OTA) report — is any technique that uses living organisms or parts of organisms to make or modify products, to improve plants or animals, or to develop microorganisms for specific uses.[2,3]

Thus the OTA definition encompasses both the old biotechnology — production of organic acids and antibiotics by fermentation, and the new biotechnology — processes involving microorganisms, plants, or animals that have been modified by recombinant DNA or by other genetic manipulation techniques.

In a wordplay, we see that the word biotechnology is actually an amalgamation of three words: bi –o' – technologies, or a combination of two older technologies.

2   Mother Nature, Father Time and Author Children

The growing arsenal of biotechnology is providing us with powerful new tools to engage in radical experiment on Mother Earth's life forms and ecosystems.[4] Biotechnology is now capable of the wholesale transfer of genes between totally unrelated species and across all biological boundaries — microorganisms, plants, animals, and humans — to create thousands of novel life forms in a brief moment in evolutionary time. That is to say, modern biotechnology can author Mother Nature by shortening Father Time.

Corporate leaders in the new life sciences industry promise a new era of history where evolution is itself a subject to human authorship.   Intrinsic in clone propagation, mass-producing replicas of new creations and use of the creations is the risk of accidental releasing them into the biosphere to propagate, mutate, proliferate, migrate, and to colonize land, water and air. This will be the great scientific and commercial experiment in this new biotechnology century.

Human beings have been remaking Earth for as long as we have had a history. Until recently, however, our ability to create our own have been tempered by the restraints imposed by species boundaries. We have been forced to work narrowly, continually crossing close relatives in the plant or animal kingdoms to create new varieties, strains and breeds. Through a long historical process of tinkering with trial and error experiments, we have redrawn the biological map, creating new

---

[2] *Commercial Biotechnology: An international analysis*, Office of Technology Assessment, U.S. Congress, Report No. OTA-BA-218, 1984.

[3] B. Brown (ed.), *Biotechnology in a Global Economy*, Office of Technology Assessment, U.S. Congress, Report No. OTA-BA-494, (U.S. Government Printing, Washington, DC, 1991).

[4] Jeremy Rifkin, *The Biotech Century*, (Tarcher/Putnam, New York, 1998).

agricultural products, new sources of energy, more durable building materials, life-saving pharmaceuticals, and other useful products. Still in all this time, nature dictated the terms of engagement.

The new gene splicing technologies allow us to break down the walls of nature, making the very innards of genome vulnerable to a new kind of human colonization. Transferring gene across all biological barriers and boundaries to create genetically modified organisms (GMO) is a technology unprecedented in the human history. We are experimenting with nature in ways never before possible, creating unfathomable new opportunities for society and concurrently, potentially generating grave new risks for the environment.

Critics worry about reseeding the biosphere with these genetically modified organisms can lead to a different future. The cause of concern is not unfounded. The Industrial Revolution brought in petrochemical products and petrochemical pollution. Information technology brought in infoware and infollution. Now genetic revolution brings in genetically modified organisms and is ushering in genetic pollution. Genetically modified organisms (GMO) differ from petrochemical products in many ways. While petrochemical products cause pollution to our environment, GMO may cause genetic pollution to our biosphere. Because they are alive, GMO are inherently more unpredictable than petrochemicals in the way they interact with other living organisms in the environment. Consequently, it is much more difficult to assess the long term potential impacts of GMO on Earth's ecosystems. GMO also reproduce and migrate. In contrast to petrochemicals, they are far more difficult to constrain to within a given geographical locale. Once released, it is virtually impossible to recall GMO back to the laboratory, especially those that are microscopic in nature. For these reasons, GMO may pose far greater long-term potential risk to the environment if they are not properly monitored or handled.

3   From Green Revolution To AgBiotech Revolution

The world population is growing rapidly, and one third of all crops are still lost to pests and diseases. The Green Revolution of the 1960s, which depended on agrochemicals, achieved a doubling of production with only a 10-20% increase in the amount of land under cultivation. But as the world population rises to 10 billion over the 21st century, the Green Revolution will soon reach its saturation point. Its impact on the environment is also increasingly unacceptable.[5]

Sir Robert May, chief scientific adviser to the British government, warned at the AgBiotech'99 Biotechnology and World Agriculture that "...we will not be able to feed tomorrow's population with today's technology. We will now need to create crops that are shaped to the environment, with biotechnology, whereas before, with

---

[5] Susan Aldridge, "The Agbiotech Revolution", *Genetic Engineering News*, 20(1), January 1, 2000.

the Green Revolution, our environment was shaped by crops that were created with the use of chemicals derived from fossil fuels." Sir May also emphasized that The Green Revolution was built with public money, whereas genetic modification is all being done with private money, with farmers and agribusiness seen by the consumer as the chief beneficiaries of the new technology. What may be required is a reorientation through public-private partnerships.

Most soldiers in the biotechnology revolution believe that the public will eventually accept genetically modified foods, thereby ending hostilities. However, science must first offer something of value, such as improved nutrition — functional food. Just making life easier for farmers with pest-resistant crops will not outweigh real or imagined risks to people.

3.1   *Genetically Modified Crops*

Chemical and agribusiness companies are introducing a new generation of transgenic crops into agriculture with the hope of making a wholesale shift into the new genetic revolution. The biotech crops contain novel genetic traits from other plants, viruses, bacteria, and animals, and are designed to perform in ways that could never have been possible by scientists working with classical breeding techniques.

3.1.1   Yield Increase via Genetics

Scientists have succeeded in inserting "antifreeze" protein genes from flounders into the genetic code of tomatoes to protect the fruit from frost damage. Chicken genes have been inserted into potatoes to increase disease resistance. Firefly genes have been inserted into the biological code of corn plants to serve as a genetic marker. Chinese hamster genes have been inserted into the genome of tobacco plants to increase sterol production.[6]

Ecologists are unsure of the impacts of bypassing natural species boundaries by introducing genes into crops from wholly unrelated plant and animal species. The fact is there is no precedent in history for this kind of experimentation. For more than ten thousand years, classical breeding techniques have been limited to the transference of genes between closely related plants and animals that can sexually interbreed, limiting the number of genetic combinations. Natural evolution appears to be circumscribed. As a result, there is little or no precedence for what might occur in the wake of a global experiment to redefine the fundamental rules of biological development to suit the needs of market-driven forces.

Much of the current effort on agricultural biotechnology is centered on the creation of herbicide-tolerant, pest-resistant, and virus-resistant transgenic plants. More than a third of the entire field releases in 1993–1994 in the Organization for Economic Cooperation and Development (OECD) nations involved herbicide-

---

[6] Jane Rissler, and Margaret Mellon, *The Ecological Risks of Engineered Crops*, (MIT Press, Cambridge, 1996).

tolerant plants, while 32% of the field trials involved pest-resistant plants, and 14% of the test releases were virus-resistant plants.[7]

Herbicide-tolerant crops are a favorite because chemical companies can increase their share of the growing global market for herbicides. By creating transgenic crops that can tolerate the company's own herbicides, the company can increase its share of both the seed and herbicide markets. An example is Monsanto's herbicide-resistant patented seeds, which are resistant to Monsanto's best-selling chemical herbicide, Roundup. Chemical companies hope to convince farmers that the new herbicide-resistant crops will allow for a more efficient eradication of weeds since farmers will be able to spray at any time without having to worry about killing their crops. Because there is no fear of damaging their own crops, farmers are also more likely to use greater quantities of herbicides to control weeds. This resulting increase use of herbicides raises the possibility of weeds developing resistance, forcing an even greater use of herbicides to control the more resistant strains.

Virtually all of the pest-resistant crops contain a gene from a naturally occurring soil bacterium, *Bacillus thuringiensis*. The bacterium produces a crystalline protein known as Bt prototoxin. When the toxin is ingested by larvae or insects, it is activated by the pest's stomach acid and destroys their digestive tracts. Unlike the naturally occurring prototoxin, a transgenic toxin has been altered so that it becomes active immediately upon production by the plant. As it does not have to be activated by stomach acids, it can destroy a wider range of insects and soil organisms. The transgene also remains toxic up to three times longer in the soil, making it far more lethal than its naturally occurring counterpart.

The unique quality of pest-resistant transgenic plants make them especially troubling to entomologists and organic farmers who worry that the widespread use of Bt crops will build resistance among affected insect species, rendering Bt useless as a bio-pesticide. Their worry is well founded for resistance to Bt bio-pesticide first showed up in the 1980s. Since that time, more than eight major species of destructive insects have developed resistance to Bt toxin.

3.1.2   Genetically Fortified Crops — Golden Rice

Biotechnology may be a solution to world hunger. Metabolic or nutritional genomics is using genes to improve nutritional value of plants. For example, rice is a relatively poor source of many essential nutrients, including vitamin A and iron, but is a staple food for half the world. An estimated 124 million children worldwide are deficient in vitamin A, including a quarter million in Southeast Asia who go blind each year because of the deficiency. Improved nutrition can prevent 1 to 2 million deaths a year.

---

[7] Ricarda A. Steinbrecher, "From green to gene revolution: the environmental risks of genetically engineered crops", *Ecologist*, November/December 1996, pp. 277.

Vitamin A and iron deficiency affect 24% and 14% of the world's population, respectively. Rice endosperm does not make chlorophyll or β-carotene, giving it its natural white color. It is naturally low in both iron and provitamin A. Furthermore, rice is high in phytate, which inhibits iron resorption in the intestine, and low in sulphur-containing proteins, which enhance iron resorption.

At a Swiss laboratory, the International Rice Research Institute (IRRI), Zurich's Federal Institute of Technology (ETH), Ingo Potrykus and his colleague, Peter Beyer of the University of Freiburg, Germany, have succeeded in splicing three genes into rice to make it iron-enriched and four other genes to make it rich in β-carotene, a source of vitamin A.[8]

The ETH team aims at both to increase the iron content of the rice by transferring a ferritin gene from *Phaseolus* and to reduce phytate by transfer of a gene for a thermotolerant phytase [note phytate vs phytase] from *A. fumigatus*. Finally, a gene for a cysteine-rich metalloprotein from *Oryza* is also transferred. The transgenic rice thus created has a two-fold increase in iron, high phytase activity, and a 25% increase in cysteine content.

The provitamin A metabolic pathway in rice terminates prematurely, with a precursor called GGPP (geranylgeranyl diphosphate). The first gene, from daffodil, encodes for phytoene synthase, which combines two 20-carbon compounds GGPP into colorless 40-carbon phytoene. A second desaturase enzyme encoded by a bacterial gene introduces four double bonds to make red-colored lycopene (plants normally need two enzymes to do this job). The third enzyme, a daffodil cyclase completes the pathway and transforms GGPP to provitamin, producing plants with a yellow endosperm. The yellow color is due to provitamin A, synthesized in levels that would provide the necessary dose of vitamin A. Rice naturally produces its own β-carotene, but it is lost in the milling process. The engineered variety will have the β-carotene right in the endosperm, the part people eat. Just 300 grams of golden rice each day should prevent vitamin A deficiency.

The iron-enriched rice and vitamin A-enriched rice are combined by crossing. The new crop, dubbed "golden rice" because of the hue the β-carotene gives to it, is not expected to be available to farmers for several years. Nevertheless, IRRI is already working on breeding the new trait into popular varieties.[9]

3.1.3   Plant Antibody Factory

AgBiotech can deliver more than just food. Biologically based plants have also been engineered to produce medicines and other products. For example, corn has been engineered to produce human mucosal antibodies used in passive

---

[8] Xudong Ye, et al., "Engineering the provitamin A (β-carotene) biosynthetic pathway into (carotenoid-free) rice endosperm", *Science* 287, 2000, pp. 303–305.

[9] Philip Brasher, "Geneticists find way to add vitamin A to rice", *San Francisco Chronicle*, January 14, 2000.

immunization. There is a market force to drive down the production cost of therapeutic antibodies as new indications are discovered. Plants are an alternative that can meet the demand. A single ear of corn can produce 280 mg of antibody at a cost of about $20–30 per gram. The corresponding cost of antibody manufactured fermentation is $200–300.

3.1.4 Petro Alternatives

There is also a growing interest in using plants as an alternative to petrochemicals. The current world oil reserves are expected to deplete in 60–140 years, peaking at around 2020. In 50 years or so, there will be a great need for seed-oil producing crops for making products like biodegradable plastics and cosmetics that are based on fatty acids stored in seeds. Most natural crop plants make only a limited range of fatty acids, whereas a wide range is needed to get the variety of products we have. However, there is a huge spectrum of seed oils present in nature. Thus there are two ways to achieve the goal. First genes for desirable fatty acids can be transferred to crop plants. The current status is that many key genes are known, but the major impediment is the lack of knowledge of how lipid biosynthesis is regulated in the plant. The key gene does not always have the expected effect when it is transferred. For example, the fatty acid may be broken down instead of going into storage.

Second, plants that naturally produce the desirable seed oils can be domesticated. Until recently, this has been a long process. For example, wheat took thousands of years to domesticate. However, advances in plant genomics may accelerate the process of domestication. It is now known that agronomics traits is governed by a few genes, and it is likely that the time scale for domestication can be reduced to a few decades.

3.2 Genetically Modified Trees

First there were herbicide-tolerant soybeans and worm-resistant corn. Despite the controversy swirling around such genetically engineered crops, they are now planted on millions of acres of U.S. farmland. And there are growing indications biotech trees will be next.[10]

Three of the world's largest paper and lumber producers have formed a joint venture, called ArborGen, that hopes to be the first group to commercialize genetically modified trees. ArborGen expects to seek approval from the U.S. Department of Agriculture and other regulatory bodies for mass planting by 2005. Based in Summerville, South Carolina, the company has access to the vast research base of its corporate parents: International Paper, Westvaco, and Fletcher Challenge Forests, plus a New Zealand-based genomics company, Genesis Research and Development.

---

[10] Kathryn Brown, "Industry hugs biotech trees", *MIT Technology Review*, March 2001.

What trees are likely first candidates? ArborGen is not saying. But the venture has access to several technologies for tree modification. Among them are ways to suppress genes that produce lignin, a plant component that must be chemically removed to make paper. Other technologies include ways to make trees herbicide-tolerant and to minimize crossbreeding risks by controlling their ability to reproduce. Trees headed for the market from ArborGen include genetically modified hardwoods like poplar and sweet gum, used for high-quality paper products. ArborGen also believes it can stock plantations of modified pine and eucalyptus trees, workhorse species that can be used for building products or ordinary paper pulp.

Table 1. A sample of U.S. research on genetically modified trees. (Source, USDA).

Organization	Species	Desired Trait
Michigan Tech, MI	Poplar	Reduced lignin content
Westvaco, NY	Pine, sweet gum	Altered plant development
Exelixis Plant Sciences, OR	Apple	Altered fruit ripening
International Paper, NY	Sweet gum	Herbicide resistance
University of Wisconsin, WI	Poplar, spruce	Resistance to moths
Cornell University, NY	Apple	Disease resistance
Oregon State University, OR	Poplar	Insect resistance, herbicide tolerance, reproductive control

Environmental groups are already concerned. There is a difference between genetically modified crop plants and trees, in the sense that trees are perennial and live a long time. The question is whether tree roots will pump out a genetically modified toxin for years? And what will happen if insect-resistant trees spread genes to relatives that flourish outside, in unmanaged ecosystems? Like in other instances of genetically modified organisms, there are many ecological issues with modified trees that need to be carefully studied.

To prepare itself for the emerging debates over biotech-tree safety, the industry has joined with the state of North Carolina — a center for both biotech research and the $3.2 billion forest-products industry — to form the Institute of Forest Biotechnology, a Research Triangle Park, North Carolina-based think tank chartered to promote the societal benefits of modified trees and address environmental concerns. Steven Burke, the institute's director, argues fast-growing, efficient stands of modified trees can better serve the needs of the paper and wood industries while reducing chemical pollution from papermaking and leaving more wild forests alone. The think tank will have a lot to think about: ArborGen is not the only group developing biotech trees. In the past decade, the USDA has given 136 approvals, most in the last three years, for small outdoor test plots, some of which include fruit trees. David Wheat of the Bowditch Group, a Boston-based agricultural technology consulting firm, says work is also progressing in laboratories and testing fields as far flung as Canada, Australia and South Africa. Genetically modified trees have already reached the point of practical application but will be phased into the commercial sphere gradually partly because of regulatory and public concerns, and partly because genetically modified trees are not planted on million of hectares the first year out. The prediction is that genetically engineered paper will probably not hit the market for another 10 years.

3.3 Genetically Engineered Animals

The risks in introducing novel genetically modified organisms (GMO) into the biosphere are reminiscent to what we have encountered in introducing exotic organisms into the North America habitat. Over the past several hundred years, thousands of non-native organisms have been brought to America from other regions of the world. While many of these organisms have adapted to the local ecosystems without severe dislocations, a small percentage of them have run wild, wreaking havoc on the flora and fauna of the continent. For example, the mongoose, introduced into Hawaii from India to control rodents that were damaging the sugar cane crop, became an environmental disaster. They devour a wide range of native animals, and in the process, destabilize the ecosystems of the island. The zebra mussel is another example. A native of Europe, it migrated to North America by attaching itself to ships and has become a formidable pest in the Great Lakes, blocking up water pipes at filtration plants and edging out native species.[11]

A number of experiments are already underway to release genetically modified animals into the environment, including predator insects that will prey on noxious insects and genetically engineered fish with growth hormone and antifreeze gene inserted into their genetic codes to allow them to grow faster and bigger and be able to tolerate colder waters. These fish are genetically engineered to increase the efficacy of food conversion, tolerate cold and salinity, and be disease resistant. In the wild, they might have selective advantage because they can potentially out-compete native fish species and create havoc. Even the accidental release of sterile males can pose unanticipated problems. If the transgenic species are larger and stronger as a result of the addition of a gene to produce increase growth hormone, they might secure easier access to female eggs, crowding out the native males, and because they are sterile, seriously deplete the indigenous fish population.[12,13]

---

[11] Bernard Rollin, *The Frankenstein Syndrome: Ethical and Social Issues in the Genetically Engineering of Animals*, (Cambridge University Press, New York, 1995).
[12] Sue Mayer, "Environmental threats of transgenic technology", In: Peter Wheale and Ruth McNally, eds., *Animal Genetic Engineering of Pigs, Oncomice, and Men*, (Pluto Press, London, 1995), pp. 128.
[13] Hu Wei, and Zhu Zaoyan, "Transgenic fish and biosafety", In: G.X. Xue, Y.B. Xue, Z.H. Xu, Roger Holmes, Graeme Hammond, and Hwa A. Lim, eds., *Genes and Gene Families: Studies of DNA, RNA, Proteins and Enzymes*, (World Scientific Publishing Co., New Jersey, 2001), Chap. 5.

Figure 1. A genetically engineered Atlantic salmon (top), compared to a same-age control salmon shows the larger salmon reached full market size in about half the time normally expected. (Photo: courtesy of Associated Press).

3.4   Genetically Modified Microbes

Scientists have been considering the possibility of producing a genetically modified enzyme (GME) that can destroy lignin, an organic substance that makes wood rigid. They believe there may be great commercial advantage in using the GME to clean up the effluent from paper mills or for decomposing biological material for energy. But if the bacteria containing the GME should stray offsite, it will likely end up destroying millions of acres of forests because the enzyme will eat away the substance that provides trees their rigidity.

The first government-approved release of a genetically engineered organism into the open environment is a modified bacterium called *Pseudomonas syringae*. This particular bacterium is found in its naturally occurring state in temperate regions all over the world. Its most unique attribute is its ability to nucleate ice crystals. In the early 1980s, using recombinant DNA technology, University of California researchers found a way to delete the genetic instruction for making ice from the bacteria. This new genetically modified *P. syringae* microbe is called ice-minus.[14] Scientists were excited about the long-term commercial possibilities of ice-minus in agriculture. Frost damage has long been a major problem for American farmers. The chief culprit has been *P. syringae*, which attaches itself to plants, creating ice crystals. It was hoped that by spraying ice-minus on agricultural crops, the naturally occurring *P. syringae* would be edged out, thus preventing frost damage. The benefits of introducing ice-minus appeared impressive. But when one weighs the potential long-term ecological costs, then one quickly realizes that the burden can be very high.

---

[14] Steven E. Lindow, "Methods of preventing frost injury caused by epiphytic ice-nucleation-active bacteria", *Plant Disease*, 1983, pp. 327–333.

Naturally occurring *P. syringae* helps shape worldwide precipitation patterns and is a key determinant in establishing climatic conditions on the planet. Thus, if ice-minus were to be released over million of acres for a sustained period of time, replacing ice-nucleating agent (INA) *P. syringae*, the effect of global climatology can be enormous.

3.5 Germ Warfare Agents

Biological weapons have never been widely used because of the danger and expense involved in processing and stockpiling large volumes of toxic materials and the difficulty in targeting the dissemination of biological agents. Advances in genetic engineering technologies over the past decade, however, have made biological warfare viable for the first time.

Breakthroughs in genetic engineering technologies provide a versatile form of weaponry that can be used for a wide variety of military purposes, ranging from terrorism and counterinsurgency operations to a large-scale warfare aimed at an entire population. Unlike nuclear technologies, genetic engineered organisms can be cheaply developed and produced. They also require far less scientific expertise, and can be effectively employed in many diverse settings. These factors rekindle military interest in biological weapons. But at the same time, it also generates grave concern that an accidental or deliberate release of harmful genetically engineered microbes can spread genetic pollution around the world, creating deadly pandemics that destroy plant, animal, and human life on a mass scale. Biological warfare (BW) involves the use of living organisms for military purposes. The weapons can be viral, bacterial, fungal, rickettsial, and protozoan. The agents can mutate, reproduce, multiply, and spread over a large geographical terrain by wind, water, and by insect, animal, and human transmission. Once released, biological pathogens are capable of developing viable niches and maintaining themselves in the environment indefinitely. Conventional biological agents include *Yersinia pestis* (plague), tularemia, rift valley fever, *Coxiella burnetii* (Q fever), eastern equine encephalitis, anthrax, and smallpox.

Recombinant DNA designer weapons can be created in various ways. The new technologies can be used to program genes into infectious microorganisms to increase their antibiotic resistance, virulence, and environmental stability. It is also possible to insert lethal genes into otherwise harmless microorganisms, resulting in biological agents that the body recognizes as friendly and does not resist. It is even possible to insert genes into organisms that affect regulatory functions that control mood, behavior, and body temperature. It is also possible to clone selective toxins to eliminate selective racial or ethnic groups whose genotypical makeup predisposes them to certain disease patterns. Genetic engineering can also be used to destroy specific strain or species of agricultural plants or domestic animals, if the intents are to cripple the economy of an adversarial country.

Two factors make the threat of a bioterrorist attack greater than ever before:[15]

❑ First, the unspoken taboo that previously dissuaded terrorists from using chemical or biological weapons against civilians has now been broken. On 20 March 1995, the nihilistic Japanese cult Aum Shinrikyo unleashed nerve gas on the Tokyo subway, killing 12 people and hospitalizing five thousand. Aum was also developing biological weapons.

❑ Second, with the explosive growth of basic biological research and biotechnology, what was once regarded as esoteric knowledge about how to culture and disperse infectious agents has spread among tens of thousands of people.

Many experts say that it is no longer a question of whether a major bioterrorist attack will occur, but when. Indeed, just days before the start of the air war on January 17, 1991, Iraq had moved 157 bombs filled with botulinum, anthrax, and aflatoxin to airfields in western Iraq. In addition, 25 warheads missiles filled with the same biological agents were made ready for use at additional sites.[16]

4   Biotech Backlash

At the turn of the century, protests and political posturing against biotechnology in London; Paris; and Brussels, Belgium, were written off here in the U.S. as misplaced fear and anger against globalism and technology. As biotechnology lobbyists, scientists, and industry observers anxiously watched the events unfurl, they began to undertake a campaign to defuse the reaction taking hold here in the United States. The prevailing belief among those contrarians was that it could not happen here. What is more, they believed that even if a European-style backlash established a foothold in the U.S., it would most likely focus on biotech foods and not cross over into the medical side of the industry.[17]

There is growing evidence to suggest otherwise. In 2000, there have been at least 30 acts of antibiotechnology terrorism, including one at the fabled laboratory where Nobel laureate Barbara McClintock studied Indian corn in the 1940s, and another that razed the U.S. government's largest agbio laboratory, causing $2 million in damage. Across the country, new voter initiatives banning genetically modified (GM) ingredients from school lunch programs are popping up. And a recall of taco shells because they contained bioengineered corn will cost billions of dollars. There is also concern about escalating prices of prescription drugs and the ethics of gene patents.

---

[15] Robert Taylor, "Bioterrorism report: all fall down", *New Scientist*,19 September, 1998.
[16] "A new strategic concept for NATO", *Papers on International Security*, May 20, 1997, No. 20.
[17] Stephen Herrara, "Biotech's backlash bandwagon", *Redherring*, January 5, 2001.

Figure 2. A demonstration was staged in front of the San Diego Convention Center against the 3-day BIO2001 Convention, being held at the Convention Center, (June 24, 2001, Sunday). The banner reads "BIOTECH PERVERTS GET OUT OF OUR GENES". According to a local TV channel, safety measures cost the City of San Diego more than $2.4 million.

Left unchecked, these developments will hamper the biotechnology industry's ability to raise capital for research and will inevitably lead to new regulations, higher costs, and lower profit margins. And the backlash will not come only from opponents to bioengineered foods. There is a growing resistance among consumer groups, government, and health care providers to rising drug prices. For example, a one-year drug prescription for either Procrit or Epogen is $12,000 per patient, and Neuopogen costs $17,000 per patient per year.

The tremors of a backlash began in the fall of 2000 when a small fraction of the country's taco shells were found to be produced from genetically engineered cornmeal. The cornmeal product had been approved for hog feed, but not human consumption.

5   StarLink For Hogs, Not Humans

For anyone in the business of growing corn, one of the biggest frustrations of the job is a brown inchworm-like creature that spends most of the summer and fall munching and tunneling through the corn, only to emerge as a moth that flies off to spawn a lot more inchworms. Like many adolescents, corn borers can be enormously destructive. Depending on when in the growing season they arrive, they can damage arteries that carry moisture to the corn, or even cause the entire ear to fall off before harvesting time. The borer costs American farmers and their $20 billion corn crop more than $1 billion per annum, which includes diminished yields, the price of pesticides and other measures needed to keep the borer at bay.[18]

In 1995, when scientists produced an early variety of genetically modified (GM) corn that poisoned the borer shortly after its first cornstalk casserole, farmers jumped for joy. But in the summer of 2000, right in the middle of the harvest, things got messy. Plant Genetics Systems, a company now owned by Aventis, a giant European pharmaceuticals firm, had developed another borer-killing gene that it called StarLink. However, the toxin that StarLink produced in the corn plant resembled a substance that triggers violent allergies in some people. When federal regulators threatened to ban StarLink corn until its safety in humans could be established, the developers thought they had a better idea. They promised to sell StarLink seed only to farmers using it for feed corn; and in turn, the farmers would agree not to sell the seed to anyone who would put it in human food.

Soon, the matter got out of hand. Almost everybody involved screwed up. StarLink made up about 0.5% of the 80 million acres of corn planted in the U.S. in Y2000. Even though it was on the market for just three years, it began showing up in all sorts of places it did not belong, including tacos, corn chips, breweries, and muffin mix. The promises made by StarLink's inventors proved worthless, falling prey to managerial inattention, corporate mergers, blind faith, misplaced hope, woeful ignorance, political activism, and probably greedy farmers too.

The episode hardly qualifies as a disaster, since no one seems to have gotten seriously ill from eating StarLink corn. But the StarLink incident has revealed the shortcomings of federal oversight and has pointed up the inability of the grain-handling industry to segregate subtly different products. Still, StarLink has caused no end of hassles for farmers, grain-elevator operators, railroads, and food processors. Altogether, the fiasco triggered the recall of fast-food chain's taco shells, corn chips, and 300 other products. The recall will cost Aventis alone as much as $1 billion.

The long-term consequences may be more severe. So far Americans have been much more accepting of genetically modified food than the rest of the world. If StarLink triggers hysteria among Americans, the world's biggest appetite for that promising technology will shrink, and the whole science will be retarded for years. If foreign food processors that buy U.S. agricultural commodities worry that American grain "glows in the dark", they will turn even more to Brazil and other countries for their food, and U.S. farm prices, already depressed, will fall further.

One of the more surprising revelations of the StarLink mess is not that genetically modified food has suddenly appeared in the food supply, but rather how much such food is already out there. Most of us have heard about such oddities as strawberries protected from frost damage by a gene transplanted from an arctic fish. But not too many know that genetically modified soybeans now account for 60% of all soy grown in the U.S. Called Roundup Ready, the plants were developed by

---

[18] Brian O'Reilly, "Reaping a biotech blunder", *Fortune*, February 19, 2001.

Monsanto to tolerate Roundup, one of the company's weed-killers. Before Roundup, farmers used to use a quart of herbicide per acre. Now it is just ounces. Similarly engineered soy plants, including LibertyLink from Aventis, are sold by other companies.

Close on the heels of Roundup Ready soy came another kind of genetically altered plant: one that produced its own pesticide. This is where the StarLink story begins. For nearly 30 years farmers have sprayed crops with solutions derived from a soil bacterium called *Bacillus thuringiensis*. This so-called Bt spray is harmless to humans but quite effective against a variety of pests, including corn borers. However, it does not kill all corn borers, especially those that often show up in a second wave of infestation in midsummer. In 1995 seed companies such as Pioneer Hi-Bred and DeKalb won approval to sell corn genetically altered to produce the pesticide found in soil bacteria; this seed killed nearly 99% of corn borers. About 18% of corn planted in the U.S. in Y2000 was of the Bt variety.

One bag of seed corn — a quantity enough to plant about 2 1/2 acres — costs $90; Bt corn costs an additional $15 per bag. Corn-borer infestations vary widely from year to year, depending on wind and rain. If infestations are mild, it is cheaper to fight the borer with sprays. But in broad swaths of the Cornbelt where the borer is a chronic problem, the Bt varieties of seed are more economical. Roughly a quarter of the corn grown last year in Iowa, Kansas, Nebraska, and Minnesota was of the Bt type; the corresponding figure in South Dakota was 35%.

Pioneer's and DeKalb's head start in the Bt-engineered crop business worried Aventis, a $20-billion-a-year French pharmaceuticals and agricultural sciences company formed in 2000 by the merger of Rhone-Poulenc and Hoechst. Although most of Aventis' revenues come from drugs such as Allegra, a prescription antihistamine, the company's crop-sciences division had sales of $4 billion in fiscal year 2000, making it one of the biggest ag-products operations in the world. Aventis is spending about $450 million a year on R&D in the agricultural division because the company believes the agricultural sector has gone about as far as it can in fighting weeds and pests with chemicals, and thus there is a need to make a big shift to biotechnology. Even before their companies merged, executives at Rhone-Poulenc and Hoechst worried that rivals were grabbing market share in key agricultural technologies that would be difficult to win back later.

Buried in the welter of corporate subentities created by the Rhone-Poulenc and Hoechst combination was a small Belgian company called Plant Genetics, which Hoechst had acquired in 1996. Corporate life cannot have been easy for the managers and scientists at Plant Genetics, who had been working for a decade on a Bt variety of corn. Four years before the Aventis merger, Hoechst had formed a joint venture with Schering, the U.S. drug company. Plant Genetics was acquired by the joint venture, called Agrevo, which was later folded again into a division of Aventis. There was lots of upheaval at Plant Genetics. Its tiny U.S. headquarters moved through three cities in four years. It is reasonable to assume, too, that operational details surrounding a corn gene were hardly the most important concern of senior Aventis executives trying to manage a $20 billion merger.

Although scientists at Plant Genetics were a few years behind the competition, they were excited about what they had created: a variety of the Bt protein that destroyed a different part of the corn borer's gut. This was important because an additional vulnerability would make it harder for the corn borer to develop resistance to Bt pesticides. The Bt variety created in StarLink corn was called Cry9. Bt proteins have a crystalline shape, so different varieties were called Cry1, Cry2, etc. Aventis scientists thought Cry9 was a winner that would make them significant players in the next generation of agricultural products. Federal regulators in the U.S. were more cautious.

5.1   FDA Snafu

In the early 1980s, when the prospect of bioengineered crops first emerged, people from numerous U.S. government agencies met to discuss how to regulate the products. They agreed that the Department of Agriculture would determine whether a new plant was safe to grow outdoors. Would it run amok, for example, and harm other plants or animals? If a genetically altered plant was supposed to produce a pesticide, the Environmental Protection Agency (EPA) would decide whether the plant was safe in food. The Food and Drug Administration (FDA) would enforce the food safety standards established by the EPA.

In 1997, when EPA scientists were evaluating StarLink, they saw something they had not seen in other brands of Bt corn. StarLink's Cry9 protein did not dissolve in stomach acid as quickly as proteins in other Bt varieties. Nor did it break down as rapidly during cooking or processing. This means that the Cry9 protein, unlike the others, might stay in the stomach long enough to be passed intact into the bloodstream, where it may trigger an allergic reaction. In other tests, however, the Cry9 protein seemed fine.

StarLink's developers, eager to market their product, invoked a little-known EPA rule that allows some pesticides and herbicides to be used on feed for animals but not on food destined for humans. This "split registration" had never been sought for genetically modified products. However, the EPA approved the registration but impose restrictions on it. The restrictions on StarLink corn were stringent. It could be grown only for animal feed or for nonfood use, such as conversion to ethanol. Because regulators worried that windblown pollen from StarLink stalks could pass the Cry9 gene to ordinary corn, farmers had to leave 660-foot buffer strips around their StarLink fields. Farmers bringing the corn to market had to notify grain elevators that it could not be used in human food. The EPA ordered StarLink's developers to require all farmers who bought the seed to sign a form affirming that they understood the restrictions and would abide by them. The company also promised to conduct a survey of statistically confidence level of StarLink growers to ensure they were following the rules. The company agreed to accept full liability if anything went wrong.

Neither Aventis nor its predecessor companies ever produced much StarLink corn. Instead they inserted the newly spliced genes into small amounts of corn and sold the resulting sprouts to seed companies. These seed companies then planted StarLink in greenhouses, harvested the corn, and replanted it to create more seed. Eventually the seed companies contracted with farmers who grow large volumes of corn for seed under controlled conditions outdoors. Once that seed was harvested, the companies had enough StarLink seed to begin marketing.

Ultimately, about a dozen small seed companies licensed StarLink corn from Plant Genetics. The Garst Seed Co., which is near Des Moines, has one of the longest pedigrees in the seed business. It produced the vast majority of StarLink corn. Garst, as is common with smaller seed companies, relies heavily on "farmer dealers" to sell its products. These are usually farmers who use the slow winter months to convince relatives and neighbors into buying a few thousand dollars' worth of seed. In 1998, the first year StarLink was on the market, just 10,000 acres were planted. By 2000, a mere 350,000 of America's 79.6 million acres of corn were StarLink. In that year, the highest concentration of StarLink was 1.1% in Iowa, where Garst's is.

5.2   Bt Not for Butterflies

Nevertheless, the proliferation of Bt corn was causing growing concern outside the Farmbelt. In April 1999 an entomologist professor at Cornell University researching corn-borer resistance to Bt reported that he had fed a diet of corn pollen to monarch butterflies' larvae. Many of the monarchs that ate Bt pollen died. This caused a furor among environmentalists, who admire the monarch for its yearly migration between Mexico and the United States. Many environmentalists are profoundly worried about all genetically altered plants and animals, fearful that they contain health hazards that will not become apparent for years, or that the genetically-altered plants will somehow reproduce wildly and overwhelm ordinary species. For environmentalists, the monarch was about to become the poster butterfly of the anti-Frankenfood movement.

Among the environmentalists who led the charge against Bt corn was Larry Bohlen, an engineer by training and a senior official in the Washington office of Friends of the Earth. For years FOE and other greens had been trying to get the U.S. government to sign international protocols on the use of genetically modified organisms. When the Cornell study on monarch butterflies came out, they had the first tangible example of the kind of impact genetic crops could have. Bohlen wrote to President Clinton to ask that use of Bt plants be suspended until their effect on nontarget animals could be determined. He also began writing to consumer-product companies like Campbell's, Kellogg, and Frito-Lay, urging them to forswear all genetically modified food. In July 2000, the campaign began in earnest when Bohlen arranged for popular foods to be tested for genetically altered ingredients. Soon Bohlen learned about StarLink. He found out that grain-elevator operators and farmers could not properly segregate StarLink and other unapproved varieties because separation was difficult. Indeed, very little segregation was actually being conducted. There was a good chance that StarLink had made it into the food supply. In late July 2000, Bohlen went to the Safeway near his home in Silver Spring, Maryland, and filled his grocery cart with all the corn products to be sent to Genetic ID, an Iowa laboratory that routinely checks commodity shipments bound for Europe to make sure they comply with European Union standards. Genetic ID is run by John Fagan, an opponent of GM foods, and a microbiologist at Maharishi International University.

In September the news that StarLink corn had been found in tacos made by Kraft and sold under the Taco Bell brand was splashed across the front page of the Washington Post. The corn in suspect was from Aventis CropScience in Triangle Park, North Carolina.

A biotechnology industry organization immediately questioned the reliability of Genetic ID. But when Kraft ordered its own tests of the tacos; it found StarLink and recalled more than a million boxes. Other taco makers did the same. Kellogg shut down one of its mills because it feared StarLink contamination. Grain elevators, in the midst of gathering the fall harvest, scrambled for ways to test arriving truckloads for StarLink contamination. In many ways it was too late, since most of the StarLink in the nation's food had come from the 1999 corn crop. And because 1999 had been a bumper year, there were more than a billion bushels of unsold corn still sitting in silos. No one knew how much of it was mixed with StarLink.

But how did this happen? Every farmer who had bought StarLink was to have signed an agreement to keep it out of the human food supply. But in reality, many of the 2,500 StarLink farmers appear to have been clueless about the agreement. They claimed their seed salesmen never told them they were buying StarLink, and did not pass on any precautions about how to plant it. The vast majority of farmers actually did not sign any forms acknowledging planting and marketing limits until a few weeks after the StarLink news had broken. It was then that farmers who planted the seed received a letter asking them to sign and return some forms; the forms appear to have been backdated to before the spring planting. Aventis executives vigorously deny having anything to do with the letter. Garst CEO Witherspoon insists that Garst provided information to all its salesmen about StarLink. It seems unlikely that Garst's farmer salesmen would have knowingly deceived customers. The seed business relies heavily on the trust that exists when farmers sell seed to relatives and neighbors.

Garst is one of the oldest companies in the business; it began in 1930 by marketing hybrid seeds developed by Henry A. Wallace, the founder of Pioneer Hi-Bred. Wallace was such a successful entrepreneur that he was later appointed

Vice President under President Franklin Roosevelt. Garst was so well known that the former Soviet Premier Nikita Krushchev visited its founder, Roswell Garst, on his Iowa farm in 1959. But the company ran into trouble in the early 1980s, when Pioneer severed its relationship with Garst to market its own seed. Garst lost the bitter lawsuit that ensued. The family sold the business to ICI, the British chemical company, in 1985. Du Pont bought Pioneer in 1999. ICI later spun off its U.S. seed business to Zeneca, a British drug company. Garst is now a part of Advanta, a joint venture between Zeneca and Royal VanderHave Group in the Netherlands. Ironically, Advanta made headlines in Europe in 2000 when canola seeds it had sold there were found to contain small amounts of genetically altered material forbidden by the EU. The seeds, grown in Canada, may have been contaminated by windblown pollen from other canola nearby.

Aventis eventually took responsibility for the StarLink mess; the company is spending millions to locate the rogue corn so that it can be put into animal feed. Aventis executives claim that they thought Garst was spelling out the restrictions on StarLink to farmers, but hint that they did not monitor Garst carefully. There is every indication that Garst was not really motivated to be very explicit about how StarLink had to be grown and sold. Otherwise, which farmer would buy a variety of seed if (s)he was told (s)he had to allocate a 660-foot buffer strip around it, and would have to go through all sorts of special separation and storage after the harvest? Garst disagrees claiming the company sent 15 mailings to StarLink farmers. As for the "statistically significant" survey of farmer compliance that Aventis had promised the EPA, Garst conducted the survey, but did it right after the harvest when most corn was still stored on farms.

Both Garst and Aventis implied that if they failed to live up to all their agreements with the EPA, it was because they were convinced StarLink would soon get full approval for use in food and that the special conditions would be lifted.

Even after giving Aventis and Garst their share of the blame, there is plenty more to go around. The EPA concedes that a split registration for StarLink, allowing it in feed but not food, was a dumb idea. Critics point accusingly at the FDA, which was supposed to enforce food standards established by the EPA. Friends of the Earth says the FDA did not even have a way of testing for StarLink in food and that the agency moved slowly when news of the contamination first came out.

To its belated credit, Aventis has been aggressively trying to locate StarLink seed. It requested Garst's list of StarLink customers and met with all of them within days. Aventis is paying farmers up to 25 cents for each bushel of StarLink seed fed to animals. When grain-elevator owners discover that a batch of StarLink has contaminated a million-bushel silo, Aventis negotiates compensation for their added efforts and expense. The company has also paid for millions of test kits used by farmers, food processors, and grain handlers to identify traces of StarLink. Just how much is out there is anybody's guess. Because many farmers failed to plant buffer strips, pollen sometimes drifted into neighbors' fields, causing that corn to test positive. Moreover, some Garst seed varieties that were not supposed to contain StarLink turn out to have been contaminated, and that adds to the difficulty of finding it.

Jerry Rowe manages the Farmers Grain Cooperative, a four-million-bushel grain elevator in Dalton City, Illinios. He says StarLink has greatly complicated his life. At peak times he unloads a truck every two minutes. The StarLink test takes five minutes per truck. And a sampling probe could miss StarLink lurking in a far corner of a truck. A worry is that if one finds StarLink in his bins a year later, Aventis may not compensate him. Aventis claims it will.

5.3   Corn Dogs

After the StarLink taco incidence, it is expected that many more corn products will be under intense scrutiny. In March 2001, the environmental group Greenpeace commissioned tests to detect StarLink corn in corn dogs purchased in a Baltimore Safeway, and the results came back positive.

Worthington Foods Inc., a vegetarian-foods subsidiary of the Kellogg Co., which produced the corn dog, subsequently had an independent laboratory test the product.[19] It had to recall all its meatless corn dogs after its tests confirmed some of them contained a variety of genetically engineered corn that is not approved for human consumption. Worthington Foods, which was acquired by Kellogg in 1999, ordered the full recall of corn dogs to reassure customers and consumers that their trust and confidence in Worthington products was of utmost importance to the company.

5.4   The Extent of Commingling

More than 430 million bushels of corn in storage nationwide contain some of the genetically engineered variety that prompted a massive recall of corn products in the Fall of 2000, the company that developed it reported on March 18, 2001. The announcement greatly increases estimates of the amount of corn that was inadvertently mixed with StarLink, which is not approved for human consumption.[20]

Most of the commingled corn was from the 1999 crop and is in grain elevators, according to John Wichtrich, general manager for Aventis CropScience, which developed the corn. The affected corn, which is more than 4 percent of that year's U.S. corn production, will have to be rerouted to animal feed and ethanol production. The 430 million-bushel estimate dwarfs the amount of corn reported

---

[19] "Corn dog recalled after biotech corn finding", *The Associated Press*, March 13, 2001.
[20] Marc Kaufman, "Biotech grain is in 430 million bushels of corn", *The Washington Post*, March 18, 2001.

earlier from the 2000 crop as containing StarLink, about 50 million bushels grown by farmers licensed to use it and 20 million bushels from neighboring fields. Wichtrich said 99 percent of the 2000 StarLink corn has been identified and redirected.

The genetically modified protein in StarLink corn, called Cry9C, was approved only for animal consumption because of concerns it might cause allergic reactions in people. But StarLink corn was discovered in the Fall of 2000 to be widespread in the nation's corn supply, and more than 300 corn products were recalled after tested positive for StarLink.

The engineered corn apparently was mixed with other corn by farmers who inadvertently delivered StarLink to buyers without notifying them. It also could have occurred by pollen from StarLink fields blowing onto nearby plants.

Aventis CropScience will offer to set up small laboratories in mills that produce corn meal, grits and flour to ensure that the processed corn does not contain any of the genetically engineered protein in StarLink. Effort had been planned in conjunction with the Food and Drug Administration. The Environmental Protection Agency has concluded the protein does not survive the "wet milling" process that makes corn syrup and oil, and Wichtrich said on-site testing of those mills is not required.

Corn is considered unfit for human use if one kernel out of 2,400 contains the Cry9C protein.

6   Organic Food

Biotechnology observers say the cornmeal incident has ensnared other food companies in the recall fever. And it is likely to lead to more recalls, ingredient reformulations, and reassuring advertisements to consumers. H.J. Heinz and Gerber, two of the world's biggest baby-food makers, switched to more expensive organic ingredients purely out of fear of a consumer boycott. The move led to price increases. For example, a six-ounce jar of organic baby food costs twice as much as a 16-ounce can of Campbell's chicken soup.

The reverberations from a public outcry over GM foods and gene therapies have resulted in scaling back businesses or dropping them altogether. Monsanto's plastics-from-plants research was scrapped. And research into various stem-cell therapies has been delayed for at least five years because, like other tissue and cell engineering research, it relies on tissue from aborted fetuses and cadavers. The Biotechnology Industry Organization (BIO), a trade group in Washington, D.C., wrote in its October/November 2000 newsletter that antibiotechnology measures are growing across the country. Even those ballot initiatives that fell short because their supporters failed to collect enough signatures are not dead yet.

6.1 Long Term Ramifications

The long-term consequences of StarLink seed are hard to predict. No serious health problems have emerged so far. A pediatric allergist from Duke University told a scientific advisory panel convened by the EPA that unless someone has an anaphylactic reaction to StarLink, he or she does not have a food allergy. Anaphylaxis, also called anaphylactic shock, is in immunology a severe, immediate, potentially fatal systemic allergic reaction to contact with a foreign substance, or antigen, to which an individual has become sensitized.[21] Symptoms of anaphylaxis include an itching of the scalp and tongue, difficulty in breathing because of swelling or spasm of the bronchi, skin flush of the whole body, an abrupt fall in blood pressure, vomiting or abdominal cramping, and unconsciousness. In milder cases hives may spread over the whole body, and often there is a severe headache. Treatment, which must begin within a few minutes of attack, involves the injection of epinephrine (adrenaline), followed by the administration of antihistamines, corticosteroids, bronchodilators, and fluids.

The panel decided that StarLink does indeed walk and talk like a potential allergen, and advised the EPA to turn down a request by Aventis that small amounts of it be allowed in the food supply.

StarLink has not triggered widespread hysteria about genetically modified food in the U.S., to the disappointment, no doubt, of some environmental groups. But the EPA is worried that the episode may slow the acceptance of genetically modified products. The EPA is not alone. Pierre Deloffre, head of a large French vegetable-processing company, told a seed trade convention in Chicago in December 2000 that Europeans turned abruptly away from genetically modified foods during the 1990s. Deloffre blames government regulators and scientists who failed to respond properly to Chernobyl, AIDS in the blood supply, and mad cow disease for eroding Europeans' confidence in technology. "Five years ago the first boatloads of genetically modified soybeans arrived here without the slightest reaction," says Deloffre. Now the EU barely touches them.

Although Europe has not imported much American corn for years, Japan is a large customer. The Japanese have been fairly tolerant of bioengineered food, but they, too, are growing cautious. New rules that took effect in Japan the spring of 2000 would require labels on food to state if it contains genetically modified ingredients. This has led to orders from Japan for unmodified corn and soy to rise in anticipation of the new labels.

Since it caused no serious illnesses, the StarLink debacle will probably be a footnote in future agronomy textbooks. In reality, though, this was a disturbingly close brush with disaster. StarLink was probably circulating in the food supply for a year before it was found. If it had been slow acting but truly dangerous, like mad cow disease, the damage could have been enormous. Critical links in the food

---

[21] Encyclopædia Britannica, britannica.com.

chain, from Aventis and Garst to thousands of small farmers, turned out to be either unconcerned about or oblivious to what they were selling and growing.

If we are lucky, maybe StarLink will also be a wake-up call, reminding us that tinkering with Mother Nature is risky business, and that it is not just white-coated laboratory technicians who must be careful. Solving the problem of hunger and malnutrition may ultimately depend not so much on science as on our faith in science and all its stewards. And if we cannot trust a farmer, whom can we trust?

6.2   From Taco to Genes

What is happening with biotechnology legislation on a national level? In Y2000, Charlene Barshefsky, the U.S. trade representative, quietly signed a new protocol that allows any nation to ban the import or export of GM foods, most of which are produced in the United States, Canada, and Argentina, even in the absence of scientific or anecdotal evidence of harm. There is also a "Genetically Engineered Food Right to Know" bill making its way through Congress that would mandate a label on anything that is at least 10 percent GM. That kind of law would make the labels on genetically modified food in the U.S. even more restrictive than many existing European labels. The law would require farmers, grain elevator operators, wholesalers, retailers, and food processors to sequester GM grain from non-GM foodstuffs. Opponents contend the mandate will increase costs by 30 percent. As if American farmers are not scared enough by European bans, there was news in 2000 that the $11 billion grain export market in South Korea and Japan is about to go GM-free.

New regulations also have been conceived to address gene therapy. The U.S. Office of Human Research Protections recently drew up a set of costly standards for human trials in the wake of a single death in a gene therapy trial. The office has threatened to push Congress for even stricter standards if hospitals and universities do not comply. Officials shut down gene therapy trials everywhere after the fatality, stranding patients and hampering research.

Despite the small but growing surge of opposition, analysts characterize the biotechnology industry as healthy. A recent Ernst & Young study found that biotechnology has been a strong and positive influence on the economy. There are now 1,273 biotechnology firms in the United States; and they have created nearly half a million jobs in the past few years alone. And Y2000 was the first year the biotechnology industry spent as much as $11 billion on research and development.

6.3   Warranty of Edibility

But the U.S. biotechnology industry ignores the growing signs of an American backlash at its own peril. This is not just about fear of tainted taco shells. The angst is likely to spread. Consumers often do not distinguish among the different biotechnology sectors. This should give the biotechnology industry pause that liability issues (bioliability) are just starting to take hold. Just like the implied warranty of merchantability and implied warranty of fitness in product liability, there may soon be an implied warranty of edibility for food products, or implied warranty of genetizability in genetic engineering.

Spectators of the legal profession believe that gene therapy and GM foods are ripe for product liability suits. They write that allegations would likely include "failure to warn, defective design, defective manufacture, antitrust, fraud, misrepresentation, negligence and breach of express and implied warranty." Some suggest that the public education effort has to be intensive and has to start immediately for people have to realize that backlashes do not just appear out of the blue in some easily recognized form. They creep up slowly.

Chapter 4

AQUA BIOTECHNOLOGY

"Whether served as raw sushi, grilled steak or in thin smoked slices, most of the salmon you eat these days is not the sleek sport fish that has been a favorite of anglers since Izaak Walton but rather a chunky, sluggish creature raised in captivity. Indeed, salmon caught in the wild accounts for less than half of all salmon sold in the U.S."
Frederic Golden, *Time*, March 6, 2000.

1   The Fish Tale

Fish are at the lower stage of evolution but are the most diverse among the vertebrates. They first appeared around 500 million years ago, and today there are about 21,700 to 28,000 species, accounting for over half of the total number of vertebrata.[1] Fish are generally the most fecund, some producing thousands of eggs on a periodic basis. These eggs are usually large and transparent. Fertilization occurs externally and the embryonic development can be easily followed. These advantages make fish excellent animal models suitable for both field and laboratory studies.[2] Additionally, fish has enormous potential to fulfill the human requirement for protein.

In fact, fish culture is one of the earliest activities of human civilization. A vast body of animal husbandry information is accumulated through thousands of years of practical experience from fish farmers and hobbyists. The first record of pond culture in the world can be dated back 2500 years to the Handbook of Fish Culture, in which the author, Fan Li, described the domestication and cultivation of common carp in ponds. Advancement in the techniques of fish culture has increased but has always fallen short of the demands of the expanding human population. Development of high technology has become a pressing matter for the aquaculture industry. Fortunately, the advancement of modern biology, especially molecular genetics, makes fish breeding very efficient. Using gene transfer, transgenic fish with traits of growth enhancement, high food utilization efficiency and disease resistance can be produced. Thus, gene transfer technology will revolutionize traditional fish breeding. Now transgenic fish is on the verge of commercialization in China, Canada, U.S., and other countries. However, as a genetically modified organism (GMO), biosafety has been a big concern.

---

[1] J.S. Nelson, *Fishes of the World*, 2nd Ed. (Addison-Wiley Interscience Publication, New York, 1984)
[2] D.A. Powers, "Fish as model systems", *Science*, 246, 1989, pp. 352–358.

79

2   Need For Aquaculture

A 1997 estimate by the United Nations indicate that the supply of seafood products would have to increase seven-fold if we were to meet the worldwide requirement for fish and other seafood by the year 2020.[3] Given the rapid decline in world fish stocks, caused mainly by over fishing, it is clear that demand can only be met by aquaculture.

Worldwide aquacultural production of many fish species is challenged or handicapped by several factors. These include the selection and supply of suitable broodstock, growth rate and feed conversion efficiency of the fish, the feeding costs, control of the reproductive cycle, and disease protection.

Traditionally, the broodstock is selected based on cross breeding to enhance the fishes' beneficial or desired traits. However, these traits are generally slow to emerge and unpredictable, and often the fish genome might not contain the gene mediating the effects. Identifying, isolating and constructing the genes responsible for desirable traits using molecular biology and then transferring them to the broodstock are an improvement over traditional selection and breeding methods. In addition, new traits not present in a genome can be genetically transferred into the genome from an unrelated species, enabling the production of new and beneficial or desirable phenotypes.

In addition to barnyard animals, gene transfer technology is also making major inroads in aquaculture and related industries. Transgenesis, therefore, holds promises for producing genetic improvements in fish, such as enhanced growth rate, increased production efficiency, disease resistance and expanded ecological ranges. There is every indication that transgenic fish is likely to be the first marketable transgenic animal for human consumption.

2.1   Gene Transfer Protocol in Fish

As an experimental model, fish actually have several advantages over mammals. Compared with mice, which produce few eggs, a spawning female fish can produce several dozen to several thousand eggs, providing a large number of genetically uniform materials for experimentation. For example, the zebrafish (*Brachydanio rerio*) produces 150–400 eggs, the Atlantic salmon (*Salmon salar*) produces 5000–15,000 eggs, and the common carp (*Cyprinus carpio*) produces more than 100,000 eggs.[4] More importantly, once the gene transfer has been carried out using fish eggs, no further manipulation is necessary and the maintenance for the fish hatchery is relatively inexpensive. In contrast, fertilized mammalian eggs must be implanted

---

[3] Choy Y. Hew, and Garth Fletcher, "Transgenic fish for aquaculture", *Chemistry & Industry*, April 21, 1997.

[4] G.L. Fletcher, and P.L. Davies, in *Genetic Engineering Principles and Methods*, J.K. Setlow (ed.), Volume 13, Plenum Press, 1991, pp. 331–70

into appropriately prepared recipient or surrogate mothers, thus increasing the preparatory work and the operating cost.

Microinjection procedures commonly used to deliver DNA into mammalian eggs have been used with considerable success in fish. However the eggs' mortality in some species can be very high and the percentages of integration in fish are often low, which produces relatively small numbers of positives containing the transgene. Attempts to overcome this have taken advantage of the large numbers of eggs that are produced by most fish species and explored the use of several mass gene transfer techniques with varying degrees of success.

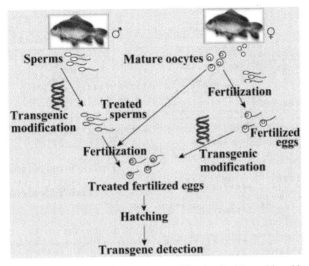

Figure 1. Gene transfer protocol in fish. Transgenic modification is either achieved by electroporation, microinjection, or using a gene gun. (Sketch: Adapted from C.L. Hew and G. Fletcher).

Electroporation, which uses short electric pulses to permeate the cell membranes allowing the DNA to enter, has been tried successfully in zebrafish, salmon, loach and others. One variation electroporates sperms, which are then used to fertilize the eggs. Other methods include the gene gun and using sperm incubated in a solution containing DNA.

2.2 Fish As Medical Tools

Though when we speak of aquaculture, we immediately think of culturing fish for food. Fish are also cultured for research purposes, just like mice are cloned for research purposes, not as bioreactors.

While it is feasible to generate transgenic mice with deleted or knockout genes using early precursor cells from embryos — embryonic stem cells (ES) — no such cells are currently available in fish. Nevertheless, promising results on ES-like cell cultures have been reported in the Japanese medaka, *Oryzias latipes* and zebrafish.

In the post-genomic world, the lowly zebrafish may be king. Scientists at various institutions have developed a technique that allows zebrafish to pass genetic modifications to its offspring. The discovery will lead to researchers being able to study genes and proteins in a less expensive way.

The tinier than two-inch, black-striped hardy zebrafish — known primarily as the last fish living in a kid's aquarium — is quickly becoming famous in the scientific world as the best animal to use when studying genetics — even better than the mouse. Zebrafish are relatively inexpensive and easy to maintain compared to genetically modified mice, this discovery could greatly accelerate new genetic experiments in vertebrates.[5]

The zebrafish is an essential model organism to a new branch of science called proteomics, also sometimes known as post-genomics. Proteomics refers to the study of an organism's proteins, just as genomics refers to the study of an organism's genetic material. Proteomics is a natural follow-up to the mapping of various organisms' genomes, including the human genome.

The human genome project are about sequencing genes, and each of the genes causes the body to produce various proteins at different times. To understand what the genes actually do, the function of the proteins have to be ascertained, and the zebrafish is an ideal model for that.

Brachydanio rerio or zebrafish is native to the Ganges river of India and Burma. Scientists describe the animal as a model vertebrate for embryo research development.

The fish small size (full grown size is 3 cm) makes it possible to house large numbers of the animal in small space, i.e., up to five per liter. Time between generations is only about three months.[6]

Spawning is virtually continuous by this active, warm water fish. They breed daily by egg-scattering with active males following active females at first light. Spawning is stimulated by the presence of light because they are a photoperiod spawner. The fish then turn around and eat eggs as they float down through the water, to replace proteins lost through activity. By spawning the fish over a bed of marbles, researchers can siphon out the eggs for studies.

The egg lends itself readily to microinjection. Genetic tagging is generally regarded as easier than with *D. melanogaster* or *C. elegans*. The fish can be spawned in water with pH ranging from 6.0 to 8.0 and can withstand a temperature range of 60°F–90°F. The average progeny per female exceeds 200 per week.

The externally fertilized eggs initially look like transparent bubbles. Cell division can be studied under a microscope from the first hour. Zebrafish eggs

---

[5] "Zebrafish could become genetic "lab rat" of choice", *Purdue News*, April 2001.
[6] Henry D. McCoy, II, "Zebrafish and genomics", *Genetic Engineering News*, 21(4), 2001.

complete embryogenesis in about 120 hours. The gut, liver, and kidney are developed in the first 48–72 hours.

The degree of conservation of synteny between zebra fish and human genome structures is high, about 75% homology. However, the zebrafish is still some years behind in terms of the development of sufficient scientific knowledge of the animal to reach the level of human, mouse, or fruit fly genome databases. But this is just a matter of time since maps and sequencing project of the fish are in progress at Stanford University, the University of Oregon, the Sanger Centre, Children's Hospital (Boston), and Max Planck Institute.

2.2.1   Knock Out The Fish That Roars

Scientists study proteins and gene function by disabling a single gene, and then raising clones of that test subject to see how they develop without the gene. Such experiments are called knockout experiments, because the gene is turned off, or knocked out.

Over the past decade, plant scientists have used a small mustard plant, called *Arabadopisis*, to conduct gene knockout experiments in crops and plants. But until now it has been difficult to conduct knockout experiments in animals.

In mice — until now the only animal in which the gene knockout technique works — a genetically modified embryo cell, called an embryonic stem cell, is inserted into a developing mouse embryo. An embryonic stem cell is an early embryo cell that has not yet begun to differentiate into various tissues. The embryo is then transferred into the womb of a female mouse using surgical techniques. After the embryo develops into an adult, some of the modified stem cells will give rise to eggs and sperm that contain the modification. Such a process is laborious and expensive. As a result, transgenic mice can cost thousands of dollars each.

Figure 2. Zebrafish eggs are transparent and are fertilized outside the body, so researchers can observe their embryos growing in laboratory dishes. Those characteristics, along with their hardiness and rapid life cycle, have made zebrafish one of the most popular creatures for studying what happens in the earliest stages of development. (Photo: Adapted from University of California at San Francisco).

With mice a researcher may have a dozen embryos to work with, and to do surgery to transplant the embryos back into a mother. In contrast, with zebrafish, a researcher can modify 100 embryos an hour, and, because the embryos develop outside the mother, no surgery is required. The entire developmental phase takes only about four days.

Because of the expense and effort required to produce a transgenic mouse, scientists have been searching for another vertebrate animal that would allow these genetic experiments. The technique has been tried in chickens, cows, pigs, sheep and other species of fish without success. The problem has been that, in species other than the mouse, once the embryo cells containing the gene are transferred into a developing embryo and that embryo develops into adulthood, the animal does not produce functional eggs and sperm. Therefore, the gene cannot be passed on to subsequent generations. Getting embryonic stem cells to develop into functional eggs or sperm once they are placed in an embryo is crucial.

In early 2001, scientists succeeded in growing stem cells in the laboratory and then transferring them into a zebrafish embryo. In other words, the stem cells contribute to the germline of the embryo. The next step is to extend the length of time cells are kept growing in the laboratory so that gene can be inserted. Scientists have been able to grow the embryo cells in the laboratory for a few days. To make a knockout, scientists have to keep the cells growing in the laboratory for several weeks.

From an agricultural point of view, transgenics and gene knockouts can be used to control reproduction, disease rate, growth rate and many things that are very valuable in livestock. This technology could be used to make sterile animals so that transgenic animals cannot breed if they escape to the environment. These new experiments could also provide information into Alzheimer's, heart disease, certain types of cancer and other diseases.

2.2.2   Zebrafish in Development Studies

How does the small glob of undifferentiated cells in an embryo multiply and specialize to form every organ of the body — bone, nerves, blood vessels and so on? Odd though it may seem, scientists have discovered over the past 15 years that the development of all creatures, from the humblest worm to the proudest *Homo sapiens*, is controlled by a remarkably similar set of genes. The completions of the first draft of the human genome and other organisms confirm this discovery. The fruit fly *D. Melanogaster*, the model organism of laboratory genetics, possesses 13,601 genes. The roundworm *C. elegans*, the staple of laboratory studies in development, contains only 959 cells with virtually no complex anatomy beyond its genitalia, possesses 19,098 genes. The general estimate for the number of genes of *Homo sapiens*, with 10 trillion cells in the body, is 30,000–40,000.

Scientists have long studied fruitflies, which, though separated by 600 million years of evolution, use the same genes, for example, to divide their bodies between front and back. Humans and zebrafish, both vertebrates that evolved from a common ancestor 420 million years ago, share even more genetic similarities.

Having backbones, zebrafish model human skeletal structure more closely than invertebrate fruitflies.

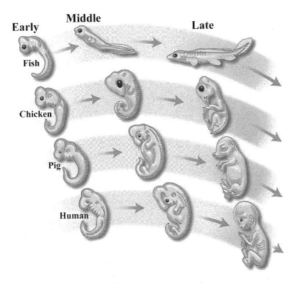

Figure 3. Scientists have long known that most creatures bear an uncanny resemblance in their embryonic stage. But it is only within the last 15 years that DNA research has shown that a few genes seem to control embryonic development. This commonality means biologists can better understand human genes by studying the DNA of fish and other simpler organisms. (Photo: Adapted from John Blanchard, *San Francisco Chronicle*).

Mice, another favorite study animal, bear even greater genetic resemblance to humans because they are both vertebrates and mammals. But zebrafish have some advantages. Cost, for instance. Though not as cheap as flies, zebrafish can be kept for a penny a day, whereas room and board for a mouse runs 15 to 25 cents per day. Zebrafish eggs are also easy to obtain and study because they are fertilized externally, allowing researchers to harvest them from the bottom of the tank. Mice develop *in utero*, which means scientists may need to kill the mother to study the embryo, and even then they would only catch the mouse embryo at one point in time.

In contrast, zebrafish eggs are transparent, researchers can see through their outer membranes, using microscopes to watch the embryo develop its heart and other organs in a matter of hours.

Regardless of which creature they study, flies, zebrafish, or mice, what scientists are looking for is genetic anomalies. Scientists try to knock out genes or cause mutations by dosing fish tanks with carcinogenic chemicals. In the latter, they then breed these potentially mutant fish, scanning the embryos under microscopes, looking for something out of the ordinary. When they spot a defect, researchers use the tools of molecular biology to cut up the fish's DNA, then compare it to the DNA of a normal embryo. The objective is to find the gene, or group of genes, that seems to set the sick fish apart.

Fertilized zebrafish eggs develop with astonishing rapidity. Within 24 hours, researchers can see the heart beating through the transparent membrane. By comparison, it takes three weeks for the human embryo to develop a heart. Five days after fertilization, the tiny fish is swimming and feeding under its own power.

These zebrafish attributes — their hardiness, quick life cycle, transparent embryos and vertebrate development — were among the traits that prompted University of Oregon molecular biologist George Streisinger to study zebrafish in the 1970s. Other scientists, including Nobel laureates Walter Gilbert at Harvard University and German geneticist Christine Nuesslein-Volhard, expanded and popularized zebra fish studies.

2.2.3   The Zebrafish — Closer to Heart than You Know

Studying embryos and the genes that control their development is more than a scientific curiosity. Researchers have discovered that our bodies contain certain cells, called stem cells, which retain some of the properties of embryonic cells. Unlike a fertilized egg which is totipotent in the sense that its potential is total or it can develop into a fetus, stem cells are pluripotent rather than totipotent. They cannot form an organism because they cannot develop into supporting tissues necessary for development. However, in the presence of the proper gene or genes, stem cells can divide indefinitely and specialize to form blood, skin, bone cells, etc.

Learning more about these stem cells, discovering which genes trigger them to grow into which types of tissues, is one of the hottest areas in bioscience. And studies of primitive organisms are helping to disclose which genes would be useful in human stem cell research.

Didier Stainier at the University of California, San Francisco, whose work is supported by research grants, has focused on studying the heart. He recently led a scientific team that discovered a zebrafish gene, dubbed *gata5*, which seems to be the master switch that tells embryonic stem cells to become heart cells.[7]

If scientists can find a human equivalent to *gata5*, it will offer the possibility of stimulating pluripotent stem cells into heart cells and thus provide a renewable source of replacement heart cells. These grown-to-order cells could be grafted onto weak hearts to strengthen the muscle — something similar to putting a rubber patch over a tire to keep it from blowing.

In July 2000, Didier Stainier, Erik Kupperman and their colleagues seeking clues to the formation of the human heart in the embryo found a chemical molecule that directs two microscopic "buds" of the zebrafish's tissue to join together and begin to beat. The molecule, called S1P, short for sphingosine 1-phosphate, is

---

[7] Tom Abate, "Window on life: transparent eggs of the zebrafish let researchers see how DNA direct embryonic development", *San Francisco Chronicle*, February 7, 2000.

actually a fatty lipid. It is found in organisms as simple as yeast, and it is the same responsible for the first steps in the development of the human heart.[8]

Stainier's group has already altered zebrafish genes to produce mutations in heart development — many of them the same kinds of mutations that result in human heart defects. And scientists say that the work could lead to new treatments for such ailments.

In the earliest stages of the zebrafish embryo's development, two tiny tube-shaped groups of cells form like buds on opposite sides of the embryo. Twenty-two hours into the embryo's life, the S1P molecule causes the separate tubes, known as "primordial", to join together into a true heart.

In embryonic mice, the same crucial union of the two heart buds develops in the eighth day after conception, while in human embryos the heart forms and begins beating at three weeks.

When a mutation prevents the S1P molecule from functioning in the zebrafish, and presumably in human embryos too, the heart buds fail to come together, and no fully developed heart can form.

This is just an example. Presumably, many more discoveries will come from zebrafish research, particularly when an international effort is already underway.

2.3   The Rule of the Game — Growth Enhancement

Since roughly half of the operating costs in fish farming, such as for the salmonids, are related to feed, the major concerns for the industry are the growth rate and food conversion efficiency of the cultured fish species. The pioneering work of Robert Brinster and co-workers at the University of Pennsylvania in creating a "supermouse" roughly twice the size of its littermates has provided the impetus for generating faster growing fish. Zhoyan Zhu and colleagues at the Institute of Hydrobiology, Wuhan, China, used the mouse metallothionein promoter linked to the human growth hormone gene to produce the first transgenic fish. Since then, several laboratories have reported the successful production of faster growing fish in several species.

Table 1. Growth enhancements (GE) using growth hormone (GH) genes in commercially important transgenic fish.

Fish Species	Promoters	Source of GH Genes	GE
Common carp	MMT	Human GH gene	1.1
	RSV	Rainbow trout GH cDNA	1.3
Crucian carp	MMT	Human GH gene	1.7
Catfish	RSV	Coho GH cDNA	1.2
Loach	MMT	Human GH gene	2.0

---

[8] David Perlman, "A small fish unlocks a secret of the heart", *San Francisco Chronicle*, July 17, 2000.

Table 1. (Continued)

	op-AFP	Chinook GH cDNA	2.5
Tilapia	CMV	Tilapia GH cDNA	1.8
	op-AFP	Chinook GH cDNA	2.0
Pike	RSV	bGH cDNA	0–1.1
Atlantic salmon	op-AFP	Salmon GH cDNA	3–10
		GH minigene	3–10
Pacific salmon	op-AFP	Salmon GH cDNA	3–10
	Sockeye MT	Salmon GH gene	6–11
	Sockeye histone	Salmon GH gene	

One of the major concerns in producing transgenic fish for food consumption is the consumer's perception of food safety. A way to ensure this is to only use DNA and genes from fish species to make the DNA constructs for gene transfer. These "all fish" gene constructs, based on the antifreeze protein gene promoter from the ocean pout linked to the Chinook salmon growth hormone gene, have been used to produce transgenic Atlantic and Pacific salmon. They showed a dramatic increase in growth enhancement, on average 3 to 5 folds, with some individual fish being 10 to 30 times larger in the early phase of growth. Although some of the largest fish had some growth abnormalities analogous to acromegaly in which the bones become excessively enlarged, most of the moderate growing fish were normal and fertile.

Subsequent generation transgenic fish have been produced. These fish are fast growing, and are fully capable of entering full strength seawater as smolts almost a year earlier than their non-transgenic siblings. Recently, transgenic tilapia with a two-fold increase in growth rate have also been produced. It is obvious that transgenesis appears to be a common and viable approach in generating faster growing fish. The magnitude might, however, depend on the species, the nature and strength of gene constructs, and the number and location where the transgene is incorporated in the host DNA.

2.4   Antifreeze

All animals contain the growth hormone gene, thus the strategy needed to enhance growth by gene transfer involves a relatively minor modification of the gene to let it be expressed by tissues other than the pituitary gland, in the liver for example. Nevertheless, it is the enhanced expression of a common gene that is mediating the biological function.

A different example is the use of the antifreeze protein (AFP) gene to confer freeze resistance. Antifreeze proteins, which are produced in several cold water marine teleosts — such as the winter flounder, ocean pout, sea raven and shorthorn sculpin — inhibit ice crystal formation in the blood and protect the fish from freezing. Not all commercially important fish contain AFPs and the gene. For example, the Atlantic salmon lacks the gene and cannot survive sub-zero temperatures. This inability to tolerate freezing temperatures is a major limitation to their sea cage farming off the Atlantic coast of Canada. In contrast, the winter flounder (*Pleuronectes americanus*) thrives in these icy environments.

To enhance the freeze tolerance of Atlantic salmon, the flounder AFP gene may be incorporated into the salmon's genome. The successful development of a freeze tolerant transgenic salmon will surely extend the range of Atlantic salmon farming in many coastal regions in Atlantic Canada.

2.5   Disease Resistance

High density culture conditions mean that many fish are under stress and are generally more susceptible to infection. A major disease outbreak translates into loss of revenue and possibly the complete shutdown of operations. So more robust strains of fish capable of resisting a wide variety of diseases are urgently needed by the industry.

Several approaches are feasible using transgenic technology. Antisense and ribozyme technologies could be used to neutralize or destroy the viral RNA such as the infectious hematopoetic necrosis virus (IHNV) which causes extensive mortality in salmonids. Another possibility is to express the viral coat proteins such as the 66kDa G protein of IHNV in the host membranes. The expression of this viral protein might titrate out the receptors for the virus, thus minimizing viral penetration.

These two methods are effective but are restricted to one or related pathogens. Alternative methods include boosting the host's own immune surveillance and expressing antimicrobial or antibacterial substances. Since immunology in fish has not yet been thoroughly investigated, several laboratories have focused on producing antibacterial substances as a way of conferring disease resistance against a wide range of pathogens.

3   Gene Transfer In Fish

Gene transfer is based on two major advances in molecular and developmental biology. The first is molecular cloning and the second is embryonic micro-manipulation. Molecular cloning enables the isolation and cloning of a single gene that codes for a unique protein while micro-manipulation enables foreign genes to be introduced into an organism's genome. In 1982, recombinant genes were cloned and transferred into host animals, of which the transgenic "super mouse" was the most exciting achievement in transgenic studies. These first transgenic animals, mice that grew to twice their normal size, thanks to rat and human genes that produce growth hormone. The mouse experiment prompted other scientists to start manipulating traits in a range of species. Many researchers saw the new technology as a way to help farmers produce more food with less resources. The idea is to grow more fish in less space to decrease pressure on the environment. This is particularly true when we factor in the fact we will never be able to catch more fish than we do now from the natural environment, and yet the world demand for fish is increasing.

Based on the technique of microinjection, the recombinant gene *MThGH*, a mouse metallothionein-I gene promoter triggering a DNA sequence coding for human growth hormone, was introduced into fish eggs. The principal advantage of microinjection is the efficiency of generating transgenic lines that express most genes in a predictable manner, and the method has been most widely and successfully used for generating transgenic animals. The art of microinjection, however, is both labor- and time-consuming. To overcome the disadvantage, much more convenient techniques, e.g., electroporation or sperm-mediated for DNA transfer, were subsequently developed. Using the new methods, transgenics have been created in a number of species of fish such as salmon (*Oncorhynchus tshawytscha*), goldfish (*Carassius auratus*), loach (*Misgurnus anguillicaudatua*) and zebrafish (*Danio rerio*), common carp (*Cyprinus carpio*) and shellfish (*Pinctada maxima Jameson*). And in recent years, more than 10 laboratories throughout the world have been successful in generating transgenic fish in a variety of species using different foreign gene constructs and transgenic techniques.

When a novel gene is transferred into the fertilized eggs of fish, there are several possible outcomes: (1) the gene may replicate and its descendants persist for several cell divisions, (2) it may integrate into the chromosomal DNA of some host cells and generate transgenic somatic cells, (3) it may integrate into the chromosomal DNA of the host progenitor germ cells and the founder will pass the transgene onto the F1 progeny, (4) alternatively, it may be lost in a few of the founder embryos.

Growth and feed utilization by *MThGH*-transgenic F4 fish fed with diets containing different protein levels has been carried out. For example, in comparison with controls, intakes of protein and energy are significantly higher in the transgenics fed with 20% protein diet, and recovered energy, as a proportion of protein intake, was significantly higher in the transgenics fed with 40% protein diet. This reveals that at a lower dietary protein level, transgenics achieve higher growth rates mainly by increasing feed intake; and at a higher dietary protein level, transgenics achieve higher growth rates mainly through higher energy conversion efficiency. That is to say, transgenics are more efficient in utilizing dietary protein than the controls, which lead to transgenics getting a significantly higher specific growth rate. Transgenic fish have significantly higher body contents of dry matter and protein, but lower contents of lipid than the controls. The apparent digestibility of amino acids tend to be higher in the transgenics than in the controls, especially in fish fed diets with lower protein levels. There is no difference in the proportion of amino acids in the transgenics and controls. Thus, the transgenic fish have higher nutrition value than the control fish. It is reasonable to assume that this kind of fish with "fast growing and less eating" and "higher content of protein and lower content of lipid" traits may fulfill humanity's increasing requirement for protein food source.[9]

4 Transgenic Fish

Americans already eat modified corn, potatoes and other crops. Soon to come are the first such animals: disease-resistant shrimp, meatier chickens and fast-growing salmon. Thanks to mouse DNA, a new pig produces a less harmful manure. New crops include a rice, mixed with daffodil DNA, that includes more nutrients.

4.1 Carp — Food and See Food

Cyprinus Carpio, Carp, are a member of the *Cyprinade* family originated in Asia, primarily China and were bred as early as 1000 BC for food. It was from China that they were introduced to Japan and later to Europe both as ornamental fish and as a food and are still eaten today in some of the eastern European countries. Carp were introduced into the U.S. by the Carp Commission of President U.S. Grant around 1877. They were brought in as a food fish. Other food species were in decline and carp were thought to be a good solution.

Carp were seen by many U.S. anglers to be a nuisance fish but are now gaining much more respect among U.S. anglers who finally realize their true potential as the great hard fighting sport fish they really are. They are a large omnivorous fish, yellow-green and brown in color. They have two barbells on each upper jaw. While the average carp is perhaps 20 pounds in weight they can grow to 80 pounds in weight. They can live alone or in schools in mud bottomed lakes, ponds, and rivers.

Carp reach sexual maturity around their third year and can live to 50 years or more. The female deposits her eggs on plants or alike in shallow water usually during the spring. It generally takes a good two weeks of sunshine and good temperatures to encourage carp to spawn. Eggs take between 4 and 8 days to hatch.

The common carp is the original strain with the mirrors being genetically bred out of the stock. In China, carp were the first fish to be bred for food with several domesticated varieties emerging afterwards. Chinese domesticated varieties include:

1. Common carp — The most "common" carp with lots of small uniform scales all over.
2. Mirror carp — A variety with small numbers of large scales scattered randomly.

---

[9] Wei Hu, and Zaoyan Zhu, "Transgenic fish and biosafety", In *Gene Families: Studies of DNA, RNA, Enzymes and Proteins*, Guoxiong Xue, Yongbiao Xiong, Zhihong Xu, Roger Holmes, G. Hammond, and Hwa A. Lim (eds.), (World Scientific Publisher, New Jersey, 2001), chapter 5.

3. Leather carp — A variety with no scales at all, but with a leathery appearance, hence the name "leather carp".
4. Grass carp — A variety similar to the common carp but with a long slender body and upturned mouth.
5. Linear carp — A variety like a mirror but with scales along the lateral line only.
6. Crucian carp — A variety that is barbell-less and is a relative of the gold fish.

The Japanese were responsible for breeding Koi Carp, which are large ornamental variety with a mixture of colors including white, orange gold, and black.

4.1.1   Genetically Modified Carp

Nevertheless, in view of the biosafety considerations, both the mouse *MT-1* gene promoter and the *hGH* structural gene are not suitable for the purpose of fish breeding. For this reason, researchers cloned common carp β-actin (CA) gene, grass carp growth hormone (*gcGH*) gene, and generated a new construct of *pCAgcGH*, an "all-fish" genomic construct cloned in pUC118. The construct of *pCAgcGH* consists of β-actin gene promoter of the common carp and the whole transcription unit of GH gene from grass carp. The β-actin gene promoter of common carp is a powerful promoter and the grass carp is a fast growing farm species. It is reasonable to believe that the *pCAgcGH* gene will be a strong "generator" to promote fish growth rates. In the spring of 1997, this construct was microinjected into the fertilized eggs of Yellow River Carp and a batch of *CAgcGH*-transgenic fish was produced. The transgenic fish weighed 2.75kg in 5 months while the largest of the non-transgenic controls was 1.1kg. The heaviest body weight of 17-month-old transgenics, 7.65kg, was more than double of their non-transgenic siblings. To date, there are more than ten "all-fish" recombinant genes constructed in different laboratories since the first "all-fish" expression vector was created in 1990.[10] The most spectacular "super fish" was produced by R.H. Delvin and colleagues who inserted an all-salmon gene construct (*pOnMTGH1*) into Coho salmon (*Oncorhynchus kisutch Walbaum*). On average, the transgenic salmon were more than 11-fold heavier than the controls, ranging from no growth stimulation to one individual 37 times larger than the controls.[11]

4.1.2   Safety Considerations of Transgenic Carp for Human Consumption

The "all-fish" gene construct *CAgcGH*-transgenic Yellow River carp has huge potential in the aquaculture industry. However, transgenic fish are genetically

---

[10] *Safety Evaluation of Food Produced by Modern Biotechnology: Concepts and principles*, (Organization for Economic Cooperation and Development, Paris, 1993).
[11] R.H. Devlin, T.Y. Tesaki, C.A. Biagy, E.M. Donaldson, P. Swanson, W.K. Chan, "Extraordinary salmon growth", *Nature* 371, 1994, pp. 209–210.

modified organisms (GMO). There are measures and regulations to guide GMO before they can be accepted for human consumption. The food safety, genetic safety and ecological safety are being strictly evaluated. At present, the widely accepted principle on safety evaluation of foods produced by modern biotechnology is the "substantial equivalence principle" put forward by the European OECD (Organization for Economic Cooperation and Development) in 1993. According to this principle, the safety class of "all-fish" gene construct *CAgcGH* has been assessed.

The construct of the *CAgcGH*-transgene consists of β-actin gene promoter from common carp and the whole transcription unit of GH gene from grass carp. Both grass carp and common carp are farmed species. They belong to the same family of *Syprinedae*. Between the two species, the similarity of the exons of the *GH* gene is 84.1–93.2% and the homology of the amino acid sequence of *GH* polypeptide is 97%. As a result of the transgene's expression, the transgenic Yellow River carp contains grass carp GH of 2–10ng/mL serum.

A study has been conducted in mice to see if this level is safe for human consumption? 120 mice of each group were fed with fresh meat juice of the *CAgcGH*-transgenic Yellow River carp and the non-transgenic controls, respectively. The feeding dosage per day was 10g/Kg body weight and 5g/Kg body weight, respectively. In comparison with the control, the test group of mice did not show any significant differences in growth, reproduction, general appearance of blood, biochemical indicators on blood, and histochemical analysis of tissues. The same results have also been ascertained for the F1 generation of the two groups. In addition, the polypeptide of GHs is unstable and were degraded rapidly in acid, alkali and heat. Thus, both the transgene construct and the expression product in the transgenics are as safe in the "parental" ones as in food resources.

4.1.3   Environmental Safety — Gene Flow

To evaluate the genetic and ecological safety of *GH*-transgenic fish, preliminary studies in a polyculture system have been conducted. The *GH*-transgenic red common carp with crucian carp (*Carassius auratus* L.), grass carp, big-head carp (*Aristichthys nobilis*) and silver carp (*Hypophthalmichthys molitrix*) are stocked in very well isolated ponds. PCR analyses show that the *GH*-transgene can only flow among individuals within species but not between species by natural reproduction. In addition, the transgenics and the controls show no observable effects in the growth of crucian carp, grass carp, bighead carp and silver carp. Furthermore, the total yield in polyculture system with the transgenic common carp is higher than that with the non-transgenic counterparts. The preliminary results suggest that stocking the transgenics in ponds would gain fish productivity without transgene flowing between species.

A way to solve the genetic and ecological challenges of transgenic fish in aquaculture is to make them sterile. Polyploid-breeding in fish has become very popular. For example, an artificial tetraploid fish have been obtained via hybridization of common carp against goldfish. By crossing the tetraploid fish with the haploid transgenics, the resulting transgenics are sterile. It is reasonable to assume that stocking infertile transgenic fish will lessen their impact on the water ecosystem.

4.1.4 No Carping with Carp — Natural Gene Flow

In fact, gene flow between fish species occurs in water body from time to time. For example, crossbreeding between closely related species happens naturally. And artificial crossbreeding is even more frequently carried out in fish farms and when no special precautionary measures are taken to ensure the offspring from escaping to natural water body. Each fish genome has about 105 genes. There are 105 genes added from one species to the other when crossbreeding occurs.

In producing transgenics, there is only one gene or a few genes from one species transferred to the other. For example, in the *CAgcGH*-transgenic Yellow River carp, one *GH* gene from grass carp is transferred into the common carp genome; the mass of gene flow is 1/105 as the two species crossbreeding. In other words, it may be argued that the risk of stocking "all-fish" gene-transferred fish is likely to be much less, or at most, substantially equivalent to that of stocking cross fish. In other words, according to the substantial equivalence principle, the safety class of "all-fish" transgenic Yellow River carp is evaluated to belong to the safest level or "level I". Technically speaking, the *CAgcGH*-transgenic common carp, as well as another transgenic fish (salmon) in Canada, are likely hopefuls to be the first cases of commercial transgenics being adopted by the market.

While the researchers are very optimistic, some critics are carping (carp, a verb, not the fish) safety concerns.

4.2   The Tale of Catfish — Transgenic Creole

A few miles outside the college town of Auburn, Alabama, down a gravel road that runs through rolling woodlands, Rex Dunham has turned a set of muddy ponds into a high-security prison, for catfish that is. Electric wire keeps raccoons at bay. Netting blocks the herons from swooping in. Filters stop the fish from slipping out with the waste water. Federal officials asked Dunham to protect the local environment from the catfish he grows here because nothing like them has ever cut the waters of Earth. These catfish have been laced with DNA from salmon, carp and zebrafish, which makes them grow as much as 60% faster than normal. That could help farmers feed more people for less money and boost efforts to end world hunger.[12]

---

[12] Aaron Zitner, "Gene-altered catfish raise environmental and legal issues", *Los Angeles Times*, January 2, 2001.

Figure 4. The genetically modified catfish (top) reaches market size in much shorter time than the control. (Photo: Courtesy of *Los Angeles Times*).

But there also is a chance that fast-growing fish might touch off an environmental disaster, according to scientists who have studied the matter. Their greatest fear is that Dunham's catfish will escape and wipe out other fish species, as well as the plants and animals that depend on those fish to survive. And now, some scientists and government officials are raising a second and equally troubling concern: that the federal government has limited legal authority to protect the environment from Dunham's catfish, or from some of the dozens of other genetically modified plants and animals now being readied for market.

Though we are on the brink of remaking life on Earth through genetic engineering, we still do not have a thorough process for reviewing the environmental impacts. William Brown, science advisor to Interior Secretary Bruce Babbitt remarked that the current system is full of holes. Bill Knapp, a senior fisheries official with the U.S. Fish and Wildlife Service feels that the current system is not going to be okay and that there are going to have to be changes, or a whole new system put in. This view is far from universal. But concerns about the government's legal authority are significant enough that President Clinton ordered federal agencies in May of 2000 to review the relevant laws and probe for holes.

Dunham, an Auburn University researcher, already has started seeking federal approvals to sell his fish. And he could be among the first to win approvals to sell a genetically modified animal to American consumers. Dunham had spent years using traditional breeding techniques to modify the channel catfish, which is by far the most farmed fish in the United States.

Normally, catfish stop growing in the winter, when the genes that produce growth hormone all but shut down. Dunham and his team began producing catfish that had an extra copy of a growth hormone gene. They also added a piece of DNA from salmon, carp or other species that acts like a year-round "on" switch for the gene. The resulting catfish grow to their market size of about 2 pounds within 12 to 18 months, rather than the normal 18 to 24 months.

Dunham and his research partner, Zhanjiang "John" Liu, hope to turn the fish into a commercial product. Several fish geneticists believe the Auburn catfish could be the second genetically modified animal to reach American consumers. A/F Protein Inc., a Massachusetts firm, is expected to be first. It is seeking approval for a fast-growing salmon that it is developing in indoor tanks in Canada.

Dunham and Liu also have begun researching how their fish would behave in the wild. So far, they have found no cause for concern. One published study found that the fish have slightly less ability to avoid predators than do native catfish. Two other studies determined that the Auburn catfish do not have a competitive edge over native fish for food and have equal reproductive ability.

What these studies point to is that these fish have no environmental advantage, or maybe are a little handicapped in the natural environment. But the principal point is that more research is needed to determine what the environmental risk is.

4.3   Salmon for Sushi

Whether served as raw sushi, grilled steak or in thin smoked slices, most of the salmon one eats these days is not the sleek sport fish that has been a favorite of anglers but rather a chunky, sluggish creature raised in captivity. Indeed, salmon caught in the wild accounts for less than half of all salmon sold in the U.S.

Now gene splicers have cooked up a replacement that sounds like a fish tale: a veritable superfish, one that can grow at least twice as fast, resist disease and potentially outmate competitors. If approved, it could provide protein to millions of people at a time when fish stocks are perilously low. But, as one may have expected, some critics are carping (carp, a verb, not the fish carp). They consider the supersalmon a biological time bomb that could destroy the remaining natural salmon populations and wreak other environmental havoc. To them, the supersalmon is nothing less than a "Frankenfish".

Unlike other genetically modified foods or so-called Frankenfoods — the supersalmon was born almost accidentally. About 20 years ago, a fish researcher in Newfoundland found that even though his saltwater tank had frozen, the flounder in it survived. Adapted to icy Canadian waters, the fish turned out to have a gene, known in other polar fishes, that produces an anti-freeze protein. While trying to splice this gene into salmon so it too could be grown in colder waters, scientists made a second accidental discovery: they found that while the gene did not keep the salmon from freezing, a portion of it, when stitched onto a salmon's growth-hormone gene, greatly speeded development, up to five or six times as fast as in the early months and about twice as fast overall. Patenting their discovery, the scientists started a company in Waltham, Massachusetts, called A/F Protein (A/F stands for antifreeze).[13]

As of Y2001, the company has 10,000 to 20,000 Atlantic supersalmon swimming in endless circles in 136 tanks at three locations in Canada's maritime provinces. The hope is that these fish will soon be producing eggs for commercial aquaculture not just in Canada but in New Zealand, Chile and the U.S. as well. By turning to the supersalmon, according to Elliot Entis, A/F's president, fish farmers can double production without doubling costs because the fish converts food into

---

[13] Frederic Golden, "Make way for Frankenfish", *Time Magazine*, March 30, 2001.

body mass so much more efficiently than ordinary salmon. That will mean "more fish for more people at a lower price".

Figure 5. These salmons are siblings, but the genetically modified one is spectacularly larger. (Photo: Courtesy of R. Devlin, Fisheries and Oceans, Canada).

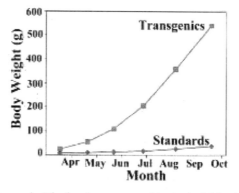

Figure 6. Growth of transgenic Atlantic salmon compared to standard Atlantic salmon between April through October 1995. The salmon were first fed on March 2, 1995 and were held at approximately 16°C. (Figure: Adapted from Aqua Bounty, Canada).

But this so-called blue revolution may not reach U.S. shores for a while. Although gene scientists in the U.S. have been tinkering with a variety of marine creatures — not only salmon and trout but also carp, catfish, tilapia and shrimp — these efforts are drawing criticism similar to that directed at genetically modified foods. Opponents, who complain about the fertilizers and other pollutants released into coastal waters by the fish farms, are especially concerned about the potential impact on the gene pool. They note that domesticated fish, transgenic fish inclusive, regularly escape from their pens into the wild and breed with native stocks, upsetting the balance of nature.

No one knows what ripple effects might occur if the new supersalmon escaped into the wild. One of the few studies done by U.S. researchers found a lower survival rate for eggs produced by transgenic fish. Still other studies show that despite their name, the so-called superfish have diminished muscle structure and swimming performance. Canadian fish geneticist Robert Devlin finds that science at the moment is unable to provide a reliable risk assessment.

Entis and others are confident that whatever the risk, it could be lowered to almost zero by raising the fish in closed tanks rather than in storm-exposed pens. Still another tactic under consideration is shocking the fertilized eggs so they create fish that cannot reproduce — a marine equivalent of the self-destructing terminator gene that Monsanto once considered putting in its patented plant seeds.

Fearing a consumer backlash, New Zealand King Salmon, a major producer of Chinook salmon — the largest Pacific salmon — announced in March 2001 that it was suspending its gene-modification experiments. Entis of A/F Protein, by contrast, believes he can win acceptance of his supersalmon through public education.

But supersalmon lox are probably not going to be available for public consumption anytime soon. The Food and Drug Administration must first approve introduction of the fish into the U.S., something that probably will not happen before 2001 ends.

4.4   Shrimps

Increasing consumer demand for shrimp products in the U.S., combined with static shrimp fishery yields, has led to rising shrimp imports resulting in a $2.9 billion annual trade deficit. Ocean Institute (OI) and its consortium partners, through funding from the U.S. Department of Agriculture, are developing and evaluating environmentally sensitive technologies to expand domestic shrimp aquaculture production.

Through the U.S. Marine Shrimp Farming Program, OI has been a leader in the selective breeding of high health, genetically improved shrimp and currently maintains the world's most genetically diverse population of Pacific white shrimp *(Litopenaeus vannamei)*. These shrimp have been bred for rapid growth and improved disease resistance, providing significant advancement for the U.S. shrimp aquaculture industry.

To date, over 500 full-sibling families have been evaluated for harvest weight in different growout environments and for resistance to Taura Syndrome Virus, a pathogen that has plagued shrimp farmers in the Western Hemisphere for almost a decade.

OI has been a leader in the design and evaluation of biosecure shrimp production systems in response to problems of virulent shrimp viruses and growing concerns about environmental pollution from shrimp pond effluent. These systems rely on minimal water exchange and provide U.S. shrimp farmers with the technology to move production away from the coastline. In 1998, the prototype biosecure shrimp production system at OI produced more than 3.8 pounds of shrimp per square meter, using recycled water. Current research efforts are directed at making this system cost-effective by increasing production through the development of genetically improved shrimp and non-polluting feeds that support rapid shrimp growth.

5 The Prospect Of Transgenic Fish

The transgenic fish that have been produced at present are far from a genetically homogenous strain. Scientists are pursuing transgenics with transgenes of site-specific integration, controllable expression and stable germline-transmission. The constructs of foreign genes being transferred in fish are usually less than 10 kilobase in size. These smaller constructs might miss some important elements for regulation of gene activity. With the development of artificial chromosomes capable of cloning DNA fragments up to 2 megabases long, it is now possible to use intact genomic loci as transgenes. Consequently, the foreign genes would be similar to the context of their native sequence environment in the transgenics, enabling the transferred genes to be expressed at levels comparable to that of the corresponding endogenous gene, as well as in a tissue-specific and developmentally correct manner. In 1998, a simple method for modification of bacterial artificial chromosomes (BACs) through Chi-stimulated homologous recombination was developed and used for zebrafish transgenesis. The DNA constructs microinjected into zebrafish embryos with the modified BAC can display the correct spatio-temporal gene expression pattern. More importantly, those embryos show less mosaic and had improved the foreign gene expression in special cells compared with the smaller constructs.

It is suggested that the artificial chromosomes have potential for producing transgenics of site-specific integration and controllable expression. The "gene targeting" technique may be an alternative way for producing a pure line of transgenic fish. Gene targeting — homologous recombination between DNA sequence residing in the chromosome and a newly introduced cloned DNA sequence — allows the transfer of any modified gene into the host genome of living cells and is hopeful to be carried out to gain embryonic cell lines carrying artificially modified and site-specific integrated gene.[14] Fish are at lower evolutionary stage and the totipotency of fish cells (both embryonic and somatic cells) is much higher than that of other vertebrate cells. Additionally, embryonic stem-like-cell line has been partially achieved in zebrafish and medaka (*Oryzias latipes*).[15,16] Meanwhile, nuclear-cytoplasm hybrid fish could be produced between different species and even between different genera via nuclear transplantation. Therefore, by nuclear

---

[14] M.R. Capecchi, "Altering the genome by homologous recombination", *Science* 244, 1989, pp. 1288–1292.

[15] Y. Wakamatsu, K. Ozato, and T. Sasado, "Establishment of a pluripotent cell line derived from a medaka blastula embryo", *Mol. Mar. Biol Biotech.* 3, 1994, pp. 185–191.

[16] L. Sun, C.S. Bradford, C. Ghosh, P. Cotllodi, and D.W. Barnes, "ES-like cell cultures derived from early zebrafish embryos", *Mol. Mar. Biol. Dev* 4, 1995, pp. 193–199.

transplantation with gene-targeted embryonic cells, a genetically homogenous strain of transgenic fish with desired traits could be generated.

6   Other Fish To Fry

Recent successful commercial culture of channel catfish, salmonids, and crayfish has led researchers and entrepreneurs to investigate and invest in a wide array of potential finfish, crustacean, and molluscan candidates for commercial rearing. Commercial production is essential in meeting the increasing demands for food fish and for sport fishing, the provision of aquarium fishes and bait fish, and the local production of fish to replace imports. Such rapid expansion of the industry, however, has raised many concerns. Conflicts in both inland and coastal waters can be expected to intensify as aquaculturists, recreationalists of all types, developers, environmentalists, and commercial fishers contend for use of the same bodies of water.

Disease problems, genetic pollution, escape of exotic and introduced species, and eutrophication are areas of greatest concern. There is the possibility of amplifying pathogenic organisms in an intensive culture system which might be released with or without fish into wild populations. All states and provinces should have fish health programs, but because of the diverse nature of these programs, only the federal government may be able to consistently apply equal standards throughout the country.

Biotechnologies are now providing mechanisms to genetically manipulate organisms to promote economic advantages through increased growth rates, sex reversal, etc. However, the effects of the escape or release of these genetically altered organisms into the natural environment are not known. It is imperative that aquaculturists understand the need for and that governmental agencies enforce regulations to safeguard wild populations from escaped aquaculture species, whether genetically altered or exotic.

Successful commercial aquaculture usually implies a highly intensive management system that often results in nutrient-rich effluents. Since many factors such as ratio of volume of receiving water to effluent, frequency of discharge, nutrient load, geographic location, etc, are involved in each aquaculture operation, acceptable standards should be set and enforced by regulatory agencies to avoid eutrophication of receiving waters.

The growth and profitability of worldwide aquaculture depends on efficient, economical, and environmentally sound hatchery and growout methods. Whether working with foodfish or ornamentals, for example, Ocean Institute's Marine Finfish Program yields benefits for Hawaii's own aquaculture industry while also addressing issues of national and global relevance. OI is working with more than eight species, representing all ecological zones, for purposes of fisheries restoration and reef conservation, as well as to encourage commercial foodfish and ornamental markets.

Although there has been great attention paid to whether these foods are safe to eat, the potential risk to the environment could be an even bigger concern. And, the government is stretching outdated laws to cover the genetic revolution, as if using 19th century railroad laws to regulate airlines.

Some warn that genetically modified plants and animals could move into the wild and breed disruptive traits into local species, similar to the way African "killer bees" escaped a Brazilian research facility in 1957 and spread their aggressive traits. Others fear an opposite scenario: that instead of thriving, the modified plant or animal could interbreed with its natural cousins in ways that would destroy the species entirely.

Scientists call this the "Trojan gene" effect, because the modified organism is undermined by the new genes that it takes in. William M. Muir, a geneticist at Purdue University, has used a mix of laboratory observation and computer modeling to show that it could happen with gene-altered fish. Fast-growing fish might enjoy a mating advantage in the wild, yet produce young that are ill-equipped to survive. This could locally take a population to extinction.

And yet, federal officials say that no law requires people who alter fish genes to keep the fish isolated and away from local waters. The Agriculture Department was able to ask researchers to build "fish prisons" only because their research is backed by federal funds. Moreover, it is unclear whether any federal law penalizes a person who releases genetically modified animals into the wild.

More troubling to some critics is that certain species may escape federal regulation entirely. For example, at least one company is altering the genes in creeping bentgrass, a common golf course turf, so that it is more resistant to weed killers. That would allow lawn managers to use herbicides without harming the turf. But it could also make the grass, which already invades lawns and gardens, harder for homeowners to control. Officials are divided over whether the government has the authority to regulate genetic changes to the grass. The Agriculture Department claims authority over all "plant pests" and potential pests, and it is using that authority to supervise the company working on creeping bentgrass genes. But others disagree, saying that the legal definition of plant pests clearly excludes the grass. The department has overstepped its legal authority, according to these people.

Similarly, several teams are working to modify algae as a food and laboratory substance, according to Anne Kapuscinski, a fish geneticist at the University of Minnesota. Algae is not a plant pest, and consequently, who is going to have authority over it?

The confusion arises because the government, starting with the Reagan administration, decided that decades-old food and agriculture laws could be stretched to cover genetically altered species. For example, some corn and potato varieties already on the market have been genetically modified to produce their own insecticide. Because the Environmental Protection Agency has jurisdiction over insecticides, it takes a lead role in regulating these crops.

For other crops, the Agriculture Department claims a leading role because scientists commonly use bacteria and viruses to modify the crop genes. The agency already regulates those bacteria and viruses as plant pests, and it claims jurisdiction over the crops as well.

Jane Rissler, senior staff scientist with the Union of Concerned Scientists, called this rationale "an awkward stretch of the laws" that does not cast a broad net over all gene-altered plants. She is of the opinion that the mere fact genes have been engineered should be enough to bring a plant or animal under federal scrutiny. Besides, scientists now are modifying genes in ways that do not rely on bacteria or viruses but that should not release them from federal regulation.

In regulating fish, some people believe the laws are being stretched in equally awkward ways. For example, in the transgenic catfish case, the lead regulator would be the Food and Drug Administration(FDA), but not because the fish would be a food. Instead, the agency considers the fish's extra growth hormone to be a drug! But some wildlife experts say that, although the FDA is well-equipped to assess drugs, it is the wrong agency to rule on whether genetically modified fish pose a risk to the environment.

FDA officials say they are routinely called on to consider environmental effects. John Matheson, senior review scientist for veterinary medicine, noted that, when the agency recently reviewed a growth hormone for cows, it studied potential changes in land-use patterns, soil erosion and methane levels.

Critics of the system raise another complaint about the FDA's role: It operates under a federal law that aggressively protects company trade secrets, and an often anxious public cannot learn what genetically modified plants and animals are on the road to winning federal approvals.

Chapter 5

CLONOLOGY AND CLONETECHNOLOGY

"One egg, one embryo, one adult-normality. But a bokanovskified egg will bud, will proliferate, will divide. From eight to ninety-six buds, and every bud will grow into a perfectly formed embryo, and every embryo into a full-sized adult. Making ninety-six human beings grow where only one grew before. Progress."
Aldous Huxley, *Brave New World*, Chapter 1, 1932.

1   The Genetic Age

Throughout the Industrial Age, we have radically altered our society and environment by extracting and transforming massive amount of inanimate material. Numerous metals, minerals, and fossil fuels are mined, filtered, dug, and pumped from Earth. They are burned, forged, soldered, melted, reconstructed, and recombined to create the machines, structures, and artifacts of the modern world.

We are now adding living materials to the inanimate matter being transformed in our system of production. Advances in gene technology have enabled us to begin the engineering and commodification of living materials.

But what is the use or profitability of isolating and patenting human and other valuable genetic material unless it can be copied and reproduced in industrial quantities? Why create and patent transgenic livestock and research animals unless they can be duplicated at will? And finally, why genetically engineer "superior" humans, or perfect body parts, if one cannot then make endless copies? In order to establish an efficient living matter production line process, biological engineers need to invent a process for the industrial-scale reproduction of life forms. Just like manufacturing, biofacturing is a procedure that will systematically reproduce exact duplicates of genetic material, endless "carbon copies" of engineered microbes, plants, and animals, including humans and human subparts.

From the onset of the genetic revolution, the challenge was clear: If we can manufacture millions of identical machines, books, clothing, or computers, why can we not biofacture millions of identical life-forms, including the human body and all its subparts?[1] Natural procreation, as in sexual reproduction, clearly will not do the trick of biofacturing. Creation through intercourse is too slow, unreliable and unpredictable. It simply does not have the efficiency or speed required for a life-form body shop, be it microbes, plants, animals or humans.

Thus for over four decades, genetic engineers have been attempting to meet the challenge of artificially and efficiently producing life. Having overcome the challenge, now they are in the process of perfecting biofacturing.

---

[1] Andrew Kimbrell, *The Human Body Shop*, (Harper, San Francisco, 1993).

2   Clonolgy

Cloning, or photocopying of life forms, involves the production of genetically duplicate organisms from the biological information contained in a single cell. The chronology of cloning — clonology) — started almost seven thousand years ago.[2] In 5000 B.C., early humans discovered that if they planted seeds produced by the heartiest plants, the successive crops would be a stronger one. This is generally regarded as the first instance of manipulating life to suite human needs.

2.1   Parthenogenesis — Virgin Birth

Over a century ago, researchers began to manipulate a natural form of asexual reproduction called parthenogenesis. Parthenogenesis is a strange and unusual form of reproduction that occasionally occurs in plants and animals. It is derived from a Greek word. *Parthenos* meaning virgin to signify that in parthenogenesis there is no mixing of parental genes because all the genes come from one parental organism. Parthenogenesis produces offspring by certain actions within a single cell. These actions include cell division in bacteria, cell budding in yeasts, and vegetative duplication in plant cutting.

Parthenogenesis has been observed in certain insects and animals. By the end of the nineteenth century, pioneer experimenters with parthenogenesis use chemicals on the eggs of various species to artificially induce parthenogenic creation of offspring. For example, in 1896, German embryologist Oskar Hertwig added strychnine or chloroform to seawater containing sea urchin eggs. Remarkably, the process fertilized the eggs without any contact with sperm. Three years later, another German, Jacques Loeb, successfully duplicated the sea urchin experiments.

The postal child of parthenogenesis was on the cover of the 1939 *Life* magazine. Despite its normal appearance, the rabbit had been created inadvertently by Gregory Pincus by administering thermal shock treatment to a female rabbit.[3] Interestingly enough, Pincus would later conduct pioneering birth control pill work.

2.2   Photocopying Life Forms

Parthenogenesis was the first important step toward artificially reproducing life forms. Cloning has to wait for more significant advances in the understanding of cells and genes and the technique did not become practical until 1952 when Robert Briggs and Thomas J. King performed a landmark experiment replacing the nuclei of freshly fertilized frog eggs with foreign nuclei from another frog of the same species. All the resulting tadpoles were duplicates of the tissue cell donor. In other words, the frog had been cloned.

---

[2] MSNBC Interactive, www.msnbc.com/news/49963.asp#BODY

[3] D.S. Halacy, Jr., *Genetic Revolution, Shaping Life for Tomorrow*, (Harper Row, New York, 1974).

Other successful experiments in creating frog clones followed, but researchers would struggled for decades to achieve the far more difficult feat of cloning mammals. Nevertheless, the transition from human parthenogenesis to cloning has been made. The egg has been proven to be essentially an environment and that it is the implanted nucleus with its full set of chromosomes that determines the genetic makeup of a clone individual.

Table 1. The chronology of cloning (pre 1960). This table is adapted from Thinkquest.[4]

Pre-1800
    Early humans learned about plant breeding by trial and error.
1885
    August Weissman states that genetic information of a cell diminishes with each cell division.
1902
    Walter Sutton proves that chromosomes hold genetic information.
    Hans Spemann divides a Salamander embryo in two and shows early embryo cells retain all the genetic information necessary to create a new organism.
1928
    Hans Spemann performs the first nuclear transfer experiment.
1938
    Hans Spemann proposes "fantastical experiment" of cloning higher organisms.
1944
    Oswald Avery discovers genetic information is carried by the nucleic acids of cells.
1952
    Robert Briggs and Thomas King clone tadpoles.
1953
    Watson and Crick discover the structure of DNA.
1958
    F.C. Steward grows whole carrot plants from carrot root cells.

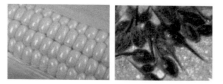

Figure 1. Left: Early humans learned that if they planted seeds from heartiest plants, subsequent harvest would be a better one. Right: In 1952, a tiny tadpole was the first cloned animal. (Photos: Courtesy of Yoav Levy).

By the 1970s, Paul Berg and Stanley Cohen invented the techniques of genetic manipulations. Cloning stepped down to the minute level with the first cloning of a

---

[4] http://library.thinkquest.org

gene. Scientists isolated the gene, then bound it to an organism (in this instance a yeast) that incorporated the gene into its own DNA and multiplied, producing many copies of the desired gene in the organism's multiplication process. In 1976 Rudolf Jaenisch of the Salk Institute for Biological Studies in La Jolla, California, injected human DNA into newly fertilized mouse eggs to produce mice that were part human. When the mice reproduced, they passed their human genetic material to their offspring, creating a slew of so-called transgenic mice. Different human diseases can be studied by creating mice with the appropriate genetic composition

Two years later, in 1978, the world clamored for a glimpse of Baby Louise, the first child conceived through *in vitro* fertilization. Using the husband's sperm, British doctors fertilized an egg in a petri dish, and then implanted the embryo in the uterus of a healthy woman.

In 1987, the first mammals, sheep and cows, were cloned from embryonic cells. But animals cloned from embryonic cells contain the genetic material of both parents because the embryos are sexually fertilized. Clones from embryonic cells from the same parents fertilized at different times are as different as brothers and sisters.

Table 2. The chronology of cloning (1960–1980). This table is adapted from Thinkquest.

1962
    John Gurdon claims to have cloned frogs from adult intestine cells.
1963
    JBS Haldane coins the term "clone".
1966
    Establishment of the complete genetic code.
1967
    Enzyme DNA ligase isolated.
1969
    Shapiro and Beckwith isolate the first gene.
1970
    First restricted enzyme isolated.
1972
    Paul Berg creates the first recombinant DNA molecules.
1973
    Cohen and Boyer create the first recombinant DNA organisms.
1977
    Karl Illmensee makes claims to have created mice with only one parent.
1978
    The release of David Rorvik's book "In his image: the cloning of man" sparks a worldwide debate on cloning ethics.
1979
    Karl Illmensee makes claim to have cloned three mice.

Figure 2. In 1972, cloning steps down to the minute level with the first cloning of a gene. Scientists isolate the gene, then bind it to an organism (in this instance a yeast) that incorporates the gene into its own DNA and multiplies, producing many copies of the desired gene. In 1976, Rudolf Jaenisch of the Salk Institute for Biological Studies in La Jolla, Calif., injects human DNA into newly fertilized mouse eggs to produce mice that are part human. When the mice reproduce, they pass their human genetic material to their offspring, creating a slew of so-called transgenic mice. In 1978, the world clamors for a glimpse of Baby Louise, the first child conceived through in-vitro fertilization. Using the husband's sperm, British doctors fertilize an egg in a petri dish, then implant the embryo in the uterus of the healthy woman. (Photos: DNA, courtesy of Dan McCoy; in vitro fertilization, courtesy of Owen Franken).

2.3   Win a Clone at the Diary Expo

In February 1988, a front-page *New York Times* story featured Grenada's newfound ability to bring factory-like efficiency to animal reproduction.[5]   The article described how Grenada had successfully used a modified version of the nuclear transfer technique utilized in the Briggs and King frog experiment to clone cattle embryos. Grenada scientists inserted the nuclei taken from a prize cow into the fertilized eggs of normal cows, whose nucleic genetic material had been removed. The altered fertilized eggs were transplanted into surrogate mother cows for gestation. The resulting calves were genetic replicas of the prize cow nucleus donor.

Subsequently in the 1990 Dairy Expo in Madison, Wisconsin, the following advertisement appeared in a special issue of *Holstein World*,[6]

Genetic cloning is the newest, and one of the fastest ways to upgrade the genetics in your herd. To develop greater awareness of "cloning as a method of genetic advancement," Grenada Biosciences, Inc. and Holstein World are offering you an opportunity to win a "clone with outstanding genetics for milk and protein"... Remember come to Holstein World's booth during the World Dairy Expo... view the video on "cloning and other industry advancements"... and maybe there will be A CLONE JUST FOR YOU!"

Table 3. The chronology of cloning (1980s).

1980
    US Supreme Court rules live, human made organisms are patentable material.

---

[5] Keith Schneider, "Better farm animals duplicated by cloning", *New York Times*, February 17, 1988.
[6] *Holstein World*, special Diary Expo edition, June 1990.

Table 3. (Continued)

1983
Kary B. Mullis develops the polymerase chain reaction for rapid DNA synthesis.
Solter and McGrath fuse a mouse embryo cell with an egg without a nucleus, but fail to clone using their technique.
1984
Steen Willadsen clones sheep from embryo cells.
1985
Steen Willadsen joins Grenada Genetics to commercially clone cattle.
1986
Steen Willadsen clones cattle from differentiated cells.
Neal First, Prather, and Eyestone clone a cow from embryo cell.

Figure 3. In 1987, the first mammals, sheep and cows, are cloned from embryonic cells. But animals cloned from embryonic cells contain the genetic material of both parents because the embryos are sexually fertilized. Clones from embryonic cells from the same parents fertilized at different times are as different as brothers and sisters.

Table 4. The chronology of cloning (1990s).

1990
Human Genome Project begins in earnest.
1995
Ian Wilmut and Campbell clone sheep from differentiated cells.
1996
Dolly, the first animal cloned from adult cells, is born.
1997
President Bill Clinton proposes a five-year moratorium on cloning.
1997
Richard Seed announces his plan to clone a human.
Ian Wilmut and Keith Campbell create Polly, a cloned sheep with an inserted human gene.
1998
Teruhiko Wakayama develops the Honolulu technique and creates three generations of genetically identical cloned mice.

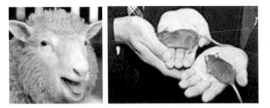

Figure 4. Left: Dolly the sheep, the world's first mammal cloned from a cell of an adult animal, was born in 1996. Right: Scientists at the University of Hawaii cloned more than 50 mice from adult cells, creating three generations of identical laboratory animals. (Photos: Courtesy of Jeff Mitchell, and Peter Morgan).

3   Clonetechnology

The pace of cloning picked up in the second half of the 1990s. Clonetechnology entered another epic.

Dolly the sheep, the world's first mammal cloned from a cell of an adult animal, was born in 1996, but her existence was not revealed to the world until February 1997. Embryologist Ian Wilmut and colleagues at the Roslin Institute in Scotland cloned Dolly from a cell taken from the udder of an adult ewe. In response to public concern, President Bill Clinton issued a moratorium on the use of federal funds for human cloning research.

In 1998, scientists at the University of Hawaii cloned more than 50 mice from adult cells, creating three generations of identical laboratory animals. Meanwhile, several independent teams of researchers successfully cloned calves using differing techniques. Most notably, Japanese researchers produce eight genetically identical calves from the biopsied cells of an adult cow with a success rate of 80 percent, making it the most efficient cloning endeavor to date.

Sheep and cows were followed soon afterwards by monkeys. In 2000, Oregon researchers revealed the existence of Tetra the cloned monkey. The rhesus macaque was cloned using a very different method than Dolly. Tetra was made by splitting a very early embryo consisting of only eight cells into four pieces. These were then nurtured into new embryos, but only one survived. Thus unlike Dolly, Tetra has both a mother and father and is a clone of neither, but is rather an artificial quadruplet. Additionally, the company that had helped produce Dolly unveiled a litter of five cloned piglets.

3.1   The Celebrated Dolly

Scientists who asserted that a mammal could not be cloned from an adult cell must be feeling a bit sheepish. Dolly is the first viable offspring ever derived from adult mammalian cells. The procedure used was deceptively simple — the researchers,

Ian Wilmut and colleagues of the Roslin Institute, Edinburgh, Scotland, removed an unfertilized oocyte (egg cell) from an adult ewe and replaced its nucleus with the nucleus of an adult sheep mammary gland cell. This egg was then implanted in another ewe, and Dolly was the result. By February 1997 when Dolly was announced to the world, she was already seven months old and appeared normal in every respect.[7,8,9]

The clone is appropriately named Dolly because the adult cells with which Dolly was cloned came from a mammary gland. Dolly is named after the well-endowed country singer Dolly Parton.

Figure 5. Left: The young lamb Dolly (left) and her surrogate mother. Note the surrogate mother is a Scottish blackface sheep so that the mother would look different from her baby. Right: Ian Wilmut, whose team cloned the celebrated Dolly, was runners-up Times Magazine Man of the Year, 1997. (Photo: Courtesy PPL Therapeutics. Photo of Wilmut: adapted from *Times*, December 29, 1997, Man of the Year, Runners up).

3.1.1   Adult Cell Reprogramming

This remarkable experiment answers one of the biggest questions facing geneticists in the past 30 years, whether the genetic material of differentiated cells from adult animals undergoes some kind of irreversible modification, rendering it useless for cloning. The study suggests that with careful handling, the genetic material of such cells can indeed be introduced into an egg cell, and that normal cell development and differentiation culminating in live birth can occur.

Other research teams have managed to clone tadpoles and cattle from embryo tissue. Indeed, the same researchers that produced Dolly had only a year earlier raised seven other sheep from oocytes whose nuclei had been replaced with nuclei from either fetal or embryonic tissue.[10] The researchers built on this knowledge, using a similar process that had proved useful in their earlier experiments with fetal cells, but this time with adult cells. The process depends on the experimental

---

[7] Sean Henahan, "Send in the clones", *Access Excellence*, February 24, 1997.

[8] Gina Kolata, *Clone: The Road to Dolly and the Path Ahead*, (Allen Lane, The Penguin Press, London, 1997).

[9] Ian Wilmut, Keith Campbell, and Colin Tudge, *The Second Creations*, (Farrar, Straus and Giroux, 2000).

[10] K.H.S. Campbell, et al., *Nature* 380, 1996, pp. 64–66.

protocol which forces the donor cell — the cell that is to provide the nucleus — into a 'quiescent' state so that it is not replicating its DNA or dividing. This appears to make the nucleus more susceptible to re-programming by the recipient egg cell. And the result is Dolly. Dolly is a first, the only animal ever cloned from genetic material of adult cells.

In this particular experiment, the researchers managed to produce live births derived from three cell populations: 9-day old embryo, 26-day fetal and 6-year old mammary gland. Nearly two-thirds of the reconstructed embryos did not survive. However, eight ewes did give birth. Only one of these, Dolly, resulted from adult-cell nucleus cloning. The birth of this lamb shows that during the development of that mammary cell, there was no irreversible modification of genetic information required for development to term. Wilmut claims that this is consistent with the generally accepted view that mammalian differentiation is almost all achieved by systematic sequential changes in gene expression brought about by interactions between the nucleus and the changing cytoplasmic environment.

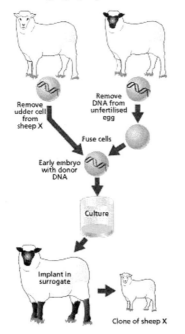

Figure 6. The technique involves several steps. First, the donor cells are grown under special conditions in culture. In this way the number of cells can be increased by several orders of magnitude. It is also possible to make genetic modifications and to select just those cells in which the desired modification has occurred. The selected cells are then fused with an unfertilized egg from which the introduced nucleus can lead to the formation of an embryo. The embryos are then transplanted into female sheep and the lambs are born naturally. (Figure: Courtesy of *The New Scientists*).

The implications of this work are far-reaching, and not limited to sheep. It seems probable that the same methods used to produce clones of sheep would work with other mammals, including humans. Cloning farm animals might allow a farmer to replicate a disease resistant cow that is an especially good milk producer. However, too many clones in a herd can potentially reduce essential genetic diversity to a dangerous level, leaving the animals vulnerable to disease and unexpected problems.

On the positive side, cloning livestock can allow the creation of animal herds that can be farmed for milk, blood and organs. Indeed, a company that supported the sheep cloning research is already farming animals genetically engineered to produce human proteins with a potential use for the treatment of cystic fibrosis. Similar methods might be used to farm treatments for hemophilia and emphysema, and can lead to a better understanding of other diseases including BSE (Mad Cow Disease). Cloning can also prove a great advance for genetic researchers studying gene expression, aging and cancer.

It may not be that farfetched to imagine that one could clone one's dead relatives, or even one's self. Aldous Huxley described such a world in his novel "Brave New World". The breakthrough is also likely to excite interest in cloning extinct species in which a compatible living host can be found. For example, elephants and mastodons, and a real life Jurassic Park.

3.2   *Monkeying with Nature*

Taking one step closer towards the potential for cloning humans, a group of Oregon researchers, led by Donald Wolf, announced in March 1997 to have cloned two Rhesus monkeys.[11] Taking one step back from the adult-cell cloning method used by Ian Wilmut and colleagues to clone a sheep, scientists at the Oregon Regional Primate Research Center cloned the monkeys from early stage embryos using the nuclear transfer method. A set of chromosomes was removed from each of the eight cells in a primitive monkey embryo and then inserting into egg cells from which the original DNA had been removed. These embryos were then implanted in the wombs of host mothers using *in vitro* fertilization techniques. The two monkey clones, one male and one female, were then six months old and appeared normal in every respect. They were being raised by their "mothers" and were expected to live as long as 20 years. The breakthrough could prove a major boon for researchers by allowing the creation of genetically identical animals.

---

[11] Sean Henahan, "Monkey business", *Access Excellence*, March 4, 1997.

Figure 7. One of the two cloned identical Rhesus monkeys. The other one, of course, is genetically identical to this one and looks just like this one. (Photo: Adapted from *Access Excellence*).

Like in mouse clones, monkey clones are aimed mainly for research purposes. Having test groups of cloned animals would remove some of the uncertainties in research that might have been attributed to genetic differences among animals. This will provide researchers with higher confidence levels in conclusions drawn from animal studies, and can allow the design of trials using far fewer animals.

What the team intends to do is to establish an immortal cell line, something like an embryonic stem cell line, where one can produce literally unlimited numbers of these things. The monkey cloning advance can prove useful in the design of studies involving cancer treatments and AIDS drugs. In addition, the technology used in the research might also be used to help infertile women, who might be able to have their DNA inserted in a donor embryo.

Wolf and associates are continuing their research and develop a line of monkey clones. However, they do not plan to clone monkeys from adult cells, and are opposed to the idea of cloning humans. One unusual aspect of the monkey cloning research is in the way it was announced. Typically, such important research would be announced in a leading peer-reviewed journal. For example, the sheep cloning announcement coincided with the publication of the research in the journal *Nature*. The monkey cloning study, in contrast, was announced at a press conference, and has not been published yet.

This effort really throws a monkey wrench into the works of Mother Nature.

3.3 Sheep Factory

Dolly is the first viable offspring ever derived from adult mammalian cells. Now we have Polly. Unlike Dolly, Polly is the first sheep cloned by nuclear transfer technology bearing a human gene. Investors hope the lamb will make them a mint because the arrival of Polly could be good news for hemophiliacs and others who rely on expensive protein therapy of their diseases.[12]

Polly was created by the same team at the Roslin Institute that gained fame with the birth of Dolly. In Polly's case, the researchers did not use adult cells. Rather, they used fibroblast cells obtained from a sheep fetus. This is considered a somewhat less tricky procedure. However, Polly is the first sheep to receive genetically altered fetal cells, in this case modified with a human gene. The human

---

[12] Sean Henahan, "Hello Polly", *Access Excellence*, July 24, 1997.

gene was introduced into the nucleus of the lamb fibroblast which was then inserted in an enucleated donor ovum.

This technique represents a significant advance over techniques currently used to produce transgenic animals. The current technology involves placing target genes from one species into the fertilized egg of another, and then waiting to see if the gene is expressed. This laborious process produces results about ten percent of the time at best. Since fertilized eggs are in short supply, the ability to use the more common fibroblast cells can increase the success and efficiency of cloning transgenic animals.

In the experiment, five lambs were produced by the new technique. The scientists have so far found marker genes in two of the five lambs. These marker genes confirm that the human gene was expressed in the sheep. While the researchers report that the gene inserted in the sheep is of "therapeutic value" they have yet to reveal what gene it is. Like the announcement of the monkey quadruplets months before, the researchers took the controversial step of announcing their findings in a press release rather than in a peer-reviewed journal.

The new cloning technology will allow scientists to create large numbers of identical, milk-producing ewes. Using genetic engineering, the sheep can be modified to produce therapeutic proteins in their milk. These expensive proteins can then be removed from the milk and used therapeutically. The Edinburgh company that sponsors the research, PPL, already produces transgenic sheep that produce α-1-antitrypsin, a protein used to treat the symptoms of cystic fibrosis. Transgenic sheep have also been genetically engineered to produce proteins used by patients with clotting disorders such as hemophilia, including fibrinogen, factor VII and factor IX.

It is more than likely that rather than producing expensive herds of cloned sheep, the researchers would create a small number founder herd and then let nature takes its course, allowing the animals to breed naturally. Finding out how well the target genes carry from generation to generation is of great interest to the investigators.

The company will not be selling anything produced by the first batch of transgenic sheep. This is because of concerns about scrapie infection, a sheep disease believed to be related to TSE, or mad-cow disease. The five experimental lambs did not come from a scrapie-free herd. This problem highlights the downside of transgenic animal cloning, since there is some concern that humans can become infected with viruses or prions from the animals.

On a more positive note, the new research may open the way for scientists not only to add desired genes, but also to remove undesired genes. This would be essential for removing antigens pig organs that might some day be used as replacement parts in humans.

3.4   Dolly, Polly and "Holly" Calf

The British company, PPL Therapeutics, that helped to create Dolly the cloned sheep used the same method to create a calf. The healthy 98 lb animal, called Mr. Jefferson, was produced by its American subsidiary in Blacksburg, Virginia, and delivered at Maryland College of Veterinary Medicine.[13]

Figure 8. Mr. Jefferson, a cloned Holstein bull calf born on February 16, 1998, stands in his stall. (Photo: Courtesy of PPL Therapeutics).

PPL has established itself as a world leader in the "transgenic" production of human proteins in the milk of livestock. While Mr. Jefferson is not transgenic, the animal's birth on February 16, 1998 opened the way to producing transgenic cows, whose milk could eventually be used to treat diseases in humans. One of PPL's products, used to treat cystic fibrosis, is already undergoing clinical trials.

Mr. Jefferson was named in honor of his birth on February 16 — President's Day in the United States. The animal was produced using a technology based on that used to clone Dolly and Polly, but whereas Dolly was cloned from an adult cell, the calf was cloned from the cell off a fetus.

3.5   Got Milk? Goat Milk

The successful cloning of transgenic goats from somatic cells adds to the herds of cloned animals that will produce valuable human proteins in their milk.[14]   A collaboration between researchers in industry and academia cloned three healthy female transgenic goats. First, fetal fibroblast cells were obtained from a 30-day old female goat fetus. Next, the researchers attached a bioengineered human gene for the production of an anti-clotting protein to a promoter gene and then injected this into the nucleus of the newly fertilized egg. In some cases the human gene was incorporated into the DNA of the goat embryo.

---

[13] "Dolly's creators clone a calf", *BBC News*, February 23, 1998.
[14] Sean Henaha, "Goats on the clone pharm", *Access Excellence*, April 29, 1999.

Figure 9. The three identical goat clones with human genes. (Photo: Courtesy of *Nature Biotechnology*).

After removing the nucleus of recipient egg cells, the researchers activated and fused the donor eggs with the fetal fibroblast cells now containing the human gene. The cloned embryo was then transferred into a recipient female goat "mother" that carries the clone to term. Female progeny will produce milk containing the human protein, which can be extracted from the milk for use in medical treatments. Moreover, half of the female offspring produced by subsequent generations of the clones will also produce the human protein.[15]

Richard Denniston, a researcher with the Louisiana State University Agricultural Center, claimed that the technology used to clone the three goats is one of the first applications of the nuclear transfer cloning procedure to produce transgenic goats for the pharmaceutical industry. The bioengineered goats carry a human gene called AT III. The gene produces a protein that helps keep the blood from clotting. The protein is being investigated as a potential treatment for victims of heart attack and stroke. In this study, the three transgenic female goats did produce the AT III protein in their milk. It is possible that this might be the first therapeutic agent derived from transgenic animals to be approved by the FDA. The market for AT III is $200 million. A herd of less than 100 bioengineered goats will produce the amount of protein to meet the market. Even with the cost being a half to $1 million per animal, simple arithmetic shows that the potential earnings from the pharmaceutical product are enormous.

While researchers are also investigating the possibility of developing transgenic cows to produce human proteins, goats offer some important advantages. First of all, it is cheaper to produce transgenic goats than cows. Second, the goat's gestation period is shorter (5 months to a cow's 9 months). The results, and others, will revolutionize the biopharmaceutical industry because we now have a method that is a faster, reliable and more cost-effective way to produce complex pharmaceuticals for humans and animals.

3.6   Second Chance — Clone of the Oldest Bull

Scientists at Texas A&M University have successfully cloned what is believed to be the first calf cloned from an adult bull, which is also the oldest animal ever cloned, a

---

[15] *Nature Biotechnology*, May 1, 1999.

21-year-old Brahman. Their research could have enormous implications in the beef cattle industry and in the future applications of cloning technology.[16]

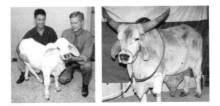

Figure 10. Left: Second Chance, clone of the oldest bull, and Jonathan Hill (left) and Mark Westhusin (right). Right: This 1998 photo shows the late bull Chance, whose tissue was used to clone Second Chance. Chance, a Brahman who died at about 21 years of age, is believed to be the oldest animal ever cloned. (Photo: Courtesy of Texas A&M).

Researchers Jonathan Hill and Mark Westhusin accomplished the cloning of the bull, fittingly named Chance, in a year-long project. Chance's offspring, Second Chance, displays identical markings as his father and has identical DNA. The owners of Chance wanted to have their prized bull cloned because of his unusually gentle nature, and they considered the cloning effort a good opportunity to see if an identical copy of Chance might also have such an easy going disposition. Chance was great around people, and he was in several TV commercials, performed in the Houston Rodeo and was even on The Late Show with David Letterman. It will be interesting to see if Second Chance lives up to his heritage. Chance was unable to reproduce naturally because of the removal of both diseased testicles two years earlier. Therefore, cloning Chance was the only option for preserving his genetics.

Second Chance is an intact male and should be able to sire offspring when he reaches puberty. But Chance died at age 21, shortly before his DNA was used to produce Second Chance. There is considerable interest in keeping track of Second Chance throughout his lifetime because of the age of the cells used to clone him. In the Spring of 1999, scientists revealed that the DNA of Dolly, the first cloned sheep, had some characteristics of the older cells that were used to generate her.

To create Second Chance, it took 189 attempts, that is, transferring 189 cells into 189 different eggs, before a pregnancy ended in the delivery of Second Chance. Because Second Chance came from the oldest animal cloned to date, he has received intensive monitoring and treatment since birth by a team of veterinarians and intensive care technicians at the Texas A&M Large Animal Hospital. Like many previously cloned calves, at birth he displayed some symptoms that resembled those seen in premature human babies.

The successful cloning effort could dramatically impact the multi-billion dollar beef cattle industry in Texas and throughout the world. The cloning of Second Chance was funded by the Texas Coordinating Board of Higher Education's

---

[16] "Texas A&M scientists clone first-ever bull", *Texas A&M Press Release*, September 2, 1999.

Advanced Research Program and by Dr. Charles Looney of Ultimate Genetics in Franklin, Texas.

By cloning Second Chance, the team has taken the bull by the horns.

3.7   Artificial Monkey Twins

The first success at creating identical twin monkeys, an alternative method of cloning, was achieved by Gerald Schatten and his colleagues at the Oregon Regional Primate Center.[17] The first living monkey born of this technique, named Tetra, was created via an artificial twinning. This involves splitting tiny embryos to make multiple copies. This process will provide scientists animals that are exact genetic copies. This is important because researchers will have identical twins for a variety of studies. In other words, the control and the test groups will be exactly the same. In contrast, the famed clone, Dolly, was created by a method in which the nucleus is scooped out of an adult cell and placed into an egg cell. In that method, a small set of auxiliary genes were left behind, leading to Dolly not quite an exact copy of her cell-donor parent.

In the artificial twinning process, Schatten and colleagues used a procedure very similar to *in vitro* fertilization by taking an egg from the mother monkey and sperm from the father monkey and then mixing them together to create a fertilized egg. Once the embryo had grown into eight cells, they divided the embryo into four identical embryos consisting of two cells each. Repeating the process numerous times, they created 368 tiny embryos by splitting apart small groups of cells that came from 107 slightly bigger embryos. They did 13 embryo transfers into surrogate mothers, got four pregnancies and one live birth, Tetra, 157 days after implantation.

Figure 11. Tetra, the singleton rhesus macaque, was cloned using a very different method than Dolly. Tetra was made by splitting a very early embryo, consisting of only eight cells, into four pieces. These were then nurtured into new embryos, but only one survived. So unlike Dolly, Tetra has both a mother and father and is a clone of neither, but is rather an artificial quadruplet. (Photo: Courtesy of *Science* magazine).

Tetra, from the Greek word for four, is not the first time this technique has been used to create mammalian twins. The same technique is already being used in cattle.

---

[17] Tom Abate, "First monkey born using new method of cloning", *San Francisco Chronicle*, January 14, 2000.

A physician also reported using the technique to clone human embryos as far back as 1993. Nor is it the first time that monkeys have been cloned. Researchers from the same Oregon research group, Donald Wolf and colleagues, reported cloning a monkey in 1997 using the nuclear transfer method. That method involves removing a set of chromosomes from each of the eight cells in a primitive monkey embryo and then inserting into egg cells from which the original DNA had been removed. These embryos were then implanted in the wombs of host mothers using *in vitro* fertilization techniques.[18]

What is new about the creation of Tetra is that this is the first time researchers have created a perfect genetic copy of a monkey by using the embryo splitting technique. Unlike the earlier monkey clone, Tetra is the first to possess both identical nuclear and cytoplasmic components. This offers researchers for the first time the opportunity to produce a line of identical primates for medical research.

Most medical therapies are now first tested in mice, but monkeys would be more reliable in developing new techniques such as gene therapies or growing new organs from stem cells. For example, research using human embryonic stem cells is controversial because to produce the cells requires the death of an embryo. Genetically identical monkeys may be used to develop treatments using embryonic stem cells, the ancestral cells from which all organs and tissue grow during gestation.

Will this monkey business eventually turn into money business? Time will tell.

3.8   Inserted DNA and ANDi

It has been four years since Dolly, the cloned sheep, and many other clones afterwards put a reassuringly face on the new era of Frankenstein biology. One would think with all the cloning, genetic engineering, and general monkeying around scientists have done since then, they would have produced at least one grotesque monster. Instead, the mutants just keep getting more adorable.[19]

Figure 12. ANDi, aka Jelly Belly, a rhesus macaque monkey with a short stretch of jelly-fish DNA. (Photo: Adapted from National Geographic World).

ANDi is the most adorable member of the biotechnology petting zoo yet. ANDi (the name stands for "inserted DNA" backwards) is mostly monkey, rhesus macaque

---

[18] Sean Henahan, "Tetra the singleton twin monkey", *Access Excellence*, January 14, 2000.
[19] Thomas Hayden, "Monkeying with nature", *US News*, January 22, 2001.

to be precise. But nestled somewhere in his chromosomes is a short stretch of DNA that makes him part jellyfish, too. His birth announcement, made in the journal *Science*, removes what little doubt was left that genetic engineering is not just for mice and soybeans. As the world's first "transgenic" monkey, ANDi is living proof that the technique can work on our closest biological relatives—and therefore on us.

As disarming as he may appear, ANDi raises more questions than he answers. Scientifically, the experiment demonstrates just how technically difficult it is to do genetic manipulation in higher animals. At the same time, even a preliminary advance with our genetic next of kin ignites all the ethical concerns about tinkering with human DNA and the prospect of designer babies.

Like all of us, ANDi started out as an unfertilized egg. A team of scientists at the Oregon Health Sciences University used a modified virus to carry the jellyfish gene into monkey eggs, which were then injected with sperm and implanted in surrogate mothers. The gene itself is just an easy-to-spot marker used by biologists to test new procedures. The team leader Gerald Schatten said they have shown the technique works and one can use the same method to insert almost any gene one desires. This would be a boon to researchers hunting for cures for everything from blindness to Parkinson's disease, who currently use genetically modified mice for their work.

As useful as those lesser critters are, sometimes the differences between mice and humans are too great to make them useful tools for researching human disease. Only primates have a monthly menstrual cycle, for example, which can have important impacts on breast and ovarian cancer. Rodents also lack a macula, the part of the eye's retina that is lost in macular degeneration, the leading cause of blindness in the United States. Perhaps most significant, rodent brains are just too simple to show the subtle effects of psychiatric and neuro-degenerative diseases such as schizophrenia and Alzheimer's.

3.9   Clone of a Clone — More Bulls

On January 24, Japanese scientists at the Kagoshima Prefectural Cattle Breeding Development Institute (KPCBDI) reported that they had succeeded in cloning a bull. Though researchers in the United States have successfully bred clones of cloned mice, this experiment marks the first time a large cloned animal has itself been cloned.

There has been speculation that cloned animals may not be as healthy or live as long as normal animals. The calf born is part of a project to study the life expectancy and aging of cloned animals. The three generations of genetically identical bulls — the original animal and the two clones — are being studied at the Institute in southern Japan.[20]

---

[20] Shihoko Goto, "Cone of cloned steer bred in Japan", *Associated Press*, January 24, 2000.

Figure 13. These four calves were cloned from the ear cells of the bull standing next to them. These clones are part of an experiment to answer the question of life expectancy of clones and acceptance by diners. (Photo: Courtesy of Associated Press).

To create the new calves, skin tissue was taken from the ear of a cloned bull in April 1999, when the animal was four months old. Those cells were fused with an unfertilized egg that had been stripped of its nucleus and placed in the womb of a cow. The resulting bull calf weighed nearly 100 pounds at birth.

According to Norio Tabara, one of the research scientists involved in the experiment, the primary objective of the Institute is to produce good cattle consistently and if there is a stud of the highest quality, the bull will be made available more widely. In other words, the Japanese are interested in producing a herd to provide tasty beef.

Cloning also reduces the amount of time needed for breeding. The tissues of an animal as young as three months can be used for cloning, while cows do not mate naturally until they are about 14 months old. Cloned beef is already on sale in Japanese supermarkets, although the government's announcement in April of 1999 that it had already allowed cloned beef to be sold unmarked for at least two years sparked a beef boycott nationwide.

3.10 Five Piglets: Oink! Oink! Oink! Oink! Oink! — Organs for Sale

The British company that helped clone Dolly the sheep is at it again when it announced that it had created the first cloned pigs, animals that can eventually be used to harvest organs for transplant to human beings. PPL Therapeutics said that five cloned piglets were born on March 5, 2000 in Blacksburg, Virginia, ushering what can be a new era in cell and organ transplant and an end to the chronic shortages of donors worldwide.[21] More than 110,000 people in the U.S. and Europe alone are on waiting lists for hearts, kidneys, and livers. The list grows larger each year while the number of donors is shrinking.

---

[21] Marjorie Miller, "New breed of cloned pigs — organs wanted for humans", *Los Angeles Times*, March 15, 2000.

Like Dolly, the five piglets, Millie, Christa, Alexis, Carrel, and Dotcom, were cloned from adult cells. They also represent the first stage in creating pigs whose organs can be transferred to humans. The first, Millie, was named after the millennium; Christa, after Christiaan Barnard, the surgeon who performed the first human heart transplant in 1967; Alexis and Carrel after transplant pioneer and Nobel prize winner Alexis Carrel; and Dotcom to reflect the growing use of the Internet.

Barnard Neethling Christiaan*, a South African doctor, had spent many years experimenting with heart transplants, mainly with dogs, before he walked into an operating room at Groote Schuur Hospital in Cape Town at about 1:00am on December 3, 1967 to replace Louis Washkansky's 53-year-old heart — crippled by diabetes and heart diseases — with the heart of a 25-year-old Denise Darvall, who had died in an automobile accident. Washkansky lived for 18 days before dying of double pneumonia attributed to his suppressed immune system. Though short-lived, the replaced heart was coaxed into beating and circulating blood, making a medical history.

Soon after Washdansky, Barnard transplanted a heart into Philip Blaiberg, who lived for more than 18 months before succumbing to chronic rejection. More than three decades later, heart transplantation, while not quite routine, has been refined to a state that 90% of patients survive, and 85% live for more than a year.

PPL hopes to help supply the $6 billion market for organs and a similar market for cellular therapies to treat different diseases. Pigs are preferred over other species for xenotransplantation or animal-to-human transplant because they can be bred quickly and their organs are the same size as those of humans.

Clinical trials of transplating harvested pig organs into humans can begin in as little as four years (year 2004).

Figure 14. The five piglets, Millie, Christa, Alexis, Carrel, and Dotcom, were born on March 5, 2000 in Blacksburg, Virginia, where PPL has a research facility. (Photo: Courtesy of Mike Theiler, Reuters).

PPL Therapeutics' next plan is to genetically alter cloned pigs to "knock out" a specific gene that is responsible for adding sugar group to pig cells that is foreign to the human immune system and therefore, provokes it into rejecting the transplant organs. Three new genes will then be introduced into the cells of the cloned pigs to

---

* Just as this book was being completed, news came in that Dr. Barnard Christiaan died on September 2, 2001, at age 78, suffering from a fatal asthma attack after going for a swim at a resort in Paphos on the southwest of Cyprus.

control the causes of organ rejection, and finally, potential transplant patients will receive a blood transfusion containing the modified cells of the pigs supplying the organ. This will serve as an immunization to help the human system absorb the new tissue or organs.

Critics are concerned about the safety and ethics of the procedure. While successfully cloning the piglets is a positive step forward for those in line for tissue or organs, there are still risk such as that pigs are harboring viruses, called retroviruses, which have not yet crossed species barriers.

3.11 Spring Holy Cows

If a spring chicken is a young person, then a spring calf must be a biologically young stud.

In the Spring of 2000 and after the pigs, news of six calves cloned using a new technique was announced in *Science* magazine.[23,24] The technique allows the animals to start life biologically younger than the aged cells from which they were derived. This is in contrast to Dolly, who prematurely ages, which led to fears that cloning will spawn creatures doomed to grow old before their time.

The six claves grew from embryos fashioned by a nuclear transfer process. The DNA needed to produce the calves was taken from cells of a calf fetus. In the laboratory, the DNA donor cells were allowed to replicate to the point of exhaustion, that is, molecular aging. This is to be contrasted with how Dolly was created in 1996 where scientists transferred "younger" cells that had been put in a state of dormancy.

Figure 15. Five Holsteins that started life younger than the cells from which they were cloned. (Photo: Courtesy of Advanced Cell Technology).

The nuclei of the biologically aged calf cells were then transferred into essentially hollow vessels, bovine egg cells from which the original genetic material had been removed.

---

[23] Carl T. Hall, "New way to clone raises hope for medical miracles", *San Francisco Chronicle*, April 28, 2000.
[24] *Science*, April 27, 2000.

In normal sexual reproduction, DNA from sperm and egg mix together, contributing equally to a genetically unique new offspring. The calves, like in Dolly, are clones of the original DNA donor, that is, exact genetic copies rather than individual mixtures of male and female DNA. But the cattle differ from the sheep in one subtle but fundamental way. Dolly turned out to have cellular characteristics of an older animal, showing at age 3 the signs of a 6 year-old sheep. The cells of the calves appear to be every bit as young as they look. In particular, the calves have lengthy telomeres, even though the donor genetic material had been aged to exhaustion. Telomeres are caps at the ends of DNA strands that ordinarily wear away each time cells divide until further replication becomes impossible. This cellular senescence is believed to be at the root of age-related disorders, including problems affecting the skin, eyes and internal organs. But this is hardly the only factor that causes aging in animals.

Although the calves look to be ordinary, they raise the tantalizing possibility of a new era in regenerative medicine, in which any worn-out body part may be replaced as easily as in replacing automobile parts.

The lead author of the report, Robert Lanza of Advanced Cell Technology (ACT), Inc. in Massachusetts, which sponsors the research, said that they had shown for the first time that cloning can take old cells back to a youthful state. The animals have cells that appear to be younger than their chronological age. The goal is to fashion replacement parts from a patient's own rejuvenated cells to overcome the problem of organ shortages and transplant rejection while ensuring the grafts will last long enough to make the procedure worthwhile.

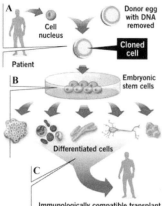

Figure 16. Figure to show how the technique for cloning the calves may be adapted to replace aging tissues. (A) A cell nucleus from a patient is transferred into a donor egg cell that has been enucleated. (B) Stem cells capable of becoming various types of tissues and taken from the resulting clone are genetically identical to the patient, but have youthful traits. (C) The stem cells are differentiated in the laboratory into various genetically matched human cells and tissues for transplantation. (Source: Advanced Cell Technology. Figure adapted from Todd Trumbull, *San Francisco Chronicle*).

There are still unanswered questions surrounding the biological aging of clones, most notably the connection between the cellular signs of aging and the actual lifespan of an animal. Despite her old-looking cells, Dolly has been leading what appears to be a normal ewe's life and has enjoyed two reproductive cycles. The oldest calf, a Holstein named Persephone, is at this writing one year old, and along with her five cohorts, Lily, Daffodil, Crocus, Forsythia, and Rose, appear to be completely normal in all aspects.

Figure 17. Cloning seems to have no ill effects so far. Dolly, the first mammal ever to be cloned from adult cells, gave birth in 1998 to Bonnie, who by all accounts is normal. This past year Dolly delivered a healthy set of triplets. (Photo: Adapted from Roddy Field, Roslin Institute).

3.11.1 Telomerase — A Vignette of the Fountain of Youth

In recent years, telomerase, the enzyme for lengthening telomeres, has been hailed in the press as a fountain of youth. Telomerase has caused a stir among researchers because of its implicated role in aging and cancer.

Discovered in the 1930s by Barbara McClintock, telomeres are linear nucleoprotein complexes at ends of eukaryotic chromosomes. They are implicated in cell cycle, replication, and senescence. Capped at chromosome ends for protection from division and damage, and for ensuring complete replication, telomeres are TTAGGG DNA repeats of about 1,000–15,000 base pairs in length.

Telomere loss occurs during every cell division by some 200–2,000 nucleotides, and throughout the life cycle of a normal cell. Somehow, telomere loss eventually limits the proliferative capacity of cells. A hypothesis is telomere shortening functions as a mitotic or telomere clock for senescence.

Telomerase is a multi-subunit ribonucleoprotein that adds the oligonucleotide repeats to the end of DNA strand to be replicated. It has an RNA component, and RNA binding protein (TEP1), and a catalytic component (TRT) belonging to the class of reverse transcriptases, which use RNA as a template for replication like the retrovirus human immunodeficiency virus.

Telomerase is expressed in embryonic cells, some germline cells, and regenerative somatic cells like epidermal basal cells. The common denominator of these cell types is their high proliferative capacity.

If uncontrolled growth and proliferation are characteristics of cancer, then there may be an association between tumors and telomerase. Research at Amgen shows that 85% of all cancer cells are telomerase-positive. Mice lacking telomerase RNA show adverse effects in highly proliferative tissues, coinciding with substantial erosion of telomeres, and therefore knock-out mice experiments support the aging hypothesis.

Despite the link supporting cancer hypothesis of telomerase, in somatic cells, where TRT is normally undetected, researchers at Geron found expression of human TRT, introduced by genetic manipulations, does not cause oncogenic transformation. Thus, in humans, there are still 10–15% of cancer cell types in which telomerase is undetected.

3.12   Xena — No Warrior Princess but a Piglet

Xena is not a warrior, but a scrappy, black-coated piglet born from a white-coated sow. As a swine produced by cloned genetic material from fetal pig cells, Xena represents a living milestone in the rapidly evolving field of genetic cloning. She is named after the field of research that scientists hope her birth might advance — xenotransplantation — the use of genetically modified animal organs for transplant into humans.[25]

Figure 18. Xena, a piglet cloned in Japan, represents the second successful pig cloning. Her white-coated surrogate mother is in the background. (Photo: Adapted from *Science* magazine).

Pig tissue and organs, particularly the liver, are thought to be similar to human's and hold the best possibility for human transplant. However, even with the successful cloning of a pig, scientists still must find methods of ensuring pig's organs are not rejected in the human body and are able to function in human hosts.

Xena is not the first cloned pig to enter the world. That honor belongs to a litter of five cloned piglets born in March of 2000 at the Scotland-based PPL Therapeutics, the same laboratory that produced the famous female, Dolly the cloned sheep.

The procedures both groups used are different. According to Akira Onishi of the National Institute of Animal Industry in Japan, his team used the same techniques developed by researchers at the University of Hawaii to clone mice. To create Xena, scientists used a needle-like pipette to inject genetic material from fetal pig skin into eggs that had been stripped of their own genetic material. Next, the

---

[25] Amanda Onion, "Xena: a new generation", *ABC News*, August 16, 2000.

team stimulated the injected eggs with an electrical pulse that triggered them to develop into embryos. Those embryos were then transplanted into four surrogate sows. Healthy and black-coated, unlike her white-coated surrogate mother, Xena was the one successful birth of 110 transplanted embryos. When researchers cloned Dolly the sheep, they fused entire cloned cells into empty host eggs. Researchers on Xena's team believe their technique of injecting only genetic material will allow scientists more flexibility in the future to manipulate the genetic material of pig cells and effectively engineer transplant-safe organs because with microinjection, researchers can be quite selective about genetic transfer. The chromosomes can be separated out to avoid contaminating the egg with the rest of the material from the donor cell nucleus.

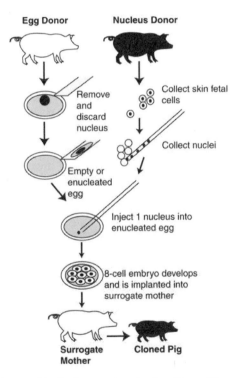

Figure 19. The technique used by the Japanese team to clone Xena. (Diagram: Adapted from Mary S. Gibbs, *Genome News Network*).

Scientists at PPL Therapeutics contend the success rate for the procedure that created Xena remains too low. To ensure more successful fertilization, scientists at PPL Therapeutics added a step to the cloning process. Rather than electrically stimulating the cloned egg to develop, they inserted the cloned material into the shell of a fertilized cell, or zygote. Though a bit more labor intensive, PPL scientists claim their approach is a more efficient method by pointing out that their method created five healthy piglets in March of 2000, as opposed to one by the Japanese team. Efficiency in cloning may become an important factor in the future if and when scientists are able to use pig organs for human transplant. About 180,000 people around the world are estimated to be waiting for organ transplants, but fewer than one in three will receive one because of the widening gap between supply and demand. Many see organs derived from pigs as a solution to the problem.

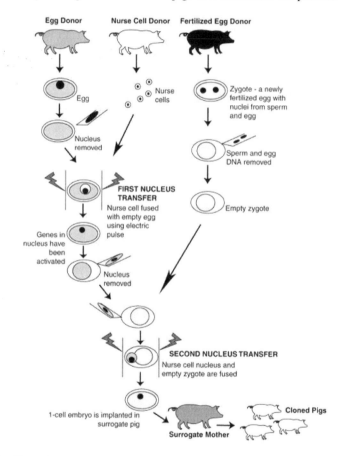

Figure 20. The technique used by the PPL Therapeutics team to clone the five piglets. (Diagram: Adapted from Mary S. Gibbs, *Genome News Network*).

Research is already under way to manipulate genes responsible for organ rejection. Once scientists find a way to turn off organ-rejecting genes, they hope to clone pigs with genetic material that is missing those genes. Organ rejection is not the only concern. Transferring pig organs may also introduce contagious viruses that exist in all pig cells, called porcine endogenous retroviruses or PERVs. Xenotransplantation falls into the category of hazard where, although the risk is probably low and the benefit to individuals undoubtedly substantial, the public consequences could be catastrophic.

Xena may represent a big step, but there are many more to come before any animal's organ is made "user friendly" for people. John Brems, the director of transplantation at Layola University Medical Center in Chicago, points out pigs walk on all fours, and if their organs are put in an animal that walks standing up, there are a lot of differences, and these obstacles have to be overcome to ensure the organs will work.

3.13 Cow Pregnant with Clone — Jurassic Amusement Park

In what could represent a new way to save endangered species, scientists at a Massachusetts-based biotechnology company said that they had cloned an endangered Asian gaur and implanted the resulting embryo into a cow in Iowa. The gaur, an ox-like creature that is native to India and Southeast Asia, is expected to be born in another month.[26] The gaur is usually brown or black with white or yellow socks on each leg and a humplike ridge in its back. Gaur population has dwindled to about 36,000 because of hunting and because of degradation in forests, bamboo jungles, and grasslands in India and Southeast Asia.

Figure 21. Left: Noah weighs 36 kilos (80 lbs) at birth. Right: Noah receiving oxygen shortly after birth. (Photo: Courtesy of BBC News).

To create the gaur, scientists took a skin cell from a recently deceased gaur and fused it with a cow's egg from which the chromosomes, containing the cow's genetic material, had been removed. The DNA of the gaur commandeered the egg, which grew into a gaur embryo. The embryo was implanted into the womb of a cow serving as a surrogate mother. The result should be an exact genetic copy of the gaur from which the skin cells were obtained.

Previously, many scientists thought such cross-species cloning would be impossible because the DNA of the cloned animal would not be able to interact properly with the rest of the egg cell.

---

[26] Andrew Pollack, "Cow pregnant with cloned ox", *New York Times*, October 9, 2000.

The technique currently has a low rate of success. The scientists created several hundred embryos, but only 81 grew to the stage where they could be implanted. Some 42 were planted into 32 cows, but only eight cows became pregnant. The fetuses were extracted from two cows for examination and five cows suffered spontaneous abortions, leaving only one cow that is still pregnant. The work is described in a paper in the current issue of the journal *Cloning* and in an article in the November issue of *Scientific American*.

If the gaur is born, it would represent the first cloning of an endangered species and the first cloned animal to use another species as a surrogate mother. Scientists say the technique could not only help preserve endangered species but also possibly even revive species that have been extinct. In fact, the company, Advanced Cell Technology (ACT), of Worcester, Massachusetts, said that it had received permission from the government of Spain to clone the already extinct bucardo mountain goat, using cells collected from the last goat before she died earlier in 2000.

As expected, the technique is raising ethical questions. Kent Redford, director of biodiversity analysis and coordination at the Wildlife Conservation Society in New York said the product would be more like an amusement park version of the species rather than the wild species. He is of the opinion that species should be preserved in their natural environments.

4   No Jurassic Amusement Yet

In October 2000, Advanced Cell Technology (ACT), Worcester, Massachusetts, announced that a cloned gaur bull, a nearly extinct wild ox, had made it to late fetal development, and they were anticipating its birth in a few months. On January 15, 2001, the company reported that the baby gaur was born at 7:30 p.m. on Monday, January 8, 2001. The birth of the baby bull gaur, named Noah, was the first successful birth of a cloned animal that is a member of an endangered species.[27]

Noah was created by a process known as cross-species cloning, in which somatic cells from a frozen sample of gaur cells were fused with an enucleated cow egg. After implantation into a surrogate domestic cow, Noah was carried to full-term, and the 80-pound baby delivered at Trans Ova Genetics, Genetic Advancement Center in Iowa.

While healthy at birth, Noah died within 48 hours of a common type of dysentery believed unrelated to cloning. As explained by Philip Damiani, a researcher with ACT, the data collected clearly indicate that cross-species cloning worked. Despite the unexpected loss, Noah's birth brightens the prospects that the technology can be applied to many species on the verge of extinction.

---

[27] Laura DeFrancesco, "Clone of first endangered animal is born and dies", *Bioresearch Online*, January 15, 2001.

Robert Lanza, vice president of medical and scientific development at ACT explained the loss. Noah died from clostridial enteritis, a bacterial infection that is almost universally fatal in newborn animals.

Collaborating with ACT were Trans-Ova Genetics of Sioux Center, Iowa; Jonathan Hill of Cornell University; and The San Diego Frozen Zoo and the Reproduction of Endangered Species of Captive Ungulates (RESCU) International that supplied the gaur cells.

The birth of any cloned animal is a very tense period, when careful observations are needed to determine how well the animal has made the transition to the outside world. Immediately after Noah's birth, veterinarians and technicians under the direction of Jonathan Hill from Cornell University intensively monitored Noah and administered a variety of treatments. Hill was also involved in cloning Second Chance, the clone of the oldest adult bull, at the University of Texas A&M. Hill explained that the first 48 hours is a particularly vulnerable time for newborn animals similar to a premature human baby. Within 12 hours of birth, Noah was able to stand unaided and began an inquisitive search of his new surroundings. But at 1 day old, Noah began to exhibit symptoms of a common infection, and succumbed to it despite treatment efforts. The good news is his mother, Bessie, the domestic cow, remains healthy.

Commenting on the experiment, Kurt Benirschke, former president of the Zoological Society of San Diego and founder of the Frozen Zoo, expressed optimism and said science has advanced to the point of being able to successfully create a healthy trans-species gaur clone. Though the Zoological Society was saddened by the news of Noah's death, they were encouraged that scientists are learning to perfect the process and have continued hope for its inevitable role in the conservation of endangered species.

"We are gratified by the hard work and vision of many people on an effort carried out almost flawlessly," Michael West, president and CEO of ACT, summed it up nicely, "While we set the 'bar' at long-term survival, this was clearly a huge step forward."

And hopefully, we will not have to wait till the cows come home to see the first successful attempt.

Chapter 6

PET CLONING AND BIOFACTORY

"Which came first: The chicken? Or the egg? Or the therapeutic protein?"
TranXenoGen advertisement.

1 Attempts At Cloning

Scientists have long been intrigued by the possibility of artificially cloning animals. In fact, people have known since ancient times that some invertebrates — animals without backbones — such as earthworms and starfish, can be cloned simply by dividing them into two pieces. Each piece regenerates into a complete organism. The cloning of vertebrates, however, was much more difficult.[1]

The first leap forward in the cloning of these more complex organisms came in the 1950s with work done on frogs. Beginning in 1952, Robert Briggs and Thomas King, developmental biologists at the Institute for Cancer Research (present name: the Fox Chase Cancer Center) in Philadelphia, developed a cloning method called nuclear transplantation, or nuclear transfer, which was first proposed in 1938 by the German scientist Hans Spemann. In this method, the nucleus — the cellular structure that contains most of the genetic material that controls growth and development — is removed from an egg cell of an organism, a procedure known as enucleation. The nucleus from a body cell of another organism of the same species is then placed into the enucleated egg cell. Nurtured by the nutrients in the remaining part of the egg cell, an embryo — an organism prior to birth — begins growing. Because the embryo's genes come from the body cell's nucleus, the embryo is genetically identical to the organism from which the body cell is obtained. In their experiments, Briggs and King used body cells from frog embryos. From these cells, they were able to produce several tadpoles.

1.1 Cell Specialization

Briggs and King used embryos consisting of only a few thousand cells as the source for body cells and nuclei, because at that stage of development an embryo's cells are still relatively unspecialized. As an embryo develops into a completely formed organism consisting of billions of cells, its cells become increasingly specialized. Some cells become skin cells, for example, while others become blood cells. Skin cells can normally make only more skin cells. Likewise, blood cells can normally

---

[1] World Book, www.worldbook.com.

133

make only blood cells. By contrast, each of the unspecialized cells of an early embryo is capable of producing an entire body. At the time of Briggs' and King's experiment, researchers were not sure whether specialization occurs because different cells get different assortments of genes or because genes that are not needed in a particular kind of cell become inactive.

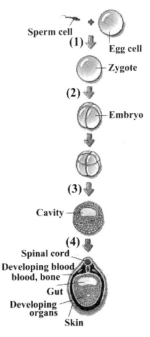

Figure 1. (1) A sperm cell and an egg cell combine to form a zygote cell. (2) The zygote begins dividing, forming an embryo. At this early stage of development, the embryo's cells are still unspecialized, and each has the ability to develop into a complete organism. (3) Cell specialization begins when a cavity forms in the embryo. This cavity will eventually develops into the organism's gut. (4) As the embryo continues to develop, its cells become increasingly specialized as they begin to form the organism's various parts, such as the spinal cord and skin. By this point, many of the embryo's cells have lost their ability to develop into complete organism, and eventually all the body cells will lose that ability. (Photo: Adapted from Roberta Polfus, World Book).

Additional research on nuclear transplantation was conducted in the 1960s and 1970s by John Gurdon, a molecular biologist at Oxford University in England. In 1966, Gurdon produced adult frogs using nuclei from tadpole intestine cells. This experiment proved that even cells that have undergone a great amount of specialization remain totipotent, that is, they are capable, under certain circumstances, of directing the development of a complete organism. Totipotency implied that all of a fully developed organism's body cells contain a complete set of genes and that specialization occurs because certain genes are active in some cells and inactive in other cells.

Despite the demonstrated totipotency of body cells, scientists were repeatedly frustrated in their attempts to use nuclear transplantation with nuclei taken from the cells of adult vertebrates. In the rare cases in which offspring resulted from such experiments, the young never survived to adulthood.

1.2 Embryo Splitting

A different and simpler cloning procedure, called embryo splitting, or artificial twinning, was developed in the 1980s and was adopted by livestock breeders. In this procedure, an early embryo is simply split into individual cells or groups of cells, as happens naturally with twins, triplets, and other multiplet births. Each cell or collection of cells develops into a new embryo, which is then placed into the womb of a host mother animal, which carries it to a full term. Although this technique permits the production of multiple clones, the clones are derived from an embryo whose physical characteristics are not completely known rather than from an adult animal with known characteristics. This imposes a serious limitation for practical applications of the procedure. By the early 1990s, embryo splitting and nuclear transplantation using cells from embryos had been used to clone a number of animals, including mice, cows, pigs, rabbits, and sheep.

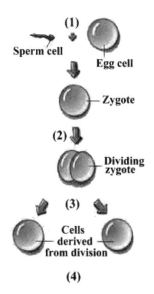

Figure 2. (1) A sperm cell combines with an egg cell to form a zygote. (2) The zygote divides into two cells. (3) The cells split apart from each other. (4) The two cells develop into identical embryos, which grow into natural identical twins who are clones of each other. If the embryos are split artificially like in artificial twining, the embryos are implanted into surrogate mothers to complete the development. (Photo: Adapted from Roberta Polfus, World Book).

1.3   Gaps in Cell Cycle

A doctoral student at Roslin Institute, Lawrence Smith, while trying a few cloning experiments in 1986, noticed that the success of cloning seemed to be related to what is called the cell cycle. Growing cells follow a certain pattern that involves proofreading their DNA for mistakes in its genetic code. Immediately after a cell has divided, each daughter cell enters a phase called G1 or Gap 1 during which it checks to make sure that its DNA is intact. The cell also begins to enlarge, adding to its bulk. Then the cell enters another phase, called S or synthesis, during which it copies its DNA in preparation for dividing. This phase is followed by G2 or Gap 2. During this phase, the cell checks its DNA for mistakes that might have occurred during the copying process. The cell also grows larger in this phase. The final phase is M or mitosis. In this final phase, the cell divides into two.

The process repeats. However, cells have a natural protective mechanism. When they are starved to the verge of death, they enter a resting phase, or G0 for Gap 0.

2   Cloning Techniques

All the above properties — cell specialization, embryo splitting and cell cycles — are being exploited in cloning. In summary, there are three known techniques, or variations therefore, used for cloning.[2]

2.1   Embryo Splitting Technique

In the embryo splitting procedure, such as the one used in cloning Tetra, an egg from a mother and sperm from a father are used to create a fertilized egg. After the embryo grows into eight cells (Panel 1), researchers split it into four identical embryos, each consisting of just two cells (Panel 4). The four embryos are then implanted into surrogate mothers. In effect, a single embryo becomes four embryos, all genetically identical.

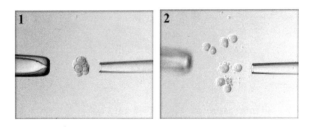

---

[2] Most of this section is adapted from Thinkquest, www.thinkquest.org.

Petory and Biofactory 137

Figure 3A. Embryo splitting begins with (1) An eight-celled embryo. (2) The embryo gets split into individual cells. (Photo: Adapted from Oregon Primate Research Center).

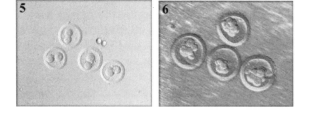

Figure 3B. (3) Two cells are injected into an empty egg. (4) A whole embryo, complete with two cells. (Photo: Adapted from Oregon Primate Research Center).

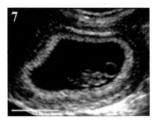

Figure 3C. (5) The resulting embryos, each with two cells. (6) The four embryos develop, gaining more cells as they divide. (Photo: Adapted from Oregon Primate Research Center).

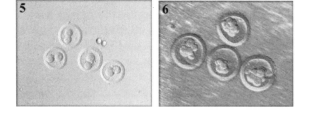

Figure 3D. (7) The sonogram shows a successful implant of the tiny embryo grows into a fetus. (Photo: Adapted from Oregon Primate Research Center).

2.2 The Nuclear Transfer Technique

First explored by Hans Spemann in the 1920s to conduct genetics research, nuclear transfer is the technique currently used in the cloning of adult animals. A technique known as twinning exists, but can only be used before an organism's cell differentiates.

All cloning experiments of adult mammals, like the Roslin or Honolulu techniques described below, have used a variation of nuclear transfer. Nuclear transfer requires two cells, a donor cell and an oocyte, or egg cell. Research has proven that the egg cell works best if it is unfertilized because it is more likely to accept the donor nucleus as its own. The egg cell must be enucleated. This eliminates the majority of its genetic information. The donor cell is then forced into the Gap Zero, or G0 cell stage, a dormant phase, in different ways depending on the technique. This dormant phase causes the cell to shut down but not die. In this state, the nucleus is ready to be accepted by the egg cell. The donor cell's nucleus is then placed inside the egg cell, either through cell fusion or transplantation. The egg cell is then prompted to begin forming an embryo. When this happens, the embryo is then transplanted into a surrogate mother. If all is done correctly, occasionally a perfect replica of the donor animal will be born.

2.2.1   The Roslin Technique

The cloning of Dolly is one of the most important breakthroughs in the cloning history. Not only did it spark public interest in the subject, but it also proved that the cloning of adult animals can be accomplished. Previously, it was not known if an adult nucleus was still able to produce a completely new animal. Genetic damage and the simple deactivation of genes in cells were both considered possibly irreversible.

The realization that this was not the case came after the discovery by Ian Wilmut and Keith Campbell of a method with which to synchronize the cell cycles of the donor cell and the egg cell. Without synchronized cell cycles, the nucleus would not be in the correct state for the embryo to accept it. Somehow the donor cell had to be forced into the Gap Zero, or G0 cell stage, or the dormant cell stage.

First, a cell (the donor cell) was selected from the udder cells of a Finn Dorset sheep to provide the genetic information for the clone (STEP 1). For this experiment, the researchers allowed the cell to divide and form a culture *in vitro*, or outside of an animal. This produced multiple copies of the same nucleus. This step only becomes useful when the DNA is altered, such as in the case of Polly, because then the changes can be studied to make sure that they have taken effect.

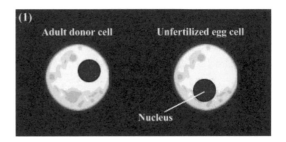

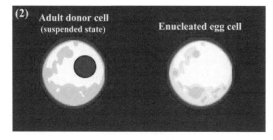

Figure 4A. (1) The nucleus of an egg is removed – enucleation process. (2) The donor cell is starved. The absence of nutrients causes the cell to enter a suspended state, matching the state of the enucleated egg cell. (Photo: Adapted from Thinkquest).

A donor cell is taken from the culture and then starved in a mixture, which has only enough nutrients to keep the cell alive. This causes the cell to begin shutting down all active genes and enter the G0 stage. The egg cell of a Blackface ewe is then enucleated and placed next to the donor cell (STEP 2). One to eight hours after the removal of the egg cell, an electric pulse is applied to fuse the two cells together and, at the same time, activate the development of an embryo (STEP 3). This technique for mimicking the activation provided by sperm is not completely reliable, since only a few electrically activated cells survive long enough to produce an embryo.

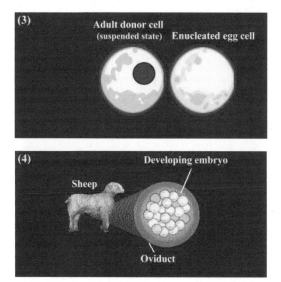

Figure 4B. (3) The donor cell is placed near the egg cell, and an electric current is used to fuse the cells together and stimulate development. The donor cell's nucleus directs the development. (4) The developing egg is placed in a sheep's oviduct for about six days. The oviduct acts as an incubator as the egg grows into an embryo and continues to develop. (Photo: Adapted from Thinkquest).

If the embryo survives, it is allowed to grow for about six days, incubating in a sheep's oviduct (STEP 4). It has been found that cells placed in oviducts early in their development are much more likely to survive than those incubated in a laboratory. Finally, the embryo is placed into the uterus of a surrogate mother ewe (STEP 5). That ewe then carries the clone until it is ready to give birth. If everything goes as planned, an exact copy of the donor animal is born.

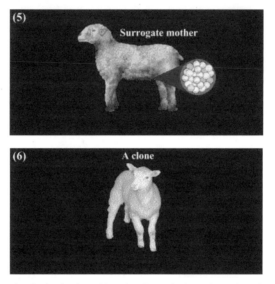

Figure 4C. (5) After developing in the oviduct for about six days, the embryo is transplanted into the uterus of a surrogate mother ewe. This ewe will carry the developing sheep to term. (6) After a normal pregnancy, a clone is born. The clone is an exact genetic duplicate of the animal which provided the adult donor cells. (Photo: Adapted from Thinkquest).

This newborn sheep has all of the same characteristics of a normal newborn sheep (STEP 6). It has yet to be seen if any adverse effects, such as a higher risk of cancer or other genetic diseases that occur with the gradual damage to DNA over time, are present in Dolly or other animals cloned with this method.

2.2.2   The Honolulu Technique

In July of 1998, a team of scientists at the University of Hawaii announced that they had produced three generations of genetically identical cloned mice. The technique is accredited to Teruhiko Wakayama and Ryuzo Yanagimachi of the University of Hawaii. Mice have long been held to be one of the most difficult mammals to clone due to the fact that almost immediately after a mouse egg is fertilized, it begins dividing. Sheep were used in the Roslin technique because their eggs wait several hours before dividing, possibly giving the egg time to reprogram its new nucleus. Even without this luxury, Wakayama and Yanagimachi were able to clone with a much higher success rate — three clones out of every one hundred attempts — than Ian Wilmut (one in 277).

Wakayama approached the problem of synchronizing cell cycles differently from Wilmut. Wilmut used udder cells, which had to be forced into the G0 stage. Wakayama initially used three types of cells, Sertoli cells, brain cells, and cumulus cells. Sertoli and brain cells both remain in the G0 state naturally and cumulus cells are almost always in either the G0 or G1 state (STEP 1).

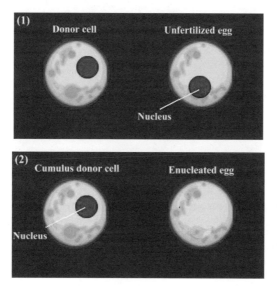

Figure 5A. (1) The nucleus of the egg cell is removed. (2) Cumulus cells are used as donor cells because they remain naturally in a suspended state. The nucleus of a cumulus cell is inserted into the enucleated egg. (Photo: Adapted from Thinkquest).

Unfertilized mouse egg cells are used as the recipients of the donor nuclei. After being enucleated, the egg cells have donor nuclei inserted into them (STEP 2). The donor nuclei are taken from cells within minutes of each cell's extraction from a mouse. Unlike the process used to create Dolly, no *in vitro*, or outside of an animal, culturing is done on the cells. After one hour, the cells have accepted the new nucleus (STEP 3). After an additional five hours, the egg cell is then placed in a chemical culture to jumpstart the cell's growth, just as fertilization does in nature.

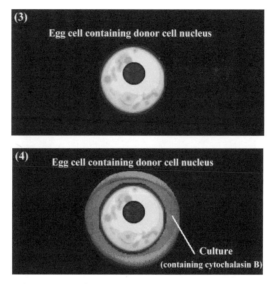

Figure 5B. (3) It takes about one hour for the egg to accept the new nucleus. Then the egg is allowed to sit for an additional five to six hours. During this time, it undergoes no development. (4) After sitting for about six hours, the cell is placed in a culture to jumpstart cell development. The culture functions in the same way as an electric shock, but is less strenuous on the cell. (Photo: Adapted from Thinkquest).

In the culture is a substance (cytochalasin B) which stops the formation of a polar body, a second cell which normally forms before fertilization. The polar body would take half of the genes of the cell, preparing the other cell to receive genes from sperm.

After being jumpstarted, the cells develop into embryos (STEP 4). These embryos can then be transplanted into surrogate mothers and carried to term (STEP 5). The most successful of the cells for the process are cumulus cells, so research is concentrated on cells of this type.

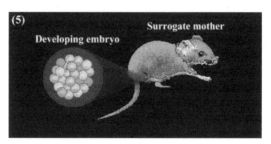

Figure 5C. (5) The egg cell develops into an embryo which is implanted into the uterus of a surrogate mother mouse. After a normal gestation period, a cloned mouse is born. (Photo: Adapted from Thinkquest).

After proving that the technique was viable, Wakayama also made clones of clones and allowed the original clones to give birth normally to prove that they had full reproductive functions. At the time he released his results, Wakayama had created fifty clones.

This new technique allows for further research into exactly how an egg reprograms a nucleus, since the cell functions and genomes of mice are some of the best understood. Mice also reproduce within months, much more rapidly than sheep. This aids in researching long-term results.

3   From Barnyard To PerPETually Yours

With the refinements in cloning techniques that led to the successes of sheep and mouse cloning, it is conceivable that the techniques, with modifications, may be applied to cloning of other species, particularly, pets. This creates a potentially huge market for petory or pet factory.

All across the United States, more people are preserving genetic material from their pets, even though pet cloning is still in the testing stages. In fact, it may be many years before canines, felines or other family pets are cloned with any regularity, if at all.

Canine sperm banks, meanwhile, once used chiefly to breed show dogs, are freezing more samples from the more ordinary family purebred or mutt. Some pet owners say that cloning or artificial insemination, though perhaps eccentric, is the best, most heartfelt tribute to a beloved family member.

At least three companies have emerged since 1998 offering people the chance, at a cost of hundreds or thousands of dollars, to take that leap of faith and freeze animal tissues now so that someday they may be among the first to clone a pet.

PerPETuate in Sturbridge, Massachusetts, which was begun in November 1998 by people with backgrounds in livestock genetics and veterinary medicine, is one of three major companies dedicated to pet cloning and its ancillary scientific benefit. Lazaron BioTechnologies in Baton Rouge, Louisiana, was started in January 1999 by Richard Denniston, who teaches animal sciences at Louisiana State University. Denniston is also known for bioengineering goats carrying a human gene called AT III in April 1999. Genetic Savings and Clone, which is affiliated with Texas A&M University, began in February under the direction of Louis Hawthorne, a San Francisco entrepreneur who views the project as 'the celebration of the mutt.'

Even if pet cloning is a success, pet cloning may initially cost $200,000 per animal before dropping to more affordable $10,000 in several years. Still, pet owners have already frozen samples from an estimated 500 to 1,000 animals, mostly dogs and cats, but also rabbits and gerbils, at more than a half-dozen companies and clinics around the country. The method for what some call 'gene banking' is a rather straightforward procedure: a cloning kit is sent to the pet owner's veterinarian, who conducts a simple biopsy on the animal's belly or neck. The tissue, roughly the size of a pencil eraser, is shipped to a laboratory to be grown in culture dishes. The results are stored in liquid nitrogen, and the owner waits for science to catch up to fantasy — from an owner's perspective, or fiction — from a science writer's perspective.[3]

The biopsy costs vary from $600 to $1,000, and there is also an annual storage fee of $50 to $75. Samples from live animals are preferred, but at Genetic Savings and Clone, about 40 percent of the samples are from animals who have been dead for less than a week.

Regardless of whether the science works, ethicists and animal welfare groups are already wary of the moral implications. Many clients learn about cloning through the Internet, and can barely contain their excitement because animals are regarded as emotional ballasts providing unconditional love in an uncertain world.

In November of 2000, Arthur Caplan, director of the Center for Bioethics at the University of Pennsylvania, surveyed audiences about their views on cloning at Duke University and at a private high school in suburban Philadelphia. The result was although few people supported cloning humans, roughly 90 percent of the audience favored cloning pets. According to Caplan and Alan Beck of the Center for the Human-Animal Bond at Purdue University, what some of the pet owners may not realize is that an animal's personality is more a function of its environment and its experiences (nurture) than of its genes (nature). So any clone, even if it is raised in the same family, will likely have a different life. That is to say, the new dog does not understand the old tricks. In addition, Beck is critical of the commercial push of the cloning effort, warning that companies may be preying on vulnerable people who have either lost a pet or are too attached to think rationally.

Sheer novelty aside, cloning might offer a variety of benefits. Owners could spay and neuter their pets, as vets and others forcefully recommend, and still breed their favorite animals. In a nation that destroys 5 million or so cats and dogs at shelters each year, cloning would produce a single pup or kitten instead of a litter.

But there are many unknowns. Some critics fear it would demean the individuality of a pet to know that DNA is already in the freezer, ready to grow into a replacement. James Serpell, director of the Center for the Interaction of Animals and Society at the University of Pennsylvania argues that it might kind of denigrate the individual to have it constantly reproduced.

Critics also say the companies are selling a fantasy that they cannot possibly fulfill: the notion that an old friend will rise from the dead. There is another caution for pet owners and cloners. In the lingo of the field, cloning is "inefficient": It produces many stillborn or deformed animals for each live cow, sheep or goat. Given the enormously high rate of miscarriages and birth defects, one wonders whether someone who really loves a pet would want to subject that pet's genetic twin to such travail.

---

[3] David W. Chen, "Pet cloning is a boon for some, raises hackles for others", *New York Times*, December 28, 2000.

Ronald M. Green, a Dartmouth College ethics professor opines that pet cloning will be a test bed for human cloning. If proved safe in pets, the practice will likely accustom the public to cloning as a form of reproduction and will make it more likely that people will accept human cloning somewhere down the line.

3.1 K-9 Cloning

If you lead a dog's life this life, you may just be luckier when you get cloned into another family.

Debbie Thieme, an emergency room nurse near Pittsburgh, paid $1,500 to preserve cells from three of her dogs, who are fighting various forms of cancer.

Jonathan Hill, a researcher involved in the Second Chance cloning project, is a veterinarian trained in Australia and at Texas A&M who used the cloning work as part of his doctor of philosophy studies in physiology with Mark Westhusin. He is also a member of another research team led by Westhusin that is involved in the Missyplicity Project. The Missyplicity Project is a 2-year effort to produce the first cloned dog. The anonymous sponsors of the project have invested $2.3 million to produce a clone of their pet dog, Missy, a mixed breed border collie. A team of about 20 researchers is working on the Missyplicity Project, and some of the knowledge gained by Second Chance is helping to advance that research.[4]

In attempting to clone Missy, Westhusin's team follows the same protocol used to create Dolly the sheep, a process known as somatic cell nuclear transfer. But before the protocol can be applied, there is a set of species-specific problems that need to be resolved. For instance, a dog has a bursa, or pouch, encasing their ovaries making egg harvesting more difficult. Sheep, on the other hand, ovulate regularly, once every 19 days. Dogs release ova randomly, once every 6 to 12 months.

Sixty bitches are serving as egg machines that supply the raw materials for the Missyplicity Project, the joint venture between Texas A&M and California-based Bio-Arts and Research Corporation (BARC). After eggs are collected, they will be retrofitted with Missy's DNA, cultured in vitro, and if the embryo is viable, surgically implanted into another dog's oviduct for gestation. If everything goes perfectly, Missy II will be born in 63 days.

To date, none of the attempts has led to a perceptible heartbeat yet.

3.2 Feline the Copy Cat

Cats have nine lives? Think again. With cloning, a cat can have as many lives as it wants!

---

[4] Charles Graeber, "How much is that doggy in the vitro?" *Wired Magazine*, March, 2000.

While no one has cloned a cat yet, three top-notch U.S. teams are racing for what is the next big trophy in the burgeoning field of cloning. Experts say the first one could be born in year 2001, with the first cloned dog probably coming later, its arrival hampered by the peculiar hurdles of the canine reproductive system.

Even before the first copied cat arrives, companies connected to each research team are already running a test of what happens when cloning is offered as a consumer product. The looming question is whether cloning, if ever perfected, will win acceptance as a way to produce children. So far, the idea has provoked more outrage than approval, with scientists and ethicists condemning an Italian doctor, Severino Antinori, for announcing plans to try to help infertile couples through cloning.

But when it comes to pet animals, Johnson and hundreds of other pet owners such as Phyllis Sherman Raschke — a retired probation officer from Sylmar in Los Angeles County — who paid Denniston $700 to preserve cells from her cat Sammy, are proving that many people will set aside any fears about the technology and embrace it wholeheartedly.

Obtaining cells is a required step in cloning, which uses the DNA within the cells to produce a new organism with the same genetic makeup as the original. The old and new animals are thought to be something like identical twins — very close in appearance but not necessarily in personality or behavior.

Lisa Johnson of New Orleans laid her cat to rest near some alder trees in the backyard, marking his grave with a ring of stones. Then she heard about Richard Denniston and his effort to create the first cat by cloning. Soon, she was back at the grave with a shovel.[5]

And so, three days after his death beneath the wheels of a car, Johnson's much-loved Cowcat underwent a resurrection of sorts. Johnson took his body from the ground, sped it to a veterinarian and had some skin removed. She sent the tissue to Denniston, who induced the cells to multiply. Now millions of Cowcat's cells live on, frozen in liquid nitrogen and waiting for scientists to do with the cat what has already been done with the cow, pig, goat, mouse — and of course, with Dolly, the famously cloned sheep.

4   Which Comes First? The Chicken Or The Egg?

Chickens or eggs? Definitely eggs. And chickens do not have to cross the road no more.

The scientists who cloned Dolly the Sheep have now created designer chickens. It is reported their eggs have been genetically altered to help produce drugs to fight

---

[5] Aaron Zitner, "Owners willing to foot cloning bill for copied cats frozen cells may produce first pet this year", *Los Angeles Times*, April 10, 2001.

cancer. The birds, Britney, are named after pop queen Britney Spears.[6] They altered the hens' genetic make-up so the whites of their eggs are rich in tailored proteins which will form the basis of new drugs that could be commercially available within two years.

Proteins in egg white are produced according to instructions encoded in the hen's genes. Altering the genetic material in a single cell nucleus can breed a chicken which will lay eggs full of the right proteins. Each chicken can lay about 250 eggs a year producing huge quantities of the necessary proteins.

Until now, producing even small quantities of usable proteins in the laboratory has proved difficult and expensive. This has hampered the development of new drugs to treat various illnesses including ovarian cancer and breast cancer. But each egg from Britney and the rest of the flock will contain 100mg or more of the proteins which can be easily extracted.

Britney was developed over two years by U.S. biotechnology company Viragen and the Roslin Institute in Edinburgh.

In this case, one can count the chickens before they hatch, since eggs are what these companies will be selling, and the financial reward will be no chicken feed.

5 White Elephant

Everybody loves elephants. In fact, people in the Southeast Asian country of Thailand adore them so much that researchers at Mahidol University in Bangkok want to clone their own elephants.[7]

Scientists in Thailand hope to clone a prized fair-skinned, but dead elephant that was once owned by a Thai king. The elephant died 100 years ago, and a museum in Bangkok has preserved its remains in alcohol-filled jars. A Thai veterinarian Chisanu Chiyacharoensri said they hope to clone the elephant since it is the best specimen ever seen in the country.

This may sound impossible. But according to Dave Evans, a molecular biologist at the University of Guelph in Ontario, Canada, in theory the whole thing is feasible. It is possible to recreate a live animal from a dead animal's genes. But the actual process involved could make it a fantasy. First, researchers would have to figure out which genes in the elephant are responsible for its white color. It could be one gene, but it is probably multigenic.

Then researchers would have to devise a way to extract the genes from the dead elephant, then to insert them into a live elephant's embryo, or fertilized egg. Even trickier, the dead elephant's genes would have to closely match the live elephant's genes in order to join with and replace them. Finally, the experiment would have to

---

[6] "Dolly scientists clone chickens", *ITN News*, December 3, 2000.
[7] Michael Cannell, "Impossible elephant", *Science World*, March 22, 1999.

be carried out on lots of elephant surrogate moms, so that one might bear a prized transgenic (carrying altered genes) white elephant.

Depending on what chemical preservatives were added to those alcohol-filled jars, an animal's genes soaked for 100 years might well be destroyed. And that preserved white elephant may turn out to be a white elephant (a useless possession) after all.

6   A Mammoth Undertaking

A University of Arizona geology professor, Paul S. Martin, wants to restock the modern-day plains with animals of the past. Martin envisions game reserves with buffalo roaming, deer and antelope playing, and elephants browsing.

This might not sound like the range that greeted Lewis and Clark. But it does represent the wilderness of 13,000 years ago that confronted the earliest settlers into North America. And he would like to see pockets of modern America that reflect the pachyderm presence once again.

Martin shared his vision recently during the 25th anniversary celebration of the discovery of Mammoth Site in Hot Springs, South Dakota. The site hosts a museum where the 100,000 visitors each year can see the continuing excavation as well as some results of the effort — including some complete skeletons of the roughly 50 mammoths preserved when they fell into a slippery sinkhole 26,000 years ago.[8]

Martin said he wanted to do honor to his country by appreciating its true nature. We have been misled into thinking this is the home of the deer and the buffalo and the moose. That is true in historical time but in evolutionary time this land is the home of elephants, camels, horses and ground sloths.

In a paper entitled "Bring Back the Elephants", published in the Spring 1999 issue of *Wild Earth*, Martin and co-author David A. Burney note that the disappearance of North American elephants about 13,000 years ago during the late Pleistocene occurred almost yesterday in the geological time frame.

As a result of the late Pleistocene extinctions we live in a continent of ghosts, their prehistoric presence hinted at by sweet-tasting bean pods of mesquite, honey locusts and monkey ear. Such fruits are the bait evolved to attract native animals that served as seed dispersers. African and Asian elephants are the only members of the order of Proboscidea that were not lost in the mega faunal crisis of the late Pleistocene.

Seven species of Proboscidea, including wooly mammoths, dwarf mammoths and mastodons, suddenly died off during this crisis. After a million years or more of successful existence, they faded into evolutionary history in perhaps a few hundred years, evidence indicates. What is more, the rapid cycle of extinctions occurred just

---

[8] "Prehistoric times", *Science Matters*, July 25, 1999.

as the Clovis people were settling North America on a southward journey that began at the Bering Straits, a now-flooded peninsula that connected Alaska and Siberia. Martin considers the timing more than coincidence. He is one of the main proponents of the theory that humans were the catalysts for the sudden wave of extinctions of large North American mammals. Although there is no smoking gun to prove the connection, there are spear tips found in fossil mammoths. For instance, a mammoth skeleton unearthed in Naco, Arizona, contained eight spear points identified as having Clovis origin.

Thanks to cave paintings in Europe by ancient artists, scientists know what mammoths looked like, with long fur making them appear superficially different from the elephants that have so far survived into modern times.

Figure 6. A painting of a mammoth. (Sketch: Courtesy of *Science Matters*).

Their behavior, too, probably differed only superficially from that of modern elephants, which are considered "super keystone species" by some conservation biologists because of their ability to transform the environment. Elephants dismantle trees, turning forest into the savannah that can support a variety of large grazing mammals and their predators.

Martin suspects that the disappearance of the North American elephants, actively hunted by our ancestors, could have altered the environment enough to precipitate the extinction of other range animals. Along with the late Pleistocene elephants, dozens of other large mammal species disappeared from North America at that time, including ground sloths, horses, the saber-tooth tiger and the dire wolf.

Gray wolves — remnants who survived the Pleistocene but were recently driven to near extinction in the United States by ranchers and farmers — are being reintroduced to Yellowstone National Park, Wyoming. Scientists abroad are repopulating an area of Siberia dubbed Pleistocene Park with horses, musk ox and bison. Reintroducing elephants into North America could be the next step in efforts to restore the wilderness of old.

Martin believes that if we want the "super keystone species", second only to our own in their capability for altering habitats and faunas, we should start with the restoration of living proboscideans — with African and Asian elephants.

7   Animal Factory — From Farming To Pharming

With all the genetic engineering tools and cloning technique at our disposal, we are now at the point where we can use animals as bioreactors and as organ factories.

In the animal industry, whether the herds swim, graze, cluck or moo, a key word is "broodstock". On land, broodstock development is as old as aquaculture in the sea. It is a fledging industry.

The most striking of the new creatures being concocted by plucking a gene from one organism and inserting it into the DNA of another are known as transgenics or pharm animals. Pharm animals are the only way to produce some of the more complex proteins needed for use as drugs. Some very simple proteins, like insulin, can be churned out by genetically modified bacteria. But some proteins require being folded in particular ways or having sugars added before they can be used. Bacterial cells cannot carry out these advanced tasks but animal cells can, making pharm animals the producers of choice.

Transgenics was initially recognized as a novel platform for the production of recombinant products for various reasons:

❑   The ability to produce significantly greater amounts of protein with a higher expression level and volume output than traditional culture systems.

❑   The ability to express novel proteins due to the unique nature of the mammary gland for production of complex molecules.

❑   The potential for significant reduction in the cost per unit protein production due to the animal being true bioreactor requiring fewer inputs and less complex monitoring support system.

Producing the initial founder transgenic animal is the first critical step. Each pharm animal have its advantages and disadvantages:[9]

❑   The cattle: The efficiency of gene integration in the cattle is among the lowest. Bovine ES (embryonic stem) cells have been isolated and this may represent a more efficient route for targeted transgenesis. Scientists have been relatively successful in cloning cattle. Since the 1980s, over 300 cows have been cloned. Complications with cattle include neonatal deformities such as "large calf syndrome". While it has a relatively low reproduction rate and long gestation period, it is a remarkable protein factory, producing up to 9,000 liters of milk per year, and the diary infrastructure to harvest the milk is rather well established.

❑   The sheep: Dolly the cloned sheep debut in 1997. The first recombinant protein (factor IX) was expressed in the milk of sheep in 1988. A transgenic sheep can express recombinant protein to levels of up to 40 g/L.

❑   The goat: The goat produces nearly two-fold more milk than the sheep. It also displays a much lower incidence of scrapie. Features like a high protein content, a short gestation period and a short time to maturation make it

---

[9] L. Heninghausen, *Nature Biotechnology* 15, 1997, pp. 945.

particularly well-suited for biopharmaceutical production. Over 60 biopharmaceuticals and proteins have been expressed in the milk of transgenic goats, 11 of which at levels higher than 1 g/L. The goat is a year round breeder and is sexually active as early as 3 months old.

❑ The pig: The pig differs from many other livestocks because it has litters. Unless there are at least four viable fetuses in the womb, pregnancies fail to go to term. Porcine embryos are notoriously fragile because they frequently fail to divide after fusion. Transgenic proteins have been expressed both in the milk and in the blood. Pig organs and tissues, genetically modified to be compatible with a human recipient, could be used to meet the rising demand for human organs. However, there is still the issue of zoonotic infection by porcine endogenous retroviruses (PERVs) to be overcome before the organs and tissues can be used in clinical application. Because of its physiology, the pig is a good model for studying human diseases.

❑ The chicken: For genetic engineers, the chicken has a tricky reproductive cycle. Following entry into the reproductive tract, an egg is almost immediately fertilized, and on the 24-hour journey down the tract, the egg acquires layers of egg white proteins and a hard shell. It will have divided 30,000 times when it is laid. While somatic cell cloning is not possible as yet, gene targeting of avian embryonic stem cells and spermatozoa-mediated transgenesis to produce genetically modified chickens are being attempted. The main advantage of the chicken is its ability to grow a large number of birds relatively cheaply and its short gestation time.

❑ The rabbit: The rabbit has been amenable to nuclear transfer since 1985. Its short gestation period, rapid onset of sexual maturity and high number of offsprings make it a good model system. Proteins can either be produced in the milk or secreted in the blood. The rabbit also has similar lipid metabolism as the human and therefore serves as a good model for atherosclerosis and restenosis. As yet, no one has succeeded in cloning a rabbit from adult cells.

7.1 Animal Bioreactors

Several biotechnology firm in the U.S. and Europe are giving new meaning to the slogan "Got Milk?" These companies have sliced a few human genes into the DNA of goats, sheep, pigs and other mammals. These human genes are designed to produce, in the animal's milk, a specific therapeutic protein to combat human diseases. The therapeutic protein will be extracted from the milk, purified and packaged as a drug. In short, these transgenic animals are being groomed as four-legged drug factories. To date, Genzyme Transgenics is in Phase III trial of an anti-clotting protein derived from a transgenic goat. PPL Therapeutics in U.K. is experimenting with transgenic sheep's milk to produce protein drugs for cystic fibrosis and hemophilia. Pharming NV, a biotechnology firm in the Netherlands, is testing whether transgenic rabbits, cows, and mice can produce a variety of human proteins for stomach ailments and other bleeding disorders.

The American Red Cross is working with Pharming to create a transgenic pig to make the protein raw material for a new surgical bandage. The human blood contains a protein called fibrinogen which forms thin tendrils across wound to trap blood and form a clot. The idea behind the bandage is to produce the fibrinogen, smear it on a bandage and stick the bandage to a wound. Early tests show that this type of bandage will stop any kind of bleeding, including gushing wound that may prove fatal, in 15 seconds.

Currently, fibrinogen is extracted from human plasma, and the Red Cross cannot get enough plasma to meet the demand. The transgenic pigs can produce large quantities of the material at a reasonable cost.

Red Cross, Pharming and Infigen are also working on cloning cows to produce milk fortified with human Factor VIII proteins used to treat hemophilia and severe bleeding. The current process of making the protein from donated human blood is incredibly expensive. Factor VIII costs a hemophiliac $20,000 to $40,000 for a one-year supply. A herd of about 100 cows will supply a large percentage of the world's supply of factor VIII, at a much lower cost than refining it from blood.[10]

While human cloning and DNA services face ethical questions, animal agriculture has quietly moved ahead into the world of biotechnology. As a result, agricultural animal breeders can access cloning and other genetic services not available for humans. This is precisely what Infigen, headquartered in DeForest, Wisconsin, does. It offers cattle, sheep, hog and other farm animal cloning to commercial farmers. The trademarked cloning service, AgriCloning, costs between $50,000 to $100,000, depending on how many animals the customer wants cloned.[11] The cost is expected to drop when efficiencies improve. Current success rates are as low as 10 to 15%. In addition, the company also offers tissue/cell storage. A customer may order a $125 kit to take a tissue sample of an animal for future use in cloning. Farmers and breeders use the service to preserve cells of high genetic merit, injured or aged animals. Once the tissue/cell is received, a cell line is made and stored for future use. The service costs $325 to $400 per animal. Storage is free for the first year, and costs $50 for up to 25 animals.

Producing protein drugs through current manufacturing process means building factories that can cost up to $150 million. Animal production will require less capital outlay. The animal process is currently not that fast and cheap yet. Researchers must first harvest the female animal's eggs, fertilize them *in vitro* and implant human genes into these embryos to produce the desired therapeutic proteins. The transgenic embryos must then be transplanted into surrogate mothers, who

---

[10] Stuart F. Brown, "From cow's milk to medicine chest", *Fortune*, September 4, 2000.
[11] Karen McMahon, "Clone-grown: Cloning promises to upgrade animal genetics", *Farm Industry News*, March 2001.

would give birth to the animals. Once they grow to adulthood, the animals that produce the highest concentration of drug in their milk are bred to make a herd.

FDA is still to consider the safety issues raised by transgenic animals. A potential problem is human allergic reaction to proteins produced by animals, though to date, there has not been a case of such allergy. Another likely problem is whether infectious agents in animals, like the mad cow disease, might somehow jump to humans.

7.1.1 Antimicrobial Peptides

Antimicrobial peptides are an exciting class of therapeutics but to realize their potential, a reliable and economical large-scale production method is needed. These natural antibiotic agents have a novel mode of action compared to conventional antibiotics and offer a potential solution to the increasingly prevalent problem of drug-resistant bacteria that limit the effectiveness of the current generation of drugs. For widespread therapeutics use, the antimicrobial agents would need to be produced in large quantities. Unlike chemical synthesis, transgenic peptide production offers the flexibility to produce kilograms to multi tons quantities of peptides at low cost. Moreover, it is the only recombinant system capable of producing naturally amidated peptides, eliminating the need for enzymatic *in vitro* amidation.

7.1.2 Transgenic Technology Development

Unlike traditional pharmaceutical drugs that use chemical synthesis to make products, most biopharmaceutical drugs requires viable biological host cells for the production of therapeutic recombinant proteins. The first biopharmaceutical products to enter clinical trials utilized microbial fermentation, which has an established history in the production of pharmaceutical drugs. However, it was soon realized that certain products require mammalian cells to produce a biologically active molecule. Although mammalian cell culture has an established history in laboratory research, it has never been scaled up for commercial quantities. As mammalian cell culture technology was scaled up, it soon became apparent that considerable cost was added to the manufacturing process, and in some cases, prohibited further development. As an alternative for the production of biopharmaceutical products, the notion of using transgenic animals as bioreactors for the production of heterologous proteins was developed.[12]

---

[12] Robert L. McKown, and Rita A. Teutonico, "Transgenic animals for production of proteins", *Genetic Engineering News*, 19(9), May 1, 1999.

Table 1. A timeline in commercial development of transgenic technology.

- ❑ Microinjection technique for gene transfer demonstrated for multiple animal species
- ❑ Efficiency of stable transgenesis improved to 10%
- ❑ Nuclear transfer technique (cloning) demonstrated for multiple animal species
- ❑ Multiple heterologous proteins expressed and localized in the mammary glands.
- ❑ Level of heterologous protein expression in milk demonstrated at 1–40 g/L.
- ❑ Development of purification protocols with 53% yield and 99.999% purity.
- ❑ Validated removal and inactivation of potential adventitious agents during processing.
- ❑ Scale up and maintenance of transgenic herds demonstrated.
- ❑ The safe use of transgenic products in human clinic trials has been established.
- ❑ Good agricultural practices have been developed and accepted for commercial production.

7.1.3   Heterologous Protein Production

The technologies needed to develop a heterologous protein production system in transgenic animals include expressing and targeting the gene product for secretion in one of the body fluids, introduction of the recombinant DNA construct into the animal genome, and purification of the therapeutic protein from the harvested fluid.

Of the three body fluids initially proposed for transgenic proteins production, milk, blood and urine, milk-specific expression is the most developed. Heterologous proteins can be expressed and targeted to the mammary gland by fusing gene downstream from the regulatory sequences of gene products normally found in milk, such as whey acidic protein, lactalbumin, lactoglobulin or the casein proteins.

Various combinations of *in vivo* and *in vitro* embryo and fertilization technologies can be utilized to create transgenic animals. The first technique to be developed was microinjection of the DNA construct into fertilized, single-cell eggs, followed by implantation into recipient females. Embryos develop to term and are tested for expression of the heterologous proteins. The successful integration of foreign genes into the target genome was initially low, but refinement of the process has increased the rate of transgenesis to 5–10%. Transgenic founders are then mated to develop transgenic herds. The lead-time needed to express transgenic proteins in milk is directly related to the gestation period and the onset of lactation of the selected animal. For example, goats can produce milk in 18 months from initiation of microinjection. With this technology, it takes approximately three years from cell transfection to the birth of a production herd of cows.

Table 2. The current status of five different systems for expressing transgenic proteins on a large scale. (Table adapted from *Nature Biotechnology*).

	Chickens	Rabbits	Cattle	Goats	Sheep	Pigs
Gestation time	20 days	1 month	9 months	5 months	5 months	4 months
No. of offsprings	250	8	1	1–2	1–2	10
Time to sexual maturity	6 months	5 months	16 months	8 months; 3–6 months in BELE	8 months	6 months
First lactation of founder		7 months	33 months	18 months	18 months	16 months
Annual milk yield		4–5 l (2–3 lactations)	8000–9000 l	800–1000 l; 365 l in BELE	500 l	300 l (2 lactations)
Raw protein per female	Up to 0.25 Kg/year	0.02 Kg/year	40–80 Kg/year	4 Kg/year	2.5 Kg/year	1.5 Kg/year
Expressed proteins	Monoclonal antibodies, lysozyme, growth hormone, insulin, human serum albumin	Calcitonin, extracellular superoxide dismutase, erythropoietin, growth hormone, insulin-like growth factor 1, interleukin 2, α-glucosidase, glucagons-like peptide	Lactoferrin, α-lactalbumin	Antithrombin III, tissue plasminogen activator, monoclonal antibody, α1-antitrypsin, growth hormone	A1-antitrypsin factor VIII, factor IX, fibrinogen, insulin-like growth factor 1	Factor VIII, Protein C, hemoglobin

Ian Wilmut and Keith Campbell first achieved nuclear transfer of an adult mammalian cell to an enucleated oocyte in 1996 at the Roslin Institute creating the celebrated Dolly. Transgenic animals can be produced with this technology by transfection of cultured cells with a construct encoding the gene of interest. The cultured cell is then introduced into an enucleated oocyte, stimulated to develop and transplanted into a surrogate mother. Genetic manipulations of the cultured cells make it possible to select for high expression of the desired gene and to specify that only females are produced.

The advantages of the technique are genetic uniformity in production animals, more efficient herd expansion and a shorter timeline for product commercialization. With nuclear transfer technology, it takes approximately two years from cell transfection to the development of a cloned production herd. Development of a new cell-line cloning technique with a high rate of transgenesis may reduce the timeline further.

7.1.4   Collecting and Purifying Product

Two additional considerations in the development of transgenic technology for the production of biopharmaceuticals are the efficiency of collecting raw product and the complexity of purifying the product to acceptable levels for human use. The mammary gland is designed to secrete proteins to feed young, and the agricultural industry has developed highly efficient technology to collect milk. It has been calculated that the average milk production for an adult cow is 10,000 liter per year.

Natural milk has the advantage of containing relatively few proteins compared to the complex mixture of cellular proteins. Ninety percent of the milk-specific proteins (caseins) are present in a micellular suspension. The remaining proteins are primarily composed of lactalbumin and lactoglobulin with trace amount of serum proteins. The micellular caseins and milk fat can be removed in a single step. A combination of column chromatography and filtration steps has been used to separate the remaining milk proteins from the heterologous proteins to give a 53% yield and 99.999% purity.

The incidence of transmissible spongiform encephalopathy (TSE) and the presence of other adventitious agents in agricultural animals have raised concern about the use of human medical products derived from animals. The USDA has established regulatory guidelines to monitor and maintain herds free of known transmissible diseases.

7.2   *Key Transgenic Players and Future Trends*

The driving force behind the development of protein production in transgenic animals is cost effectiveness. The advantages of using transgenic animals compared to cell culture technology are low production cost and ease of scale-up.

With the FDA trend of defining the product rather than the manufacturing process, certain products may be able to move from cell culture to transgenic animals without incurring the cost of clinical trials.

Ease of scale-up also favors the use of transgenic animals in the manufacturing process. When additional raw product is needed with cell culture technology, additional time must be spent in a cGMP (current Good Manufacturing Practice) facility, or additional equipment must be acquired. In the case of transgenic animals, conventional breeding can increase the size of a production herd at relatively low cost. A number of reports have cited a ten-fold reduction in the cost of raw product derived from transgenic animals when compared to mammalian cell culture.

Thus it is not surprising that a growing number of biotechnology companies are investing time and money into transgenic animals to produce biopharmaceuticals. The most noteworthy with product in clinical trials are:

❑ Genzyme Transgenics Corporation (GTC). GTC is a subsidiary of Genzyme (Framingham, Massachusetts), established in 1993. Main efforts are producing novel therapeutics proteins in mammary glands of transgenic animals, such as goats, since these animals produce large volumes of milk and have a fairly short gestation period.

❑ Pharming Group NV. Pharming Group NV was founded in 1988 and produced the first transgenic dairy cattle in the world. It uses nuclear transfer technology to produce biopharmaceuticals in transgenic cattle, pigs, and rabbits.

❑ PPL Therapeutics. PPL Therapeutics was founded in 1987 to commercialize the production of proteins derived from transgenic animals produced by the Animal Breeding Research Organization (now Roslin Institute).

❑ Nexia Biotechnologies. Nexia (Montreal, Canada) was founded in 1993 to produce pharmaceutical proteins and biofilaments in the milk of transgenic goats. One of the products is BioSteel. The company cloned the genes for silk production out of a spider and was able to generate transformed MAC-T cells that expressed the biomaterial protein. BioSteel is considerably stronger than any material ever made. It is also lightweight and biodegradable.

Table 3. Key player in transgenic therapeutic protein products and their flagship products in clinical trials. (Adapted from *Genetic engineering News*, 19(9), 1999).

Company	Product	Indication	Status
GTC	Antithrombin III	Coronary artery bypass grafting	Phase III
Pharming	Alpha-glucosidase	Pompe's disease	Phase I/II
Pharming	C-1 esterase inhibitor	Acute myocardial infarction	Preclinical
Pharming	Collagen (Type I)	Tissue repair	Preclinical
Pharming	Collagen (Type II)	Rheumatoid arthritis	Preclinical
Pharming	Factor VII	Bleeding conditions	Preclinical
Pharming	Factor IX	Hemophilia B	Preclinical
Pharming	Fibrinogen	Tissue glue for trauma and surgery	Preclinical
Pharming	Lactoferrin	Prevention of gastrointestinal tract infections	Preclinical
PPL	Alpha-1-antitrypsin	Cystic fibrosis	Phase II

Table 2. (Continued)

PPL	Alpha-1-antitrypsin	Emphysema	Phase I
PPL	Bile salt stimulated lipase	Cystic fibrosis	Preclinical
PPL	Bile salt stimulated lipase	Acute pancreatitis	Preclinical
PPL	Superoxide dismutase	Reperfusion injury	Preclinical
PPL	Superoxide dismutase	Acute respiratory distress syndrome	Preclinical
PPL	Factor VII	Bleeding conditions	Preclinical
PPL	Factor IX	Hemophilia B	Preclinical
PPL	Fibrinogen	Tissue glue for trauma and surgery	Preclinical
PPL	Protein C	Prevention of deep vein thrombosis	Preclinical
PPL	Salmon calcitonin	Osteoporosis	Preclinical

Table 4. Biofactory barnyard. Genetic engineers are creating pharm animals or animals with given ability to produce valuable substances. (Adapted from *New York Times*, May 1, 2000).

Animal	Developer	Purpose
Chicken	University of Guelph	To produce antibiotic, lysozyme in its eggs, for keeping infection rates down in chicken eggs.
Cow	Pharming Inc.	To produce lactoferrin, a human protein, in its milk, for treating infection in humans.
Goat	Genzyme Corp.	To produce antithrombin III, a human blood protein, in its milk, to prevent blood clotting in humans.
Goat	Nexia Biotechnologies	To produce spider silk in its milk, for use in for example, light-weight bullet proof vests.
Sheep	PPL Therapeutics	To produce alpha antitrypsin for treating cystic fibrosis.
Pig	University of Guelph	To produce phylase, a bacterial protein for helping pigs digest the pollutant phosphorus better, resulting in manure containing less of it.

7.3   Barnyard Organ and Tissue Shops

On August 16, 2000, two independent research teams announced the birth of genetically identical piglets. The teams used different techniques refined from past cloning experiments in sheep, cattle, mice and goats.

But the announcements were tempered by another scientific report highlighting the risks of "xenotransplantation". Hidden diseases may be transferred from pigs to humans along with life-saving organs,[13] warned Daniel Salomon of the Scripps Research Institute in La Jolla, and the senior author of the report. In the cautionary *Nature* study, Salomon and colleagues showed for the first time that an infectious agent called porcine endogenous retrovirus, or PERV, can infect mice genetically engineered with a condition similar to immune-suppressed human transplant patients. The virus was found to move around in the mouse bloodstream and infect other organs as well beyond the transplant site. However, the infection induced in the mice did not lead to any disease symptoms, and the same could be true in humans.

To signal the significance of the research, Salomon's study and one of the other cloning-experiment reports were released by the British journal *Nature* in advance of regular print publication. The second report on cloning research was originally slated to appear two days later in *Science* magazine, but editors of that journal also agreed to release the report early so that all three studies would come out concurrently.

Authors of the *Nature* cloning study include scientists affiliated with the Scotland-based company PPL Therapeutics, which was formed to commercialize the technology that gave birth to Dolly, the famous cloned sheep. Advance word about Salomon's study led to speculation in the financial markets earlier in the week that PPL and researchers affiliated with Geron Corp. of Menlo Park were backing away from xenotransplantation research. However, PPL issued a statement to reaffirm its commitment to the field, although the company said it would be at least four years before clinical trials can commence. Geron, meanwhile, said it was refocusing its investments in cloning-related research but denied it was backing off due to medical risks.

7.3.1 Organ Supply and Demand

Safety concerns already have led U.S. regulators to impose tight restrictions on xenotransplant experiments, while some experts are advocating a blanket moratorium on clinical trials. Research also has been complicated by ethical considerations involving the use of animals as organ donors. Medical experts insisted that while such concerns must be taken seriously, cloning offers one of the few practical solutions to the critical problem of organ shortages. About 75,000 people are awaiting organ transplants of various kinds, including nearly 16,000 awaiting livers and 46,000 in need of kidneys. Only a small fraction of those on the waiting lists find a donor match.

According to the United Network for Organ Sharing (UNOS), there were 22,827 organ transplants in 2000, up from 21,655 a year before. Health and Human

---

[13] Carl T. Hall, "Scientists report first cloning of pigs", *San Francisco Chronicle*, August 17, 2000.

Services (HHS) Secretary Tommy Thompson said the need for organs was growing almost twice as fast the supply. Its numbers show that in 1990, almost 15,000 organs were transplanted, while the number of patients waiting was 22,000.[14]

The shortage of donations from cadavers, along with medical advances, was increasing the occurrence of living donor donations, in which people donate a piece of a liver or a kidney. In 2000, the number of living donors grew to 5,532 donations from 4,747 in 1999, a 16.5 percent increase, the largest rise ever recorded. Donations from cadavers rose slightly to 5,984 in 2000 from 5,825 in 1999, a 2.7 percent increase.

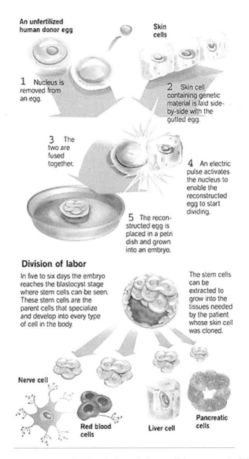

Figure 7. Using cloning techniques, scientists believe it is possible to grow in laboratory human tissues and cells for xenotransplantation. (Source: Advanced Cell Technology. Figure adapted *San Francisco Chronicle*).

---

[14] "HHS secretary launches new organ donation program", *CNN News*, April 17, 2001.

The advent of cloning, combined with genetic engineering techniques, suggests that ordinary barnyard animals might someday be transformed into walking four-legged organ factories. Alan Brownstein, president of the American Liver Foundation echoed the need is staggering, and pursuing this type of research is critically important. Pigs are considered an ideal species because they can be grown quickly to roughly human size and are relatively free of pathogens capable of causing disease in humans. However, pig reproductive biology presented some daunted technical problems for cloning pioneers. Only a single prior success in pig-cloning experiments had been reported, involving laborious methods considered impractically difficult to duplicate.

Ryutaro Hirose, a transplant surgeon at the University of California at San Francisco, said the cloning experiments amounted to a small step forward in light of other difficulties. Most pressing is the need to find ways of manipulating the pig's genetic machinery in a much more subtle fashion than can be done currently. Researchers can introduce new genes into a pig, but cannot yet knock out specific genes. A complication is pigs make all kinds of proteins that essentially are pig versions of human proteins, and those are not the kind of proteins a patient wants in the body.

7.3.2 Strides in Xenotransplant

Xenotransplant has come a long way since the first kidneys were transplanted from chimps into humans thirty or so year ago. It still, however, faces some formidable obstacles. Besides hyperacute rejection, other issues include proper functioning of the xenotransplant and risk of infection.[15] A host of strategies have been forwarded to overcome these issues.

7.3.2.1 Hyperacute rejection

Though whole-organ transplants are not yet feasible, the prospects are improving. When a regular pig organ is attached to a person or a primate, the immune system destroys it within hours. But in recent years, scientists have created pigs whose cells display antigens — a genetic flag — that human cells use to show the immune system they belong in the body. Organs from these partially humanized animals have survived up to eight weeks in baboons, and they are sometimes used externally to sustain patients for brief periods while they are waiting for human organ transplants. Despite the human camouflage, these organs still sport a pig antigen called Gal, which flags them as foreign and speeds immune rejection.[16]

To overcome hyperacute rejection, companies use nuclear transfer, transgenic animal development (adding human genes to animals), gene knockout, antibodies and thymokidney to deal with the obstacle. For whole-organ and tissue

---

[15] "Strides in xenotransplant", *Genetic Engineering New*, 20(8), April 15, 2000.
[16] Geoffrey Cowley, "The new animal farm", *Newsweek*, April 2, 2001.

xenotransplants to meet the growing needs for organs and tissues, progress has been made in inactivating specific gene to avoid rejection. For example, PPL's five cloned piglets, and Infigen Inc. of DeForest, Wisconsin, which succeeded at cloning a partially humanized pig.

Another way around hyperacute rejection is the use of cellular transplant where possible, rather than entire organs. A third strategy is genetic engineering to knockout the natural killer T cell reaction.

Pigs are thought to be the best for xenotransplantation because of the similarity of their organs to human organs in size and function. Currently, over 62,000 people in the United States are waiting for organs, with only a fraction of those organs available each year.

The three basic issues are exogenous infectious agents, endogenous viruses and transgenic animals with human genes that can act as receptors for new viruses. Transgenic animals are raised in clean environment. The environment is regularly monitored for environmental pathogens, filter air and water, carefully control feed and birth animals via caesarean section to minimize risk of infection in newborns. In other words, to start the transgenic animal germ-free, and then strategize to minimize risks throughout the animals' lives.

The animals are also tested using three methods, Western Blot, PCR and reverse transcriptase PCR for the presence of endogenous infectious agents, including porcine endogenous retrovirus (PERVs).

A problem with testing for viruses is that what may infect cells *in vitro* may or may not infect cells *in vivo*, and conversely, what may not infect cells *in vitro* may indeed infect cells *in vivo*. There are also questions about whether animals may be infected with certain viruses by other animals. Whether humans may be infected by other animals can only be determined through human testing.

In addition to PERV, other infectious agents in pigs include parvoviruses and circovirus, a newly isolated hepatitis virus that is ubiquitous in pigs and causes wasting and neurological damage.

In developing clinical trials, patient recruitment remains a major issue, due no doubt to fear of infection. Recent deaths in gene therapy trials led FDA to tighten its monitoring and oversight of xeno trials. An interagency, national xenotransplantation registry is currently under development by the FDA, NIH, and CDC to monitor patients on a life-long basis and to archive tissues for future study. The FDA has proposed a tier-review system of xenotransplantation procedure and products, which would categorize or rank risks associated with certain procedures. For example, according to the source of the transplant, whether it is encapsulated or in direct contact with tissues or only blood, whether it is a transient or lifelong transplant. Such a system is already in place for cell therapies.

7.4   Barnyard Teats as Spinnerets — <u>S</u>ilk in <u>M</u>ilk

Spider silk has long fascinated mankind for its elegant evolutionary solution — a combination of enormous tensile strength and elasticity with ultra-lightweight fiber. A class of spiders known as orb weavers, which include the usual cobweb-spinning variety, all use very similar silk thread to weave their webs. The basic silk recipe dates back to the earliest ord weavers that evolved 125 million years ago and suggests that the recipe leaves little to be improved upon.

The spider silk, that fine dragline silk-like material spiders spin to form webs, has tensiles of 400,000 lbs per square inch (2,800 MPa), 10 times stronger than steel, and 3.5 times the strength of widely used para aramid fibers. An inch thick of the silk can stop a fighter jet landing on an aircraft carrier.

Among the 35,000 species of spiders, scientists have decoded silk proteins from some 10 spider species. Of the four species of orb weavers, they have found repeating patterns that can be quite different from the repeating patterns of more primitive spiders of lineage which extends back 400 million years.

Silk, produced in the silk glands of spiders, are composed completely of silk proteins. Silk proteins are long, complex molecules, stretching hundreds to thousands of amino acids long, and are difficult to decode. Within the silk gland, the silk proteins are dissolved in highly concentrated solution that is half water, half proteins. From the silk gland, the protein solution is squeezed out of a duct like toothpaste out of a tube. The proteins, folded up in rod shapes, line up like logs flowing down a river, and they are pulled out of the duct by knobby appendages known as spinnerets. The silk proteins unwind, lock together and solidify into silk fiber.[17] How spiders unfold proteins and spin them together into silk threads is not well understood, but it is believed that the chemical conditions in the duct, like acidity and the concentration of potassium, are crucial for the silk proteins to unwind properly.

Despite its superior mechanical properties, spider silk is not used commercially because of an absolute constraint on supply. Spider farming is simply not practical. Unlike silkworms, the spider is territorial and aggressive, precluding intensive cultivation. They are cannibals, ruling out large-scale spider farming.

This is where genetic engineering comes in. By isolating the genes from the spider that code for silk protein and genetically engineering the genes into a founder herd or plant, silk can be mass produced.

To date, several spider genes have been isolated and characterized.[18,19] Bacterial and other fermentation systems, which work well with recombinant

---

[17] Kenneth Chang, "Unraveling silk's secrets, one spider species at a time", *New York Times*, Tuesday, April 3, 2001.

[18] R.V. Lewis, "Spider silk: unraveling of a mystery", *Acc. Chem. Res.*, 25, 1992, pp. 392–398.

[19] P.A. Guerette, D.G. Ginzinger, B.H.F. Weber, and J.M. Gosline, "Silk properties determined by gland-specific expression of a spider fibrion gene family", *Science* 272, 1996, pp. 112–115.

proteins, are inadequate in producing authentic silk protein. Transgenics, however, provides an elegant production method as it uses natural spider silk genes.[20]

A serendipitous find shows that the silk gland of spiders and the milk gland of goats are almost identical. The goat, domesticated some 8,000 years ago, is prized for its ability to convert plant biomass into high quality proteins in milk. As a bioreactor, the spider silk proteins can be expressed in Nigerian dwarf goat milk by genetically engineering a spider gene in the 70,000 or so goat genes. The goat is milked as in conventional dairies and the proteins are then extracted from the milk to be extruded and spun into filaments on bobbins, which are woven, braided or incorporated into industrial materials.[21] Nigerian dwarf goats are used because they have certain competitive advantages — BELE — breed early, lactate early. They come to maturity and produce milk sooner, in about 16 months or 10–20% less time than other goats and sheep, and twice as fast as cows.

The first transgenic goat, Willow, was born in 1998. Subsequently, the first pair of special dwarf goats, Peter and Webster, were born in January 2000. What is needed is for Peter and Webster to sire sufficient nannies to begin production. This is a combination of the old with the new. The old is represented by the goats and their milk. The new is the genetic engineering.

The silk finds application in lightweight and flexible body armors for military, law enforcement and peace-keeping personnel. It can be used in military and civilian aviation and space vehicles. It also finds application in high performance sports gear and apparel, and biodegradable fishing line, medical devices such as suturing and artificial tendons. It is thus not surprising that the U.S. Army is working with Ste. Anne de Bellevue, Quebec-based Nexia Biotechnologies, Inc. to develop BioSteel.

Cows are more prolific producers. They produce 7,000 to 8,000 liters of milk per year. Dwarf goats produce 250 to 500 liters per year. Each liter of milk can contain 2 to 10 grams of the protein. So for large-scale production, longer-to-production but higher-production cows are more ideal. To put things in perspectives, for ballistics alone, para aramid usage will need about 7,000 tons of the silk. It may take a goat a month to produce a lightweight vest of about a pound.

8   Ethical Issues And Regulatory Compliance

Still, one does not have to be an animal-rights activist to balk at the thought of tinkering with the genetic makeup of complex, social animals, for example, monkeys. If one considers how great the benefit to humanity will be if one can use monkeys to cure a disease, one will be helping people who have no other options.

---

[20] J. Scheller, K.H. Gührs, F. Grosse, and U. Conrad, "Production of spider silk proteins in tobacco and potato", *Nature Biotechnology* 19, 2001, pp. 573-577.
[21] F. Vollrath and D.P. Knight, "Liquid crystalline spinning of spider silk", *Nature* 410, 2001, pp. 541–548.

But one must also remember that anything one does with a monkey affects it as a being. The monkeys deserve the best life they can have.

Beyond the welfare of laboratory animals, ethicists also worry about the implications of moving genetic engineering into the world of primates, the group that includes humans. ANDi's arrival was more an inevitability than surprise, but his hybrid DNA does underscore just how possible it would now be to start engineering humans, whether to prevent inherited disease or for some less noble cause. That may not be an immediate concern, however, since the technique requires considerable refinement. Schatten's team attempted to engineer a total of 224 monkey eggs, and ANDi was the only one of three healthy babies born that carries the jellyfish gene. The researchers still are not sure if that piece of DNA is actually functional.

ANDi, at least, should not have to worry about the issues he raises, whether technical or ethical. By and large, his career as an active research subject is over. For the rest of us, a new age of experimentation is just beginning.

The use of transgenic animals for protein production does add a new layer of quality and regulatory controls not needed in cell-culture-based production.[22]

❑ A sound understanding of the health and physiology of the species is essential.

❑ Unlike cell culture production, a transgenic animal may live and produce for up to 7 to 10 years.

❑ For cell culture, each batch is made from a unique initiation or fermentation run. With transgenics, a production animal is bred and lactates annually.

❑ Each animal experiences its own physiological changes and various environments throughout its life as it develops, gives birth, and lactates.

❑ Predominantly, ruminants (cows, goats and sheep) are being used due to their high volume of milk production. Careful attention to sourcing animals must be exercised. Animals should be sourced from Transmissible Spongiform Encephalopathy (TSE) free countries such as New Zealand and Australia.

❑ Reliable and permanent identification schemes should be used. The schemes include ear tags, ear tattoos, and electronic transponders.

❑ At the point of milk collection, quality and regulatory controls should comply with traditional (cGMP) to ensure that the product is safe, pure, potent, and efficacious.

Much of the cutting edge research in animal husbandry is occurring in the field of pharming. Researchers are transforming herds and flocks into biofactories to produce pharmaceutical products, medicines and nutrients. Researchers in the field predict that by the year 2020, 95% of human body parts will be replaceable with laboratory-grown organs.[23]

---

[22] William G. Gavin, "The future of transgenics", *Regulatory Affairs Focus*, May 2001, pp. 13–17.

[23] Robert Langer, and Joseph P. Vacanti, "Artificial organs", *Scientific American*, September 1995, pp. 130.

Chapter 7

HUMAN CLONING AND STEM CELL RESEARCH

"Research on embryonic stem cells raises profound ethical questions, because extracting the stem cell destroys the embryo, and thus destroys its potential for life. Like a snowflake, each of these embryos is unique, with the unique genetic potential of an individual human being." President George Bush, United States, August 9, 2001, explaining his position on limited funding for stem cell research. 'Snowflake' is believed to have been borrowed from Snowflakes Embryo Adoption Program.

1   Human Cloning, To Be Or Not To Be?

The scene was 1992 when HAL, the author of this book, was still a tenured faculty member at a university. One day, HAL received a letter from Argentina. A mother — apparently bereaved by the sudden unexpected death of her beloved son — wrote to HAL to ask if he could help her clone the son. No doubt completely taken by surprise, but still young and early in his career, HAL made a number of phone calls to try to get some leads. He called experts, including a president's personal physician. Of the many experts, one was less polite, and spoke skeptically that "someone must have scribbled your name on a toilet wall". In spite of numerous futile attempts, HAL had a colleague who could speak the lady's native tongue to convey the disappointing news.

In the brief spanning decade since that letter from Argentina, animal cloning breakthroughs were made at an unprecedented pace. First the debut of Dolly the cloned sheep, then cows, mice, monkeys, pigs, and many others

With all these successes cloning animals, it is not surprising that various groups have been talking about cloning humans. According to an unscientific CNN/Time poll of 1005 adults in March 1997, around the time of Dolly, most Americans were wary of the use of cloning. Nearly half of those polled said cloning animals or humans was immoral and unacceptable. Half of the respondents also said they would not be willing to eat fruits, vegetables or animals resulting from cloning. The poll also revealed strong opposition to cloning research, with 66% of those surveyed saying the federal government should regulate the cloning of animals. A full 69% of respondents voiced fear regarding the prospect of cloning humans, with 89% saying this would be morally unacceptable. A contrarian 7% said they would consider cloning themselves. Today, cloning of animals has gained wider acceptance, but the current opinion is still very much against cloning of humans.

Are the polls a true barometer of the public opinion, or an opinion swayed by news media nightmarish scenarios?

Figure 1. 1) A French "cult leader", Claude Vorlihon, or better known as Rael, is a racecar driver turned prophet. His company, Clonaid, has plans to clone humans. 2) Panayiotis Zavos, formerly at University of Kentucky, has announced plans to work with Severino Antinori of Rome to clone humans. Dr. Panos Zavos, right, testifies at a hearing on Capitol Hill before the House subcommittee on Oversight and Investigations, March 28, 2001. Dr. Rydolf Jaenisch waits to testify at left. (Photo: Adapted from *ABC News*, and Stephen J. Boitano, AP Photo). 3) Italian fertility specialist Severino Antinori, shown March 9, was among the scientists testifying on a hearing in Washington on cloning. (Massimo Sambucetti/AP Photo)

The issue of human cloning, however, did not really hit home until January 28, 2001 when two infertility specialists, Panayiotis Zavos and Severino Antinori, announced that they wanted to clone humans.[1] Zavos was formerly at the Andrology Institute in Lexington, Kentucky, and an expert on male infertility. The latter, Severino Antinori from Rome has a string of reproductive firsts to his credit. He has helped a 63-year-old woman bear a child, enabled a 59-year-old woman to bear twins, and repaired the sperm of infertile men using bits of tissue taken from rat testicles. The announcement from these two experts of proven track records caused a stir all the way to policy-makers in Washington DC.

2   Politics Of Cloning

On March 28, 2001, a congressional subcommittee, Energy and Commerce Oversight and Investigation subcommittee and its chairman Republican Representative James Greenwood of Pennsylvania held a hearing testimony on whether America was ready to start cloning humans. President Bush vowed to block research, even as some scientists, both in the United States and abroad, prepare to plunge ahead with human cloning. As the hearings got underway, the White House announced that Bush planned to work with Congress on a federal statute outlawing all human cloning research in the United States. White House Press Secretary Ari Fleischer said "The president believes that the moral and ethical issues posed by human cloning are profound and cannot be ignored even in the quest for scientific discovery." Greenwood said "Congress has to make a decision about whether it is

---

[1] Aaron Zitner, "Fertility docs plan to clone humans, Italian made announcement with US cohort", *Los Angeles Times*, January 28, 2001, pp. A2.

ethical to allow the destruction of human beings as a part of a human experiment. I don't think that's a difficult question for the Congress to answer. We will probably not allow that to be legal."[2] Republican Representative Brian Kerns of Indiana introduced legislation to ban human cloning in the United States. The bill — HR 1260 or the "Ban on Human Cloning Act" — is the first to be introduced in the 107th Congress to establish a U.S. prohibition. Two other bills — HR 1644 sponsored by Representative Dave Weldon, and HR 2172 sponsored by Representative James C. Greenwood — were later introduced. Both would establish criminal penalties for any type of human cloning.[3]

3 To Be Or Not To Be, That Is The Question

From the accelerating pace with which the whole cloning field is moving forward, it is commonly believed that within the first five years of the twenty-first century, a team of scientists somewhere in the world will likely announce the birth of the first cloned human baby. Indeed, the aforementioned Italian professor, Severino Antinori, who is known worldwide for helping several post-menopausal women give birth, announced his plans to create a human clone within two years by using the same technology used to produce Dolly the sheep.[4]

The technique used by Ian Wilmut and his co-workers to clone Dolly the sheep is a technology called somatic-cell nuclear transfer. In somatic-cell nuclear transfer, researchers take the nucleus, which contains the DNA that comprises an individual's genes, of one cell and inject it into an egg, or ovum, whose own nucleus has been removed. The resulting embryo, which will carry the donor's nucleus in every one of its cells, is then implanted into the womb of a female and carried to term.

If Antinori delivers his promise, it is also likely that like Louise Brown — the first child born as the result of *in vitro* fertilization (IVF) about a quarter of a century ago — the first cloned infant will be showered with media attention. But within a few years it will just be one of hundreds or thousands of such children around the world. In other words, cloning will be routine, but only in situations where there is such a need.

As with any new technological breakthroughs, cloning has its pros and cons.[5]

---

[2] "To clone or not to clone", *ABC News*, March 28, 2001.
[3] Kristen Philipkoski, "Clone ban a life-saving ban, too?", *Wired News*, June 22, 2001.
[4] Michelle Nichols, "Angers over plans for world's first cloned human", *TheScotsman*, March 22, 2001.
[5] For an excellent review, see Ronald M. Green, "I, clone", *Scientific American*, September 1999.

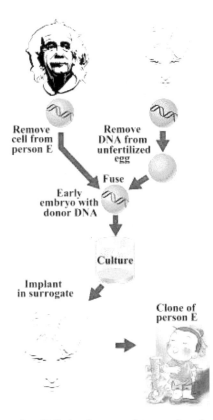

Figure 2. The technique to clone Dolly involves several steps and can be adapted for human cloning. First, the donor cells are grown under special conditions in culture. In this way the number of cells can be increased by several orders of magnitude. It is also possible to make genetic modifications and to select just those cells in which the desired modification has occurred. The selected cells are then fused with an unfertilized egg from which the introduced nucleus can lead to the formation of an embryo. The embryos are then transplanted into a surrogate mother to carry the embryo to term. The clone is born naturally. The egg donor and surrogate mother have been intentionally faded out to avoid any likeness, living or dead. The clone is a young mischievous boy to show that the development of a clone is a function of both nature (genetic) and nurture (environment). The clone of a genius may develop into a singer.

3.1   The Good — Medical Applications

The research on the basic processes of cell differentiation holds out the promise of dramatic new medical interventions and cures. Burn victims or those with spinal cord injuries might be provided with replacement skin or nerve tissue grown from their own body cells. The damage done by degenerative disorders such as diabetes, Parkinson's disease or Alzheimer's disease might be reversed. On July 16, 2001,

Former First Lady Nancy Reagan wrote to President Bush to ask that he not ban stem cell research. In the more distant future, scientists might be able to grow whole replacement organs that our bodies will not reject. When cloning is used in this manner, it is referred to as therapeutic cloning.

These important medical uses of cloning technology urge us to be careful in our efforts to restrict cloning research. In the immediate wake of Dolly, politicians around the world proposed or implemented bans on human cloning. In the U.S., President Bill Clinton instituted a moratorium on federal funding for human cloning experiments, and the National Bioethics Advisory Commission urged that the ban be extended to private-sector research as well. Congress continues to study various proposals for enacting such a total ban.

In view of the still unknown physical risks that cloning might impose on the unborn child, caution is appropriate. Of the 29 early embryos created by somatic-cell nuclear transfer and implanted into various ewes by Roslin researchers, only one, Dolly, survived, suggesting that the technique currently has a high rate of embryonic and fetal loss. Dolly herself appears to be a normal three-year-old sheep. She recently gave birth to triplets following her second pregnancy. But a recent report describes that she grows obese. This and other matters must be sorted out and substantial further animal research will need to be completed before cloning can be applied safely to humans.

Eventually animal research may indicate that human cloning can be done at no greater physical risk to the child than IVF posed when it was first introduced. One would hope that such research will be done openly in the U.S., Canada, Europe or Japan, where established government agencies exist to provide careful oversight of the implications of the studies for human subjects.

3.2   The Bad — Clandestine Efforts

Less desirably, but more probably, it might happen in clandestine fashion in some offshore laboratory where a couple desperate for a child has put their hopes in the hands of a researcher seeking instant renown. For example, a French cult leader, Claude Vorlihon, who has revealed plans to clone a human being, says his experiment will help gay couples conceive.[6] Better known as Rael, a race-car driver turned prophet, who had also performed as a singer/songwriter under the name Claude Celler, offers no proof of his claims but says they are being financed by a $1 million investment from an anonymous American couple who lost their 10-month-old baby girl as a result of an accident. The American couple is "partners in Clonaid". Another anonymous American couple has offered the sect, reportedly $200,000 to clone their dead son. The child died during a botched hospital surgery and some of his genes were preserved before he died.

---

[6] Tom Abate, "Leader emerges for disciples of human cloning", *San Francisco Chronicle*, October 12, 2000.

Given the pace of events, it is possible that a rogue researcher is already at work. For now, the technical limiting factor is the availability of a sufficient number of ripe human eggs. If Dolly is an indication, hundreds might be needed to produce only a few viable cloned embryos. Current assisted-reproduction regimens that use hormone injections to induce egg maturation produce at best only a few eggs during each female menstrual cycle. But scientists might soon resolve this problem by improving ways to store frozen eggs and by developing methods for inducing the maturation of eggs in egg follicles maintained in laboratory culture dishes.

Rael, for example, is said to have lined up a French scientist and 50 female followers to serve as egg donors. Cloning a human infant from the cells of the deceased child is only the first step in his ultimate plan. Eventually, he wants to develop an accelerated growth technique, which he witnessed among the Elohim — the alien race whom Rael claims to have created all life on Earth in a DNA experiment 25,000 years ago. This would allow a clone to be grown to adulthood quickly. With advanced computing technology, he believes people can ultimately imprint their memory patterns on the clone's blank mind.

3.3   The Ugly — Cloning Misconceptions

Once human cloning is possible, why would anyone want to have a child that way? As we consider this question, we should put aside the nightmarish scenarios much talked about in the press. These include dictators using cloning to amass an army of "perfect soldiers" or wealthy egotists seeking to produce hundreds or thousands of copies of themselves. Popular films such as *Multiplicity* — in which a man who had too much to do had multiple copies of himself made, and *Blake Runner* — in which clones were created to be drones on other planets, feed these nightmares by obscuring the fact that cloning cannot instantaneously yield a copy of an existing adult human being. What somatic-cell nuclear transfer technology produces are cloned human embryos. These require the labor- and time-intensive processes of gestation and child rearing to reach adulthood. Saddam Hussein would have to wait 20 years to realize his dream of a perfect army. And the Donald Trumps of the world would also have to enlist thousands of women to be the mothers of his clones.

Lest we forget, we emphasize that environmental factors play a significant role in human development. For all their efforts, those seeking to mass-produce children in this way, as well as others who seek an exact copy of someone else, would almost certainly be disappointed in the end. Although genes contribute to the array of abilities and limits each of us possesses, from conception forward their expression is constantly shaped by environmental factors, by the unique experiences of each individual and by purely chance factors in biological and social development. Even identical twins — natural human clones — especially those raised in geographically separated locations, show different physical and mental characteristics to some degree. How much more will this be true of cloned children raised at different times and in different environments from their nucleus-donor "parents"? Thus we have to distinguish between nature and nurture. Cloning a future Adolf Hitler might instead produce a modestly talented smooth ballroom dancer who cannot rhythm dance.

4   Louis Brown Of Cloning

So who is most likely to want or use cloning?

4.1   Infertile Parents

First are "infertile" parents, be it "infertile" naturally or by virtue of their sexual orientation.

These impotent individuals or couples who lack the gametes (eggs or sperm) needed for sexual reproduction may seek cloning. Since the birth of Louise Joy Brown on July 25, 1978, assisted-reproduction technologies have made remarkable progress in helping infertile women and men become parents. Women with blocked or missing fallopian tubes, which carry the eggs from the ovaries to the womb, can now use *in vitro* fertilization to overcome the problem, and those without a functional uterus can seek the aid of a surrogate mother. A male who produces too few viable sperm cells can become a father using the new technique of intracytoplasmic sperm injection (ICSI) — in which a usable sperm can be obtained from a man who seems to have none, or whose sperm cells are misshapen or immobile and simply unable to fertilize an egg. The procedure involves extracting an immature sperm from a man's testicle, inserting the single sperm or the progenitor of a sperm cell into a recipient egg.

Despite this progress, however, women who lack ovaries altogether and men whose testicles have failed to develop or have been removed must still use donor gametes if they wish to have a child, which means that the child will not carry any of their genes. Some of these individuals might prefer to use cloning technology to have a genetically related child. If a male totally lacks sperm or the testicular cells that make it, a nucleus from one of his body cells can be inserted into an egg from his mate that has had its nucleus removed. The child she would bear will be an identical twin of its father. For the couple's second child, the mother's nucleus could be used in the same procedure.

Same sex marriage is beginning to gain acceptance. Thus one very large category of such users of cloning might be lesbian couples. Currently if two lesbians wish to have a child, they must use donor sperm. In an era of changing laws about the rights of gamete donors, this opens their relationship to possible intervention by the sperm donor if he decides he wants to play a role in raising the child. Cloning technology avoids this problem by permitting each member of the pair to bear a child whose genes are provided by her partner. Because the egg-donor mother also supplies to each embryo a small number of mitochondria, tiny energy factories within cells that have some of their own genetic material, this approach even affords lesbian couples an approximation of sexual reproduction. Cloning might not be used as widely by gay males, because they would need to find an egg donor and a surrogate mother.

4.2   Carriers of Defective Genes

A second broad class of possible users of cloning technologies includes individuals or couples who are more advanced in age or whose genes carry mutations that might cause serious genetic disease in their offsprings.

A vast majority of genetic birth defects occur because an embryo ends up with too many or too few chromosomes, a defect known as aneuploidy. For example, as many as 40 to 50% of the eggs of women under the age of forty have one chromosome too many or one too few. In older women, the rate is even higher, as high as nine out of ten eggs. These eggs with too many or too few chromosomes may be fertilized. Similarly, a sperm with the wrong number of chromosomes may penetrate an egg and fertilize it.   The resulting embryo usually dies almost immediately and the woman miscarries often even before she knows she is pregnant. A few embryos survive to become fetuses and fewer still survive to birth, most often with an extra copy of chromosome 21.   These are Down's syndrome babies.   An extra copy of most other chromosomes is an invariably fatal condition, as is a missing copy of most chromosomes.

Couples can choose cloning as a way of avoiding this "reproductive roulette".

At present, if people with genetic defects want a child with some genetic relationship to themselves, they can substitute donated sperm or eggs for one parent's or have each embryo analyzed genetically using preimplantation genetic diagnosis so that only those embryos shown to be free of the disease-causing gene are transferred to the mother's womb.   However, the large number of genetic mutations contributing to some disorders and the uncertainty about which gene mutations cause some conditions limit this approach.

Some couples with genetic disease in their families will choose cloning as a way of avoiding what they regard as "reproductive roulette". Although the cloned child will carry the same problem genes as the parent who donates the nucleus, he or she will in all likelihood enjoy the parent's state of health and will be free of the additional risks caused by mixing both parents' genes during sexual reproduction. It is true, of course, that sex is nature's way of developing new combinations of genes that are able to resist unknown health threats in the future.   Therefore, cloning should never be allowed to become so common that it reduces the overall diversity in the human gene pool.   Only a relatively few couples are likely to use cloning in this way, however, and these couples will reasonably forgo the general advantages conveyed by sexual reproduction to reduce the immediate risks of passing on a genetic disease to their child.

Cloning also brings hope to families with inherited genetic diseases by opening the way to gene therapy. Such therapy, the actual correction or replacement of defective gene sequences in the embryo or the adult, is the holy grail of genetic medicine. To date, however, this research has been slowed by the inefficiency of the viruses that are now used as vectors to carry new genes into cells. By whatever means they are infused into the body, such vectors seem to reach and alter the DNA in only a frustratingly small number of cells.

Cloning promises an end run around this problem. With a large population of cells from one parent or from an embryo created from both parents' gametes, vectors could be created to convey the desired gene sequence. Scientists could determine which cells have taken up the correct sequence using fluorescent tags that cause those cells to glow. The nucleus of one of these cells could then be inserted into an egg whose own nucleus has been removed, and the "cloned" embryo could be transferred to the mother's womb. The resulting child and its descendants would thereafter carry the corrected gene in every cell of their bodies. In this way, age-old genetic maladies such as Tay-Sachs disease, cystic fibrosis, muscular dystrophy or Huntington's disease could be eliminated completely from family trees.

This approach is particularly pragmatic when the first draft of the human genome is already at our disposal, and an understanding of diseases at the gene level is progressing rapidly.

With the help of genetic engineering to genetically correct defective sequences of hereditary diseases, it is conceivable that we will see less wrongful birth lawsuits in which the plaintiffs sue their parents for having them with physical shortcomings.

5 Issues

Merely mentioning these beneficial uses of cloning raises difficult ethical questions. The bright hope of gene therapy — the insertion of normal or modified genes into a patient to correct genetic or acquired disorders via the synthesis in the body of missing, defective or insufficient gene products — is dimmed somewhat by the reawakening of eugenic fears. If we can manipulate embryos to prevent disease, why not go further and seek "enhancements" of human abilities? Greater disease resistance, strength and intelligence all beckon alluringly, but questions abound. Will we be tampering with the diversity that has been the mainstay of human survival in the past? Who will choose the alleged enhancements, and what will prevent a repetition of the terrible racist and coercive eugenic programs of the past? In such a scenario, will we be cloning for attributes we like, rather than what is best for the welfare of the clone?[7]

---

[7] Andrew Kimbrell, *The Human Body Shop*, (Harper, San Francisco, 1993).

5.1   The Question of Identity — Like Father Like Son

Even if it proves physically safe for the resulting children, human cloning raises its own share of ethical dilemmas. Many wonder, for example, about the psychological well-being of a cloned child. What does it mean in terms of intrafamily relations for someone to be born the identical twin of his or her parent? What pressures will a cloned child experience if, from his or her birth onward, he or she is constantly being compared to an esteemed or beloved person who has already lived? The problem may be more acute if parents seek to replace a deceased child with a cloned replica. Is there, as some ethicists have argued, a "right to one's unique genotype", or genetic code — a right that cloning violates? Will cloning lead to even more serious violations of human dignity?

In analogy with wrongful birth lawsuits in which plaintiffs sue their parents because they are physically challenged, it is conceivable that in the future we may have more wrongful identity lawsuits in which plaintiffs sue their parents for "endowing" on them identities they would rather not be associated with.

Some fear that people may use cloning to produce a subordinate class of humans created as tissue or organ donors. Some of these fears are less substantial than others. Existing laws and institutions should protect people produced by cloning from exploitation. Cloned humans could no more be "harvested" for their organs than people can be today.

The more subtle psychological and familial harms are a worry, but they are not unique to cloning. Parents have always imposed unrealistic expectations on their children and in the wake of widespread divorce and remarriage, we have grown familiar with unusual family structures and relationships. Clearly, the initial efforts at human cloning will require good counseling for the parents and careful follow-up of the children. What is needed is caution, not necessarily prohibition.

5.2   Silver Linings

As we think about these concerns, it is useful to keep a few things in mind. First, cloning will probably not be a widely employed reproductive technology. For many reasons, the vast majority of heterosexuals will still prefer the "old-fashioned", sexual way of producing children. No other method better expresses the loving union of a man and a woman seeking to make a baby.

Second, as we think about those who would use cloning, we would do well to remember that the single most important factor affecting the quality of a child's life is the love and devotion he or she receives from parents, not the methods or circumstances of the person's birth. Because children produced by cloning will probably be extremely wanted children, there is no reason to think that with good counseling support for their parents they will not experience the love and care they deserve.

What will life be like for the first generation of cloned children? Being at the center of scientific and popular attention will not be easy for them. They and their parents will also have to negotiate the worrisome problems created by genetic identity and unavoidable expectations.

All the above issues are not without precedent. *In vitro* fertilization has not become a widely employed reproductive means except in cases where it is needed. The children of IVF are not less loved than children of natural births. Louis Brown, being the first IVF baby, received a lot of media attention. Subsequent IVF babies are almost unheard of.

But with all these difficulties, there may also be some novel satisfactions. As cross-generational twins, a cloned child and his or her parent may experience some of the unique intimacy now shared by sibling twins. Indeed, it would not be surprising if, in the more distant future, some cloned individuals choose to perpetuate a family "tradition" by having a cloned child themselves when they decide to reproduce.

5.3   The Dark Cloud — All Is Fine Unless You Are The Clone

It has been four years after Dolly and evidence is mounting that creating healthy animals through cloning is more difficult than the scientists had expected. The clones that have been produced often have problems such as developmental delays, heart defects, lung problems and malfunctioning immune system.[8]

At issue is not that one particular thing goes wrong or one specific aspect of development goes awry, but rather, the cloning process seems to create random errors in the expression of individual genes. Those errors can produce any number of unpredictable problems, at any time in life.

Let us recapitulate the development to date. Before Dolly's debut in 1997, scientists thought mammals could not be cloned. But now they have cloned not only sheep but also mice, cows, pigs and goats. With mice, they have even made clones of clones on down for six generations. Dolly is apparently normal. Two infertility specialists, Panayiotis Zavos of the Andrology Institute in Lexington, Kentucky, and Severino Antinori, Rome, recently announced that they wanted to clone humans.

The initial fears that clones would age rapidly or develop cancer turned out to be unfounded. But as scientists gained more experience, and tried to understand why efforts so often ended in failure, new questions about the safety of cloning arose.

---

[8] Gina Kolata, "Researchers find big risk of defect n cloning animals", *New York Times*, March 15, 2001.

5.3.1   Current State of the Art of Cloning

In cloning, scientists slip a cell from an adult into an enucleated egg. The egg functions as a medium to reprogram the adult cell's genes so that they are ready to direct the development of an embryo, which in turn develops into a fetus into a newborn that is genetically identical to the adult whose cell was used to start the process. Our current knowledge does not understand how the egg reprograms an adult cell's genes, and that can be the source of cloning calamities. The problem seems to be that, in cloning, an egg must do a task in minutes or hours that in normal fertilization takes months or years. The difference in time scales in reprogramming is a cause for concern.

Experts such as Rudolph Jaenisch at the Whitehead Institute of the Massachusetts Institute of Technology are of the opinion that breathtakingly rapid reprogramming in cloning can introduce random errors into the clone's DNA, subtly altering individual genes with consequences that can halt embryo or fetal development, killing the clone. Or the gene alterations may be fatal soon after birth or lead to major medical problems later in life.

Scientists say they see what appear to be genetic problems almost every time they try to clone. For example, according to Ryuzo Yanagimachi, a University of Hawaii researcher who first cloned these animals and has been following consequences in them, some mouse clones grow fat, sometimes enormously obese, even though they are given exactly the same amount of food as otherwise identical mice that are not the products of cloning. The fat mice seem fine until an age that would be the equivalent of 30 for a person, when their weight starts to soar. Cloned mice also tend to have developmental abnormalities, taking longer to reach milestones like eye opening and ear twitching.

Figure 3.   Genetically identical mice.   The obese one on the right is the clone.   (Source: R. Yanagimachi, University of Hawaii).

Mark Westhusin, a cloning expert at Texas A&M University in College Station, Texas, reported cow clones are often born with enlarged hearts or lungs that do not develop properly. Dolly herself, while apparently healthy, grew fat and had to be separated from the other sheep and put on a diet. But her experience is difficult to interpret since it is hard to draw conclusions about a propensity to obesity from one animal.

The genetic effects most often seem to be fatal at the very start of life. With cattle, for example, 100 attempts to create a clone typically result in a single live calf. Cloning mice is more efficient. But even then, only 2 percent to 3 percent of the attempts succeed. Yanagimachi affirmed that cloned embryos have serious developmental and genetic problems, which usually kill them before birth, and more die right after birth usually because of lung problems. In addition, inbred strains are much harder to clone than hybrid strains of mice. Inbred animals have much less genetic diversity and so less opportunity to bypass genetic errors than hybrid animals.

Table 1. Clones and the number of attempts to produce the clones.

Date	Name	Species	Attempts	Remarks
Jul 5, 1996	Dolly	Sheep	277	Nuclear transfer from the mammary gland of a mature six- year-old ewe. Named after Dolly Parton, and not aptly named if we recall the cell was taken from a mammary gland.
Jul 1997	Polly	Sheep	425	Cells cultivated from lamb fetus. Of the 425 reconstructed eggs, 62 matured to be implanted into 22 sheep. Of the 22, 11 became pregnant to give six live births. Human Factor IX gene in only three: Polly, Molly and a third. The third was euthanized two weeks after birth because of a heart defect.
Feb 16, 1998	Mr. Jefferson	Calf		Cloned by nuclear transfer of the cell of a fetus. Named in honor of the President's Day in the US.
Apr 1999		Goat		
Aug 9, 1999	Second Chance	Cow	189	Cloned from oldest animal cell. 189 attempts (i.e., transferring 189 cells into 189 eggs), 25 embryo transfers, and 6 pregnancies.

Table 1. (Continued)

Jun 1999	Fibro	Mouse	274	Cloned from cells from the tip of a donor male tail using the Honolulu technique. First clone from male mice clones. Three male offspring from tale-tip cells. Two died shortly after birth. The survivor, dubbed Fibro because the cultured tale-tip cells resemble fibroblast.
Jan 2000	Tetra	Monkey	368	After a fertilized egg had developed into an embryo of eight cells, it was artificially twinned into four embryos of 2 cells each. Only one of the four (thus Tetra) survived, and Tetra was the outcome.
Oct 2, 2000	ANDi	Monkey	224	Jelly fish gene was inserted into 224 eggs that were then fertilized. Of 40 resulting embryos, 5 developed into fetuses, 3 were born but only ANDi carried the jelly fish gene. First transgenic monkey.
Jan 2000 Mar 14, 2000	Millie, Christa, Alexis, Carrel, and Dotcom	Bull Pig	72	Five piglets cloned from a body cell of an adult female pig after 72 double nuclear transfers. 3 are identical clones from one burse cell, and the other 2 are identical clones from another. All delivered by Caesarean section, weighed about 2.72 lbs, or 25% less than piglets from a natural mating.
Apr 2000 Jul 2 2000	Xena	Cow Pig	110	Cloned from cells of a strain of black Chinese Meishan pigs using Honolulu technique. Named after xenotransplantation.
Jan 8, 2001	Noah	Gaur	692	692 attempts to create embryos from the cells of its "father". Only 30 were sufficiently robust to be transferred into surrogate mothers, giving 8 pregnancies and resulting in 1 live birth.

5.3.2 Preliminary Evidence of Awry Clone — Problems with
Reprogramming

Scientists have long established that every cell in the body has the same genes so, in theory, all the instructions for making a new copy of an adult are present in every cell. But most of the genes in an adult cell, like a skin cell or a brain cell or a liver cell, are silenced. That is why those cells, which have reached their final stage of development, never change. In other words, a skin cell does not turn into a heart cell, a brain cell does not turn into a liver cell, and so forth. And no one expected an egg cell to be able to reprogram such an adult cell. This is what makes Dolly such a celebrity because in this cloning experiment, the enucleated egg reprogrammed the nucleus of an adult cell.

Jaenisch and Westhusin say that from preliminary molecular biology experiments they are starting to see confirmation of their belief that reprogramming can go awry. They are looking at molecular patterns of gene expression in embryos created by cloning and comparing them to the patterns in embryos created by normal fertilization. Their results so far are consistent with their hypothesis that reprogramming can result in random errors in almost any gene. But scientists say that every species is different, and it remains possible that it will be easier and safer to clone humans than it is to clone other species. For example, mouse eggs are fragile, which may complicate efforts to clone. The solutions used to bathe cattle embryos while they are being grown in the laboratory seem to create a large-calf syndrome, resulting in large placentas and huge calves that often die around the time of birth. Some growth abnormalities analogous to acromegaly in which the bones become excessively enlarged. But clinics for *in vitro* fertilization have vast experience in growing human embryos in the laboratory and have perfected the method.

So it is only a matter of time before a human clone is introduced to the world, with the bags of all the ethical and social implications.

6   Dream Things That Never Were

George Bernard Shaw is known to have said, "Some men see things as they are and say 'Why?' I dream of things that never were and say 'Why not?'"

Our society and culture, more so in the West than in the East, are based on an implicit faith in technology. Whether we are talking about transportation, navigation, space travel, electricity, telecommunication or biotechnology, advances in technology have tended to improve life spans and prosperity. Given this track record, our culture has let technology evolve at its own pace, trusting that the process would yield more blessings than drawbacks. This is not tautologous to saying that technology has been left entirely unfettered. In biotechnology, for instance, government agencies such FDA and EPA oversee medical and agricultural experiments conducted by researchers. At universities, review boards watch over the shoulders of researchers.[9]

Two trends are forcing us to question our blind faith in technology. The first trend is the suspicion that our watchdog institutions are unable to ride herd on change. Such cynicism is not new. But the explosion of knowledge and the accelerating pace of innovation would stress even a most perfect regulatory system, and we know our system is not perfect. For instance, on September 17, 1999, Jesse Gelsinger, an eighteen year-old Arizona man, who suffered from a rare metabolic disorder (deficiency of ornithine transcarbamylase or OTC), died during a gene therapy experiment. The FDA accused a professor with a financial stake in the outcome was a bit too aggressive in administering the potent medicine.[10,α] The second and newer trend, which comes through in the genetic engineering story and cloning story, is that technologies that seemed like science fiction a few years ago are now possible and within the reach of small teams of experts, respectively.

7   The Law

After 1997, when Dolly the sheep made headlines, California, Michigan, Louisiana and Rhode Island passed laws against human cloning. The federal government has banned the use of federal funds for human cloning experiments but has placed no restrictions on what private entities can do. Twelve nations worldwide have banned human cloning.[11]

The law will lumber along like an unwieldy dinosaur wending its way to extinction if it cannot keep abreast with the pace of change in technology. The law is by definition conservative, attempting to bring order out of chaos only as fast as consensus can be reached among social groups willing to conform to norms they believe are fair and workable. When the law lags behind changing conditions, conflicts arise that present new questions, ones that are not easily answered by reference to established precedents. And human cloning has no precedents in the sense that no human clone has existed yet.

The law becomes particularly hairy when at the same time two groups of conflicts began to emerge. In the first of these, a group alleges that no one owns the human life, and indeed, life is nature's creation, too inherently divine for us humans to interfere. In the second, a very different argument is made that the

---

[9] Tom Abate, "Biotech is pushing the possibilities past the breaking point", *San Francisco Chronicle*, February 5, 2001, pp. B1.

[10] "Apology for gene therapy death — researchers admit to errors and FDA violations in treating sick teen", *Washington Post*, December 10, 1999, pp. A14.

[α] To be fair, we should mention the doctor involved, Henry I. Miller, offered "post hoc, ergo propter hoc" in his defense.

[11] Mariam Falco, and Matt Smith, "House members opens hearing on human cloning", *CNN News*, March 28, 2001.

commercialization of human bodies is in conflict with established notions about the right of individuals. The question is not who has the right to traffick, but whether anyone did.

But these critics are laying the groundwork for a needed culture change. We must learn to control the inexorable growth of technology, including biotechnology. Biotechnology will be at the center stage of this change, because we are certain that science will make it possible to do things with life that our moral sensibilities will not allow. Biotechnology deals with lives and living things, i.e., ourselves. It thus has a bag of ethical and moral repercussions.

7.1 The Genie Is Out of the Bottle into The Brave New World

Should there be a law against human cloning? The issue with human cloning is this: the technology is here and is being perfected. Two known groups have already publicly announced that they will plunge ahead. It is likely that soon someone will be able to do so. The genie is out of the bottle. What we should do now is to take the necessary steps to prepare for such an eventuality. It is time for us to define civilization not by the things that our technology makes possible, but by which possibilities we choose to undertake. For example, cloning technology can be used without the end result being an embryo, such as in developing stem cells for transplant in the so-called "therapeutic cloning".

While some academicians may balk at the thought of the technology being developed by nonacademic, nontraditional entities, there is a precedent for this approach. It began back when Joseph Califano was at the helm of the Department of Health, Education, and Welfare. The decision was made then that the government was not going to fund fertility research but it was not going to stop it either. As a result, many advances in reproductive technology have come from efforts in the private sector that were often financed by the patients themselves.

Human cloning is currently privately financed. In all likelihood, a human clone will be a reality very soon. At that point, our notions of sociality and equality could be transformed. Meritocracy could give way to genetocracy, with individuals, ethnic groups, and races increasingly categorized and stereotyped by genotype, making way for the emergence of an informal biological caste system.[12]

Julian Huxley wrote that humanity — the product of evolutionary creativity — is now obligated to continue the creative process by becoming the architect for the future development of life. *Homo sapiens'* destiny is to be the sole agent of further evolutionary advances on the planet.[13] All Darwin asked of people in his "survival of the fittest" was that they compete for their own life. The new cosmology asks that people be the creator of life.

---

[12] Jeremy Rifkin, *The Biotech Century*, (Jeremy P. Tarcher/Putnam, 1998).
[13] Julian Huxley, *Evolution in Action*, (New American Library, New York, 1953).

8   Message In A Bottle From The Future

Based on the facts and discussion in the preceding paragraphs, we concur human cloning technology is within grasp and a clone is imminent. We opine the best measure is to prepare for such an eventuality.

It is best that therapeutic cloning such as stem cell research should not be banned. Reproductive cloning should be carefully curtailed and monitored until all existing physical risks have been overcome. A total ban or prohibition will not serve the purpose. It will only stifle related research, and promote clandestine activities. Restriction and regulation with caution and careful oversight will be more ideal.

Concurrently, we propose the social and ethics communities should also ready the public for the day. News media will serve the public better by educating and informing, rather than reporting unfounded facts or nightmarish scenarios.

However, what we do worry about is whether the law could ever keep pace with biotechnology, because at some point, we are certain that we will have to outlaw something. Clones differ from inanimate products (such as electronic gizmos) in various ways. Our culture and legal system still do not allow us to terminate a life like we can destroy a defective gizmo. If a clone should turn out to have unexpected traits, the clone may be a social liability. Because it is alive, it will grow, migrate and reproduce.   A clone is thus inherently more unpredictable than inanimate products in the ways it interacts with other living things in the environment. Consequently, it is much more difficult to assess all of its potential impacts.

On the issues of intrafamily relations of being born the identical twin of its parent and the pressure of it being constantly compared to an esteemed or beloved person who has already lived, we believe good counseling of parents and careful follow-up of the clone will overcome these problems.   IVF will serve as an exemplary precedent. *In vitro* fertilization has not become a widely employed reproductive means except in cases where it is needed. The children of IVF are not less loved than children of natural births. Louis Brown, being the first IVF baby, received a lot of media attention. Subsequent IVF babies are almost unheard of. If this is an indication, we anticipate human cloning will not be a widely used reproductive means except in cases where it is needed. We also expect children of cloning to be well-loved offsprings, and that with time, the media will leave them alone.

Some fear that people may use cloning to produce a subordinate class of humans for organ pharming. We regard a clone in the same status as a naturally born human being. A clone should not be harvested for organs, nor should it be subjected to any other acts that are banned from a naturally born human.

In this light, we anticipate that in the future, besides wrongful birth lawsuits, we will also have wrongful identity lawsuits in which the plaintiff clone sues its parents or creator for endowing it with an identity that it would rather not be associated with.   In addition to implied warranty of merchantability and implied warranty of fitness for inanimate stuffs, we anticipate having implied warranty of clonability and implied warranty of nurturability. Here we emphasize the interplay between nature (genetics) and nurture (environment). Although genes contribute to the array of abilities and limits each of us possesses, from conception forward their expression is constantly shaped by environmental factors, by the unique experiences of each individual and by purely chance factors in biological and social developments. Thus a clone is likely to be nurtured into an individual genetically identical (barring mutations), but socially different from its donors in a different environment. A clone of Osama bin Laden — a constant prick in the eyes of the U.S. Whitehouse — in today's society could very well develop into a proponent of U.S. democracy system.

But as we have discussed earlier, the law is always behind developments, and these warranties will probably not come about until after a precedent-setting court case.

Let us keep these in mind as we march into the "Brave New World" of Aldous Huxley.[14]

9   Sherlock Holmes — The Clone Murder Detective

A news reporter once asked HAL, given his view and position on cloning, if he would consider cloning himself. His answer is a definite NO. First, HAL is one of the most fortunate beneficiaries of advances of biomedical technologies. He is very healthy, and not impotent. He enjoys intimacy and prefers to take advantage of the genetic diversity in natural birth nature is offering him. Second, HAL is not an egoist, and he is not a very successful person either. He does not see any utilitarian reason to have more than one HAL running around.

So much for HAL II.

Some time in August 2001, while taking a break from revising this book on the picturesque Monterey Bay, California with relatives, the issue of human cloning came up. HAL was posed with the following hypothetical situation:

A clone committed a murder. To cover up the crime, the clone also got rid of his identical twin father. If the only evidence was genetic trace in the first crime scene, would the clone be able to get off the hook if he shifted the blame onto his father?

This looks like a case that Sherlock Holmes will have a hard time to crack if all that he has is a genetic magnifying glass. Do two wrongs make a right in this case?

We now return to the letter from Argentina mentioned in the opening paragraph of this chapter. If the same real-life scene could be recast in the current prevailing climate, the whole episode would have unfolded very differently. The Argentinean lady would in all likelihood not have written HAL. In all probability, she would

---

[14] Aldous Huxley, *The Brave New World*, (Penguin Modern Classic, 1932).

have called Claude Vorlihon of the Rael Group, or Panayiotis Zavos in Kentucky, or Severino Antinori in Rome. As for the skeptic expert who claimed that someone had scribbled HAL's name on a toilet wall, HAL has this to say, "You probably do not realize what you thought was impossible merely nine years ago is now within grasp!"

In this case, Sherlock Holmes will be assigned an easy task of tracking down the Argentinean lady to update her the breakthroughs. And while at it, Sherlock may also want to find out why she approached HAL, who had absolutely no expertise in cloning. As program director of a supercomputer institute, HAL used computers for his work in bioinformatics. HAL lectures and writes in English, that does not qualify him to be a linguist. HAL writes widely on diverse topics, that does not qualify him to be an expert in all these areas.

HAL talks about cloning, but he is no cloning expert.

10   Stem Cell Research

In cloning, there are reproductive cloning and therapeutic cloning. Therapeutic cloning and stem cell research are closely related.

10.1   Stem Cell 101

Gail Martin, now at University of California, San Francisco, was a post-doctoral fellow at University College London in 1974 when she figured out how to keep fragile stem cells alive in petri dish. In 1981, she succeeded in isolating for the first time stem cells from mice embryos in her laboratory at University of California at San Francisco.[15]

The issue of stem cell got pushed into public arena in 1998 when James Thomson of University of Wisconsin accomplished in isolating stem cells from human embryos, touching off the hope that such cells could be used to create cures for a host of devastating human ailments.

To understand the significance of stem cell, let us take a cursory look at human development. A fertilized egg is totipotent in the sense that its potential is total, it can develop into a fetus. Approximately four days after fertilization during which several cycles of cell divisions have taken place, the totipotent cells begins to specialize, forming a hollow sphere of about one hundred cells called the blastocyst. The blastocyst consists of an outer layer of cells and a cluster of inner cell mass.[16]

The outer layer of cells will develop into placenta and other supporting tissues needed for fetal development. The inner cell mass will develop into virtually all of

---

[15] Ulysses Torassa, "UCSF scientist opened the door — research has been built on her seminal work 20 years ago", *San Francisco Chronicle*, Friday, August 10, 2001, pp. A3.

[16] *Stem Cells: A primer*, National Institutes of Health, May 2000.

the tissues of the human body. These inner cell mass are pluripotent rather than totipotent. They cannot form an organism because they cannot develop into the placenta and supporting tissues necessary for the development in the human uterus.

The pluripotent stem cells undergo further specialization into stem cells that are capable of developing into cells of particular function. For example, blood stem cells, skin stem cells, and other human body's 200 different cell types.

While stem cells are extremely important in early human development, multipotent stem cells are also found in children and adults. Examples of multipotent stem cells include blood stem cells residing in the bone marrow or in smaller quantities circulating in the blood stream. Blood stem cells perform the life-critical role of continually replenishing red blood cells, white blood cells and platelets into the blood stream.

10.2 *Derivation of Stem Cells*

Our current state of the art technology allows human pluripotent cell lines to be derived from three sources:[17]

❑ James Thomson of University of Wisconsin isolated pluripotent stem cells directly from the inner cell mass of human embryos at the blastocyst stage. The sources of the embryos are *in vitro* fertilization clinics. This is depicted in the figure below, A→C.

❑ John D. Gearhart of John Hopkins University isolated pluripotent stem cells from fetal tissue obtained from terminated pregnancies. The stem cells are taken from the region of the fetus that is destined to develop into the testes or the ovaries. The is illustrated in the figure below, A→D.

❑ Somatic cell nuclear transfer (SCNT) is another way pluripotent cells can be isolated. In this approach, a somatic cell — any cell other than an egg or a sperm cell — is fused with a enucleated (nucleus removed) egg cell. The resulting fused cell and its immediate descendants are believed to be totipotent. When they develop into a blastocyst, inner cell mass may be extracted as pluripotent stem cells. This is diagrammed in the figure below, B→C.

---

[17] Michael Shamblott, et al, "Derivation of pluripotent stem cells from cultured human primordial germ cells." *Proceedings of National Academy of Sciences*, 95 Nov. 1998, pp. 13726–13731.

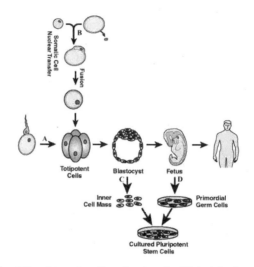

Figure 4. *In vitro* fertilization (A) and somatic cell nuclear transfer (B) both lead to totipotent cells. During development, totipotent cells become blastocyst. Pluripotent inner cell mass (C) is used for stem cell research. Cells from aborted fetuses which are committed into becoming testes or ovaries (D) can also be extracted for stem cell research. (Figure adapted from various sources, including Stem Cells: A primer, National Institutes of Health, May 2000.)

10.3   Sources of Embryos

There are many thousands of human embryos frozen at the some 300 infertility clinics in the United States. They are leftovers from *in vitro* fertilization efforts for infertile couples.

During infertility treatment, more than enough (usually more than ten) eggs are collected. Since human eggs cannot be frozen and later thawed for fertilization, all the eggs collected must be fertilized. Embryos not transferred into the surrogate mother's womb end up frozen for later use. But many end up unused and left with uncertain future — remain frozen indefinitely, donated for research, donated for other infertile couples, or being thawed and discarded.[18] Given the choice between leaving the embryos frozen and discarding them, many couples see the value of donating them for research.

---

[18] Jeffrey P. Kahn, "Making a market for human embryos?" September 4, 1000, Center for Bioethics, University of Minnesota.

10.4   Applications of Stem Cell Research

There are at least three broad categories of applications of stem cells:
❑   To delineate the complex events during human development. A primary
    goal is the identification of the factors involved in the cellular decision-
    making process that results in cell specialization. Genes are turned on and
    off but little is known about the decision-making genes or what turns them
    on and off. Some of the most serious medical conditions, such as cancer and
    birth defects, are due to aberrations in cell specialization and cell division.
    A better understanding of normal cell process will shed light on the
    fundamental errors that cause these deadly diseases.
❑   To streamline drug safety test. New medications can be initially tested using
    human cell lines. The process, however, will not replace testing in whole
    animals and testing in humans, but would streamline the process.   Only
    drugs that are both safe and appear to have beneficial effects in cell line
    would pass for further testing using laboratory animals and later human
    subjects.
❑   To use in cell and tissue therapies.   Perhaps the most far-reaching
    application of pluripotent stem cells is the generation of cells and tissues.
    Many diseases and disorders result from disruption of cellular function or
    destruction of tissues of the body. Donated organs and tissues are often used
    to replace ailing or destroyed tissues. Unfortunately, the number of people
    suffering from these disorders far outstrips the number of organs available
    for transplantation.   Pluripotent stem cells, stimulated to develop into
    specialized cells, offer the possibility of a renewable source of replacement
    cells and tissues to treat a myriad of diseases, conditions, and disabilities.
The latter, cell and tissue therapies, have stirred a lot of controversies. While
the research is holding a lot of promises, with several breakthroughs in mice, there is
still much to be done before the technology can be applied in clinical practice. The
challenges are significant, but are not insurmountable.
❑   There is still work to be done to understand the cellular events that lead to
    cell specialization in the human so that we can direct these pluripotent stem
    cells to become the types of tissue needed for transplantation.
❑   There are still a lot of problems in overcoming immune rejection. Because
    human pluripotent stem cells derived from different sources are genetically
    different from the recipient, research is still needed on modifying human
    pluripotent stem cells to minimize tissue incompatibility or to create tissue
    banks with the most common tissue-type profiles.
The second issue of tissue incompatibility is where somatic cell nuclear transfer
(SCNT) has a slight advantage.   Consider a patient with a degenerative disease.
Using SCNT, the nucleus of virtually any somatic cell from the patient could be
fused with a enucleated donor egg cell. With proper stimulation, the cell will
develop into a blastocyst. Cells from the inner cell mass of the blastocyst are then

extracted to create a culture of pluripotent cells. These cells are then stimulated to develop into the cell type required for transplanting. Since genetic information is carried in the nucleus, these cells would essentially be identical genetically to the patient. When the cell, tissue or organ is transplanted into the patient, the chances of rejection are drastically reduced.

10.5 From Degenerative Disorders to Regenerative Medicines

Limited research both in the United States and elsewhere has already shown glimmers of its enormous potential. Stem cells have allowed mice paralyzed by degenerative nerve conditions to walk again; damaged mice hearts have regenerated with stem cell injections, and stem cells could be coaxed to produce insulin in diabetic mice. In other words, stem cells can be used in regenerative medicines.

10.6 Embryonic Stem Cells versus Adult Stem Cells

As noted earlier, multipotent stem cells can be found in some types of adult tissue. Indeed, stem cells are needed to replenish worn out cells in our body. The blood cell is a very good example. However, multipotent stem cells have not been found for all types of adult tissue.

It has been shown that some adult stem cells previously thought to be committed to the development of one line of specialized cells are able to develop into other types of specialized cells. For example, recent experiments in mice suggest that when neural stem cells are placed in the bone marrow, they appear to produce a variety of blood cell types. In addition, studies in rats have indicated that stem cells found in the bone marrow were able to produce liver cells. These findings suggest that even after a stem cell has begun to specialize, the stem cell may, under certain conditions, be more flexible.

If this is the case why has there not been more aggressive pursuit of adult stem cells? While adult stem cells hold real promise, there are some significant limitations to what we may or may not be able to accomplish with them. Adult stem cells

- ❑   Have not been isolated for all tissue types.
- ❑   Are often present in minute quantities, are difficult to isolate and purify, and their numbers may decrease with age.
- ❑   May not proliferate as younger cells.
- ❑   May contain more DNA abnormalities due to exposure to external mutagens.
- ❑   May not be used for early stage cell specialization research for they have progressed further along the specialization pathways than pluripotent stem cells.
- ❑   Are multipotent, that is, they are not pluripotent.

Attempts to use stem cells from a patient's own body for treatment would require isolation of the stem cell from the patient and then grow the cells in culture in sufficient amount adequate for treatment. For some acute disorders, there may not be enough time to grow enough cells for treatment. In other genetic disorder, the genetic disorder is likely present in the patient's stem cells.

10.7 Federal Support

The National Institutes of Health (NIH) in the United States has an annual budget of $20.4 billion. In fiscal 2000, it spent $226 million on adult stem cell research, with $147 million on human adult cells, and $79 million on animal cells.

11 Private Companies In Stem Cell Business

Bush announced 64 cell lines developed by 10 laboratories in five countries around the world — United States, Sweden, Israel, Australia, and India — in his August 9, 2001 address, with each line created by harvesting stem cells from a single source: either an embryo, blood from newborns' umbilical cords or sometimes adult cells. Most scientists questioned the assertion. Fewer than ten have been reported in scientific journals, with perhaps another ten mentioned in scientific meetings. A National Institutes of Health report release in the summer of 2001 put the estimate at 30 cell lines.

Dr. Lana Skirboll, associate director for science policy at the National Institutes of Health, which will regulate the field, defended the White House claim by saying that "a lot of the research was going on behind closed doors and was being closely protected for proprietary reasons".

Nearly a month after President Bush announced that he would permit federally financed scientists to study more than 60 colonies of human embryonic stem cells, the administration acknowledged that fewer than half those colonies were fully established and ready for research. Testifying before a panel of skeptical senators on September 5, 2001, Tommy G. Thompson, Secretary of Health and Human Services, claimed that only 24 or 25 of the colonies, or lines, were ready for experiments. Noting that private research would continue despite Bush's restrictions, Thompson is confident that the private sector will fill any voids if there are any voids.

The University of Wisconsin scientist who isolated these embryonic cells patented his discovery and turned the patent over to a foundation, the Wisconsin Alumni Research Foundation. In 1999, it granted Geron exclusive commercial license to develop stem cells for certain medical uses in return for private support to continue studying embryonic cells at a time when the federal government would not pay for such research.

Geron, Inc., based in Menlo Park, California, has funded most of the embryonic stem cell work in the United States. Of the six cell lines — nerve, liver, muscle, bone, blood, and pancreatic islet cells — only two have been deemed sufficiently stable and useful to be distributed to stem cell scientists around the world. Geron is the leading owner of cell lines in the United States. It has patented the technique for extracting stem cells and has claims to any discovery that originates from the procedure.

If Bush decision holds, research firms that destroy new embryos to harvest stem cells, such as Advanced Cell Technology Inc. of Worcester, would not be eligible for federal support.

The fact that only a few companies in the world own cell lines prompts fear of price gouging and preferential treatment. The nonprofit company, WARF, in Wisconsin is selling stem cells for $5,000.

12   Reactions To Bush's Decision

Retrieved from the days-old (4–5 days) human embryos, stem cells can morphologize into virtually every kind of tissue, providing a potentially bottomless source of replacement parts. By definition, stem cells can develop into any of the more than 200 cell types in a human body.[19] The potential is enormous.

On August 9, 2001, President Bush of United States took a political tight rope of allowing stem cell research, but with restriction. The reaction can be broadly divided into four groups:

❑ Religious group — This group, which believes stem cell research should be banned completely, views the decision compromises the sanctity of life.

❑ Patient — This group, which is for stem cell research, believes the restriction will stifle or slow research into finding cures for debilitating and degenerative ailments such as Parkinson's disease, Alzheimer's disease, diabetes, spinal cord injury, and others.

❑ Scientific community — The scientific community has mixed feelings on Bush decision to limit federal funding to research on existing colonies of cells. First they are those who feel relieved that Bush had actually given in from his campaign stance of banning completely human cloning and stem cell research. Second, there are also those who believe restricting research to existing cell lines will stifle research. The restriction may block scientific progress if further research unearths problems in the existing cell lines or points to ways newly derived cells can be improved.

❑ Business community — This group believes it will slow down a potentially very commercially viable repair kits for debilitating human ailments. The

---

[19] Rick Weiss, "Scientists see growth, lament limits", *Washington Post*, Friday, August 10, 2001, pp. A01.

restriction to existing cell lines will provide windfall to the few companies who currently have access to these cell lines.

Because of limited funding, the research may go offshore to countries where the restriction is less stringent. For example (in ascending order of stringency):

- ❑ Britain — Britain is the first country to legalize cloning to allow scientists to create cloned embryos for stem cell research.
- ❑ Israel — There is no law regulating stem cell research and embryo destruction for stem cell research is legal in Israel.
- ❑ Japan — The government has established guidelines for stem cell research, a move likely to allow embryonic cells soon. The guidelines stipulate that embryonic cells used in research should come from fertility treatment that would otherwise be discarded.
- ❑ Australia — The country is drawing up laws to ban cloning but cannot reach a consensus on stem cell research. Stem cell research is legal in parts of the country.

Historically, in the 1970s, the United States did not allow cloning to be done on pathogens. As a result, the Europeans overtook the lead in studying viral infections.

13 What Next?

While it is difficult to estimate in absolute terms the commercial potential of stem cell research, we dare to say that the impact will be enormous, and the windfall to those who own the cell lines will be tremendous. There is almost no realm of medicine that will not be touched by this technology, all the way from regenerative medicine, in which ailing parts of a human body are rejuvenated using stem cells, to organ transplant. Stem cells can morphologize into virtually every kind of tissue, providing a potentially bottomless source of replacement parts.

To date, we have less than 30 viable cell lines. We opine there is still a need to generate more cell lines for the following reasons:

- ❑ Current existing cell lines may have problems. For example, of the six cell lines owned by Geron, only three are stable enough to be deemed useful.
- ❑ New ways of deriving cell lines may improve the quality of existing stock.
- ❑ More cell lines, of the order of hundred, are needed to represent the genetic diversity.

Even with a potential supply of many thousand (~100,000) spare embryos from the some 300 infertility clinics in the United States, we believe there may still be reason for making new embryos. The supply of stem cells is too limited is one reason, but a more important reason is because of the need for unique characteristics to match recipient patients to avoid rejection. Indeed, on September 6, 2001, about a month after Bush's televised announcement of federal support for stem cell research with restrictions, the United States government conceded that some of the 64 cell lines may not be ready.[20]

Cell lines exist in three phases: the proliferation phase, in which they start to grow; the characterization phase, in which scientists identify their biological characteristics; and the fully developed stage. So, for example, scientists at the University of Goteborg in Sweden have 19 cell lines. All meet the president's criteria. But only 3 are fully developed; 12 are proliferating, and 4 are in the characterization stage.

Also, to grow the lines, scientists nourish them with mouse cells that have been killed with radiation. This is a standard laboratory technique, but it raises concerns about safety, because the cells could harbor viruses that would infect people. While the Food and Drug Administration does not prohibit the use of mouse cells in human therapies, it imposes strict regulations on them, and officials do not know if the existing stem cell lines would meet the FDA's criteria.

In other words, only 24 or 25 of the 64 cell lines disclosed on August 9 are ready for experiments.

In our view, the United States policy — paying for research on stem cell but ban paying for collecting them — is an issue for concern. From a business standpoint, the United States government is effectively the market maker — a public buyer — creating a demand to be filled by other suppliers from universities or the private sector. These suppliers will need embryos to provide them stem cells. Safeguards are necessary to prevent a market from emerging in embryos trade. The potential benefits of stem cell research are staggering, but they should not come at the cost of commodifying human embryos.

In the United States, currently, there are only a handful of stem cell researchers in the public sector (University of California at San Francisco, Johns Hopkins University) and the private sector (Geron Corp, Advanced Cell Technology). With federal funding for stem cell research, though restricted only to existing cell lines, we anticipate an increase in the number of researchers in the field. Internationally, we foresee those countries with the least stringencies on stem cell research, such as the United Kingdom, or any other countries which will delve into stem cell research, will soon pull ahead of the birthplace of stem cell research — the United States — to lead the field, if the current United States position remains unchanged.

An encouraging sign is that more than 60 senators and 200 representatives have expressed support for stem cell research that would go beyond what the president has authorized. One of the strongest proponents of stem cell research is none other than Senator Edward M. Kennedy, Democrat of Massachusetts, Chairman of the Committee on Health, Education, Labor and Pensions, which held the September 5, 2001 hearing.

---

[20] Sheryl Gay Stolberg, "US concedes some cell lines are not ready", *New York Times*, September 6, 2001.

14 Hotbed For Stem Cell Research — University Of Wisconsin

Scientists at University of Wisconsin were the first to isolate and grow human stem cells, and have created rhesus macaque, common marmoset and bovine stem cell lines. In November 2001, the university announced its scientists have developed the world's first known avian stem cell line working with chickens and quail. The avian blastodermal cells were harvested from fertlized bird eggs and appear to be capable of reproduction at will and cell differentiation. The discoveries are expected to lead to the creation of transgenic birds, technology surrounding the use of eggs to grow protein-based therapeutics, and the preservation of endangered birds.[21] The Wisconsin Alumni Research Foundation (WARF), a not-for-profit technology licensing foundation, has patented the avian cell lines, adding to its already large portfolio of various stem cell lines.

14.1 6,200,806

This is not a lottery number. It is United States Patent No. 6,200,806 — granted March 13 — a number that will be on the plate of many a legal professional in the months and likely years to come.

The United States patent 6,200,806 has claims to the human embryonic stem cell. The patent, held by the Wisconsin Alumni Research Foundation (WARF), is apparently the only one of its kind in the world.[22] The patent, which covers both the method of isolating the cells and the cells themselves, gives WARF control over who may work with stem cells, and for what purpose. WARF has granted exclusive rights to the Menlo Park-based Geron Corporation to develop stem cells into six cell lines that are of great medical importance — liver, muscle, nerve, pancreas, blood and bone. Geron has vowed to be very aggressive in protecting its intellectual property position.

This patent, issued in the United States, is valid only in the United States. WARF has applied for patents in Europe. At least two foreign biotechnology companies have claimed they have cell lines that fall outside the Wisconsin's patent claim and have applied for patents of their own in the United States and elsewhere.

14.2 6,200,806 Reasons to Worry About

President Bush's decision, in principle, mandates free access for scientists to the 30 self-sustaining cell lines that currently exist in laboratories around the world. But

---

[21] *The Lifeline Report: A chronicle of life sciences trends & events*, Robert W. Baird & Co. Inc, November 2001, pp. 2.
[22] Sheryl Gay Stolberg, "Patent laws may determine shape of stem cell research", *New York Times*, August 17, 2001.

the fact is that the government does not have control over whether these cell lines get to researchers.

We argue this will only serve to impede progress of stem cell research because of lawsuits, which are expected to be both time-consuming and prohibitively expensive. The eventual winners will be no one in the stem cell research profession, nor dying patients, but those in the legal profession.

The interesting thing is that currently Geron is itself in a legal battle with Wisconsin Alumni Research Foundation, the very Foundation that granted Geron the right to develop cell lines in 1998.[23]

There are other 6,200,806 reasons to worry about. The patent, which is only valid in the United States, gives foreign countries a double advantage. They have fewer restrictions, and they may still derive and patent stem cell lines without fear of any patent infringement.

Bush's decision to limit research to only existing cell lines only serves to strengthen holders of those cell lines, particularly, WARF. By refusing funds for creation of new cell lines in the United States, we find the decision very myopic because it reduces the chances of scientists deriving and patenting new cells that might challenge Wisconsin's dominance in the field. The decision also reduces the chances of developing substitute technologies for stem cell derivations.

A silver lining in all these confusions is that WARF is eager to get its embryonic cell lines into researchers' hands with minimal restrictions. All distribution of the cells to academic researchers is handled by WARF, not by the Geron Corporation. WARF will make the cells available to academic biologists for a subsidized fee of $5,000. Any researcher who gets WARF cells and makes a patentable discovery will own that patent. But, as is routine practice, if researchers wish to commercialize any discovery they make on the basis of WARF's patent, they must negotiate a license with WARF, or with Geron for anything that falls in the scope of Geron's license.[24]

Tommy G. Thompson, Secretary of Health and Human Services announced September 5, 2001 that a registry of all 64 cell lines — only less than 30 of which are fully developed — listing their properties and biological characteristics, would be posted on the National Institutes of Health's Web site within a couple weeks. He also claimed a new intellectual property agreement had been reached between the government and WiCell Research Institute, a nonprofit subsidiary of the Wisconsin Alumni Research Foundation, or WARF.

The accord allows scientists who work at the National Institutes of Health broad access to the five cell lines created by Dr. James A. Thomson. It applies only to

---

[23] Tom Abate, "Institute sues biotech firm over stem cells: new limitations make contract hot property", *San Francisco Chronicle*, August 14, 2001, pp. A1.
[24] Nicholas Wade, "Bush's stem cell policy may streamline research", *New York Times*, August 18, 2001.

government-employed scientists and covers only basic research. If scientists want to use the cells as therapies, they will have to renegotiate.

The agreement is significant for several reasons:

❑ WiCell will make the same terms available to academic scientists in university laboratories around the country.

❑ WiCell agrees not to use its patent to block federally financed scientists from studying stem cell lines developed in other laboratories — so long as the other cell line owners do not receive more favorable treatment from the government than WiCell does.

❑ It eliminates a so-called reach-through provision that WiCell imposed in the past. Under the provision, any scientist who made a discovery using the Wisconsin cells was required to offer WiCell the first chance at commercializing it.

So what is in the crystal ball of stem cell research in the United States? Either the best of the best leave the United States for countries where there are fewer restrictions, or alternatively, we propose morphologizing a stem cell into a cell line of legal neurons and implanting these neurons into stem cell entrepreneurs. This way, they will end up making more money from a second job instigating lawsuits, while patients shall serve as their jurors.

Senator Edward M. Kennedy, Democrat of Massachusetts, the chairman of the Committee on Health, Education, Labor and Pensions, made a good point, "Millions of patients and their families expect that stem cell research will move forward as rapidly as possible. It would be unacceptable to offer these patients and families the promise of effective stem cell research but deny them the reality of it."

15 Hot Off The Press

The human cloning debate has raged for years. The U.S. House of Representatives voted to ban human cloning in the summer of 2001, but the Senate has not followed with any action, and thus no bill has been passed.

Using stem cell cultures, Chinese scientist Professor Xu Rongxiang, who had earlier made a name for himself in skin regeneration,[25] has regenerated and duplicated gastrointestinal organs. By growing mouse gastric intestinal cells in a culture fluid combined with Gastro-intestinal Capsules (GICs) — a stem cell growth factor — gastrointestinal organ clones were grown. The results prove that GICs can actively improve the regeneration process and repair gastric-intestine mucosa.[26]

A few months after the House of Representatives had voted to ban human cloning, in late November of 2001, Advanced Cell Technology, Inc. (ACT)

---

[25] "China succeeds in skin regeneration, duplication", *People's Daily*, August 9, 2000.
[26] "China succeeds in duplicating organ from stem cell", *Xinhua News Agency*, November 19, 2001.

announced they have succeeded in cloning the first human embryo.[27]   The report was featured in the online journal e-biomed, *The Journal of Regenerative Medicine.* The company cloned a six-cell human embryo by removing the nucleus of a donor egg cell and replacing it with a cumulus cell, which is responsible for nurturing egg cells while they develop.

The goal of the company is to employ the technology to create replacement cells for therapeutic purposes, and in the future, to harvest stem cells from embryos. Needless to say, the announcement generated a backlash of harsh criticism from lawmakers and critics.

John Gearhart, who led one of the two teams which in 1998 announced that human embryonic stem cells had been isolated and cultured in the laboratory for the first time, is also an editorial advisor to the online journal that published details of the "world's first human embryo clones" on November 25, 2001.   He claimed important data were missing from the online paper and that the experiment was in his judgment a failure and should not have been published.[28]

The company behind the clone research, Advanced Cell Technology, defended its work, and even released new information on work it had done on monkey embryo.

Subsequent to ACT's announcement, Clonaid, which had moved its facilities outside the U.S. after undergoing investigation from the FDA, also claimed to have cloned embryos.

Whether an embryo has been cloned, it is still debatable.

---

[27] *The Lifeline Report — A chronicle of life science trends and events,* (Robert W. Baird & Co., Inc.), December 2001, pp. 2.
[28] "Embryo clone leads to fall-out", *BBC News,* December 3, 2001.

Chapter 8

GENOMIC AND PROTEOMIC TECHNOLOGIES AND BEYOND

"If every time there's a technological change you just incorporate it into your existing process, you're a loser. Winners rebuild the process on the basis of that change. You've got to retool the entire process to take advantage of new efficiencies."
Joshua Boger, CEO, Vertex Pharmaceuticals,
on the impact of genomic technologies on drug discovery.

1 Biochip

The human genome project has been the catalyst for the development of several high throughput technologies that have made it possible to map and sequence complex genomes.[1] As of this writing, several bacterial genomes as well as the genomes of *Saccharomyces cerevisiae* (yeast), *C. elegans* (round worm), and most recently, *Drosophila melanogaster* (fruit fly)[2] have been sequenced. On June 24, 2000, the entire human genomic sequence was officially announced by the U.S. President Bill Clinton to have been completely drafted. This marked the end of a chapter in the human genome project and the beginning of many others. It is clear, however, that the identification of every gene within the genomes of the human and model organisms is only the initial step in our quest to understand what these genes do and how their expressions impact our health. Understanding the functions of the 30,000–40,000 genes comprising mammalian genomes, their roles in normal development and disease states, and their implications in variations within the population will represent a more daunting task than the mapping and sequencing efforts.

Thousands of genes and their products, i.e., proteins, in a given organism function in very complicated but orchestrated way to create the mystery of life.[3] However, traditional methods in molecular biology normally work on a "one gene in one experiment" basis, leading to a very limited throughput, and a global picture of the gene function is hard to obtain. In the past several years, a new technology, the bio-chip technology has created a lot of interest and it promises to monitor the entire genome on a single chip so that researchers can have a better picture of the interactions among the thousands of genes in the genome simultaneously.

---

[1] Hwa A. Lim, James W. Fickett, Charles R. Cantor, Robert J. Robbins, *Bioinformatics, Supercomputing and Complex Genome Analysis*, (World Scientific Publishing Co., New Jersey, 1993).
[2] "Fruit fly's genes may unlock human disorders, unraveling cryptic code called phenomenal step", *San Francisco Chronicle*, Friday, March 24, 2000, pp. A1.
[3] James D. Watson, Nancy H. Hopkins, Jeffrey W. Roberts, Joan A. Steitz, Alan M. Weiner, *Molecular Biology of the Gene*, (The Benjamin/Cummings Publishing Company, Inc., Menlo Park, California, 1987).

The term "biochip" has taken on a variety of meanings. In the most generic sense, any device or component incorporating biological (or organic) materials — either extracted from biological species or synthesized in a laboratory — on a substrate can be regarded as a biochip.[4] In practical terms, biochips often involve both miniaturization and the potential for low-cost mass production. Examples meeting these criteria include electronic nose or artificial nose chip, electronic tongue, and polymerase chain reaction chip. However, the most intensive investigations into biochip in the past few years have been in DNA microarray chip (gene chip), protein chip and lab-on-a-chip. Following commercial conventions, these biochips are categorized as follows:

❑   DNA chips — small flat surfaces on which strands of DNA probes or samples are placed.

❑   Lab chips — integrated systems which allow researchers to perform a sequence of experiments on the chips.

❑   Protein chips — similar to DNA chips, except the probes or samples are proteins.

More recently, genome chip,[5] to indicate the fact that the technology may be able to monitor the whole genome on a single chip, has been suggested. Most of the current market share and activities are in DNA and protein chips. Thus we will devote most of our discussion to DNA chip, though we will have cursory discussions of the others as well.

2   The DNA Chip Technology

DNA sequences have two very useful properties which make DNA chip a very viable technology.

❑   Complementarity — The root of DNA chip technology goes back to the discovery of base-pairing rule in the 1950s by James Watson and Francis Crick. Watson and Crick determined that the DNA molecules found in living organisms are composed of a structure of twisted double helix latticed together with pairs of nitrogenous bases: adenine (A), cytosine (C), guanine (G), and thymine (T). They also discovered that these bases always recur in the same two complementary pairs: A-T, and G-C. Thus if one knows the molecular structure of a specific genetic segment, one can deduce the other from the simple complementarity rule. These uniquely complementary strands of DNA can be sought out by using one of the strands to test for its biochemical mate. The process of one strand of DNA matching up with its counterpart strand is called hybridization. Base-pairing or hybridization is

---

[4] Arthur Chiou, "Biochips combine a triad of micro-electromechanical, biochemical, and photonic technologies", *OE Reports*, March 2000, No. 195.
[5] www.gene-chip.com

the underlining principle of the DNA chip technology: A-T, G-C are complements in DNA, A-U and G-C are complements in RNA. Thus DNA chips provide a means for matching known and unknown DNA samples based on base-pairing rules.

Figure 1. Complementarity of DNA: A-T and G-C always occur in the same pairs in a double-stranded DNA. (Figure: Adapted from Cliff Henke, DNA-chip technologies).

❑ Gene Property — Nature offers another property to make the DNA chip technology a viable tool. On average, a gene is of the order of 10,000 base pairs (bp) long. But within the gene is usually a short sequence, no more than 25 bp and is unique to that gene. By coding this short, unique chain on a chip a researcher can represent the whole gene. This genetic shorthand makes the entire process manageable.

2.1 The DNA Chip Array

The hybridization can be performed either in solution or on a solid surface. The experiment can make use of common assay systems such as glass slides or standard blotting membranes, and the spotting can be created manually. However, the spotting process is an ideal process for automation and parallelism using robotics to spot on surfaces. When this is done, the samples or probes are spotted in an array. An array is an orderly arrangement of samples or probes. It provides an automated process for matching known and unknown samples based on complementarity rules.

Figure 2. A very naïve cartoon to show hybridization process in a DNA chip. For hybridization purposes, a double stranded DNA is first denatured into single stranded DNA (A→B). One of the single strands (in the figure, the right strand) is used as a probe. Among the targets, the probe hybridizes only a complementary strand (C→D). For detection purpose, either the probe or the target (in the figure, the target) is labeled with a phosphorescent substance. (Figure: Adapted from Cliff Henke, DNA-chip technologies).

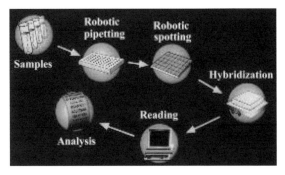

Figure 3. From sample preparation to analysis in a typical DNA microarray experiment. To rid of partial hybridizations, there is also a washing process between Hybridization and Reading. During washing, partially hybridized samples are washed away and are thus not detected.

The array can be a microarray or a macroarray depending on the size of sample spots. Macroarrays contain sample spot size of about 300 microns or larger (>300μm) and can be easily imaged by existing gel and blot scanners. Microarrays usually have thousands of sample spots with size typically less than 200 microns (<200μm) in diameter. Microarrays require specialized robotics and imaging equipment and the system is not yet generally commercially available as an integrated fully automated system.

2.2   Formats

This immediately brings us to the definition of probes and targets. In order to avoid confusion or ambiguity, in standard nomenclature,[6] a probe is the DNA with known sequence. A target is the DNA sample whose identity or abundance is being studied. Either the probe or the target can be spotted (or immobilized) on a surface, yielding different formats.

❑   Format 1 — One method that is being developed for high-throughput measurement of expression patterns of thousands of genes is the gridded cDNA microarray technology first developed by Pat Brown and colleagues at Stanford University.[7] This method utilizes a high speed, high precision robot to spot thousands of DNA samples onto glass slides. The slides are then simultaneously probed with florescent-labeled cDNAs which are generated from mRNA isolated from cells or tissues in the two different states that one wishes to compare. A different fluorescent dye is used to make the cDNAs for each physiological state to allow direct comparisons on a single chip. Current robotics technology has the precision to spot at least 100,000 PCR products onto a single glass slide.

---

[6] B. Phimister, *Nature Genetics* 21, 1999, pp. 1–60.
[7] R. Ekins and F.W. Chu, *Trends in Biotechnology*, 17, 1999, pp. 217–218.

Figure 4. In this particular experiment, DNA samples (targets) are spotted on a glass slide to make a DNA microarray. Each spot represents a different DNA molecule. mRNA from a tissue in two different states (A and B) are used to make cDNA. After cloning, 10,000 clones are obtained, each representing a specific gene expressed either in only state A or in only state B or in both. State A/B clones are tagged with red/green fluorescent molecules. The two samples (probes) are then mixed and the mixture is hybridized to the DNA microarray. If a gene is expressed in both A and B, then the spot on the microarray corresponding to that gene will bind both red and green probes and will appear yellow. If a gene is expressed in only A/B, then the spot on the microarray corresponding to that gene will only bind to red/green probe and will appear red/green. (Photo: Adapted from an experiment in Biochemistry Course 462a, University of Arizona).

❑ Format 2 — The second method uses an array of oligonucleotide (20~25-mer oligos) probes which is synthesized either *in situ* (on-chip) or by conventional synthesis followed by on-chip immobilization. The array is exposed to labeled sample DNA, hybridized, and the identity/abundance of complementary sequences are determined. This method, traditionally called genechip arrays or DNA chips, was first developed at Affymetrix Inc. It is designed to have high accuracy and low cross hybridization.

Thus in Format 2, probes are fixed and labeled targets are not. There are other combinations thereof.

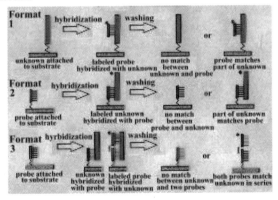

Figure 5. Either the probe or unknown (target) can be immobilized to a substrate, leading to different DNA microarray formats. Conventionally, if the unknown is immobilized, it is Format 1. If the probe is immobilized, it is Format 2. There are combinations thereof. Note that the immobilization is a salient feature for detection. Because the probe/unknown is immobilized, after washing, undesirable unknowns/probes are discarded, and the calorimetric or fluorimetric detector only detects labels associated with immobilized hybridizations.

2.3   Applications of DNA Microarray Technology

DNA microarray technology has a wide applications. It can be used, but not limited to the following areas:

2.3.1   Functional Genomics Research

Functions of genes can be deduced from their expression patterns in different tissues. This focus uses the chip to discover the expression pattern of unknown genes.

Gene expression can be used to study, for example, what the body is doing to fight against a bacterial infection, or what the bacteria are doing to survive in the host. A microarray can give data of genetic patterns in the normal and infected states 100 to 1,000 times faster than conventional methods. It can tell which of the genes are in play during the infection process and also the genes one should attack in the bacteria to inhibit the bacterial growth. It also shows the strongest response exhibited by the body of the host. In this way, the microarray will help in studying the mechanism of actions of both the host and the pathogen during the course of an infection.

2.3.2   Disease Diagnosis

This is not the traditional chip for gene expression analysis. It is usually designed to study unique disease genes. We shall describe a few popular studies.

2.3.2.1   Breast cancer

Breast cancer affect as many as two million women in North America in the 1990s. More than 500,000 of the women affected eventually succumb.

When breast cancer runs in a family, as many as 80% of the women affected show mutations in either or both BRCA1 (5,000bp) or BRCA2 (10,000bp). The mutations make a woman's chance of developing the disease as high as 85%.

BRCA1 and BRCA2 are believed to cause as many as 60% of all cases of hereditary breast and ovarian cancers. In BRCA1 alone, over 500 different mutations have already been discovered, and more mutations are being found. Using conventional methods, one tends to study only locations where mutations are expected. Using microarrays, one not only can get the entire data report of a DNA sample and easily determine the locations of variation, but also the factors that influence the variation.

2.3.2.2   Breast cancer gene region in chimps, gorillas and orangutans

Scientists at the National Human Genome Research Institute (NHGRI) report that they have used DNA chip technology to analyze the structure of significant regions in several species of animals, including three of humanity's closest relatives: the chimpanzee, the gorilla, and the orangutan.[8] The chip in the study compared the structure of 3.4 kb strand of BRCA1, a human gene known that causes hereditary breast and ovarian cancer, with corresponding regions from other animals, mostly nonhuman primates. The BRCA1 chip worked best when researchers looked at close relative of humans like chimpanzee, gorilla and orangutan, but was less accurate when researchers analyzed the DNA from other primates, including rhesus monkeys, galago, red howler monkey.

To determine whether an individual possesses a mutation for BRCA1 or BRCA2, a researcher first obtains a sample of DNA from her blood as well as a sample that does not contain a mutation in either gene. After denaturing and cutting into smaller more manageable fragments, the researcher labels the fragments with fluorescent dyes. The individual's DNA is labeled green, normal DNA is labeled red, for example. Both sets of labeled DNA is inserted into the chip and allowed to hybridize to the synthetic BRCA1 or BRCA2 DNA on the chip. If the individual does not have a mutation for the gene, both the red and green samples will hybridize with the probe. If the individual possesses a mutation, the red (normal) DNA will still hybridize perfectly, but the green (individual's) DNA will not hybridize properly in region where the mutation is located. The researcher can then examine the area more closely to confirm that a mutation is present.

2.3.3 Sequencing and Gene Discovery

Extending from the above, one can think of the chimp as a collection of sequence changes relative to the human. If a microarray works for finding disease-causing mutations, it ought to work pretty well at finding sequence differences between humans and other primates, as long as there are not too many of them. Since the international human genome project has made huge investment into getting the human sequence, one can then use that investment to get the sequences of human's closest relatives at a relatively cheap price. The chimpanzee DNA sequence differs from the human DNA sequence in about 1.5% of their sequences. Thus, one can think of a relatively cheap Chimpanzee Genome Project by using microarrays to locate the differences to sequence.

There is yet another property in the nature's tapestry. In DNA regions that are strongly conserved, which is true of the coding regions of many genes, the extension mentioned above may be used to look further away to other mammals. Sequencing projects may be extended beyond the Chimpanzee Genome Project to other more distant relatives of the human.

The same argument should apply to other organisms that have been fully sequenced. One can envision a cheap and easy way to sequence the genomes of their close relatives, leading to alternative genome projects or comparative sequencing, rather than brute force beginning-to-end approach.

---

[8] National Institutes of Health, Press Release, Monday, January 26, 1998.

2.3.4   Drug Discovery

Drug discovery has transitioned from a technology-driven process to an indication-driven process. Gene-based diagnostics are helpful for both treatment targeting and patient surveillance. High volume gene expression assays can optimize pharmaceutical therapies by targeting genome-based treatments to specific patient populations and providing time-efficient methods to study genes involved with, for example, cancer growth patterns and tumor suppression.[9]

Microarrays can be used to fine-tune drug design as well as for medical diagnosis. Drug companies test a huge number of potential drugs against a spectrum of biochemicals to determine their interactions. They also try thousands of variations of promising compounds to see which ones maximize benefits and minimize side effects.

Also every major manufacturer of microarrays has a strategic relationships with a major pharmaceutical company, taking advantage of the eagerness of pharmaceutical manufacturers to find ways to screen potential drugs faster and more accurately. This usually consists of "early access partnerships" in which partners have access to a company's technology for specific applications only, and before other companies do.

2.3.5   Toxicogenomics and Pharmacogenomics

By definition a drug interferes with body processes and therefore has toxic properties. The more potent the drug the more toxic the drug tends to be, producing a range of adverse side effects. If an illness is caused this way, the illness is known as iatrogenesis, or doctor-induced diseases.

The goal of toxicogenomics, a hybridization of functional genomics and molecular toxicology, is to find the correlation between toxic responses and the gene expression profile.[10]

Expression patterns can also be used to distinguish between people with apparently identical diseases. By studying the expression patterns of tens of thousands of genes from subgroups of patients, illnesses can be distinguished. Each subgroup normally has a different pattern of gene expression. Tiny differences can be a matter of life and death for each subgroup may respond differently to the same treatment regime. By understanding the genetic behavior of a specific form of a disease and its reaction to different treatments, therapies can be more carefully designed for each patient. This is where pharmacogenomics comes in. Pharmacogenomics or personalized medicine is the hybridization of functional genomics and molecular pharmacology. Microarrays can be used to study the

---

[9] Trina Slabiak, "Hybridization array technologies", *BioOnline*, 1999.
[10] E.F. Nuwaysir, M. Bittner, J. Trent, J.C. Barrett, and C.A. Afshari, "Microarray and toxicology: the advent of toxicogenomics", *Molecular Carconogenesis*, 24, 1999, pp. 153–159.

correlation between therapeutic responses to drugs and the genetic profiles of patients.

2.3.6 Environmental Studies

Applied to water quality controls, microarrays will screen for the presence of microorganisms in the water by matching genetic fingerprints. For example, Lyonnaise des Eaux (Paris), the world leader in water management and bioMerieux, a world leader in *in vitro* diagnostics (IVD) had formed a joint venture to use GeneChip™ of Affymetrix to do water quality control.[11] This will be a 8.5 million euros 5-year R&D partnership. The approach has three major advantages for water quality control: higher performance, decrease processing time, and significant cost savings. Affymetrix's GeneChip™ array will permit the detection of lower concentrations of microorganisms in the water and the accurate identification of many types of water contaminants. The chip is expected to reduce the current test time of 48 hours to just four hours, and the cost is expected to be about ten times cheaper.

2.3.7 Agricultural Science

Botanists and agronomists can use microarrays to understand how plants cope with stress caused by heat, drought, excess rain or the lack of certain nutrients. By charting genetic responses to specific stresses, botanists and agronomists may be able to engineer more versatile plants that thrive in more varied environments. Researchers can also read the genetic signatures of the changes in plants bred for specific advantages such as drought tolerance or cold hardiness, and then engineer the same changes in other strains.

DNA chips may even help turning plants into chemical factories. There has been some work on producing biodegradable plastics from plants. Ultimately, DNA chips will give botanists a much better understanding of the processes going on inside plant cells. That would help make better choices in the modifications to be made to the plants, while developing hardier and more productive strains.

3   Biochip Market

Enormous forces are impelling the development of DNA chips.[12] The human genome project has contributed in two ways:
- ❑ It has created a genomics market to characterize genes.
- ❑ It has helped researchers develop more and better tools for their gene-hunting work.

---

[11] "GeneChip arrays for water quality control", *Microtechnology News*, March 1999.
[12] Cliff Henke, DNA-chip technologies, *IVD Technology Magazine*, September, 1998.

3.1   Market Segments

New York-based investment bank Cowen and Co. estimates that gene chip markets will grow from US$600M in 2000 to $2 billion by 2004. BioInsight, a market research firm in Redwood City,[13] takes a more conservative view, putting today's worldwide market at about $226M, reaching almost US$1 billion by 2005. BioInsight splits the market into three categories:

❑   DNA chips — small flat surfaces on which strands of probes or targets are placed.

❑   Lab chips — an integrated system to perform a sequence of experiments on a chip.

❑   Protein chips — similar to DNA chips, except they sample proteins. The market for this segment is less clear for we are far from identifying all the 300,000–1,000,000 proteins.

FrontLine Strategic Management Consulting Inc. of Foster City estimated that the DNA chip marketplace grew at a compound rate of 77% between 1998 and 2001. Almost all of these chips were DNA microarrays but by 2001, 10% of the market are in protein chips and by 2005, 18% is expected to be lab chips, and by 2010, lab chip will be the dominant share.

In a more recent report,[14] Frontline estimates the DNA microarray market will grow to $3.6 billion over the next five years as academic institutions and pharmaceutical companies move to more aggressively adopt the instruments and as the prices of mircoarray start coming down. For example, growth in the integrated systems segment will occur at an average annual rate of 22 percent. The scanner and arrayer markets would experience the greatest growth, with sales of these systems increasing at a compound annual growth rate of 44 percent, reaching $1.64 billion and $887 million, respectively, by 2006.

On the other hand, Strategic Directions International predicts that by 2003, the microarray and associated equipment market will be $777 million. In segments, the DNA chip market will be ~$420 million, the array market will be ~$64 million, the fluid station market will be ~$96 million, the scanner market will be ~$190 million.

Whether viewed as segmented biochip markets, or as segmented microarray and associated technology markets, the combined market of biochip technology is predicted to be huge.

Affymetrix, Apogent, and GeneMachines currently control 73 percent of the arrayer market.

---

[13] Tam Harbert, "A chip off the old block?: *Electronic Business*, April 2000.
[14] *DNA Microarray: A strategic market analysis*, Frontline Strategic Consulting, November 19, 2001, Foster City, California, USA.

Table 1. Organic growth of worldwide market for biochips in US$ million, segmented into three categories.

Year	DNA Chip	Lab Chip	Protein Chip	Total
1999	158	14	4	176
2000	197	23	6	228
2001	249	35	8	292
2002	310	62	10	382
2003	395	85	17	497
2004	506	115	35	656
2005	725	157	68	950

As the demand for DNA chips and the computers that read the results grow, the cost for both should plummet. Chips now can cost as much as $2000 each to make and the equipment needed to make and read them is about ten times more expensive. In the current market, the average price is $400 for each slide with 4,000–7,000 genes. So far only a few dominant companies have marketable DNA-chip products. The barriers to market entry remain relatively high, though there are many me-to companies.

Table 2. Companies offering DNA chips and their flagship products.

Company	Product	Chip Type
Aclara BioSciences Inc.	LabCard	Lab
Affymetrix Inc.	GeneChip	DNA
Agilent Technologies Inc.	2100 Bioanalyzer	Lab
Caliper Technologies Corp.	LabChip	Lab
Clinical Micro Sensors Inc.	Not named	DNA
Hyseq Inc.	HyChip	DNA
Incyte Genomics	GEM	DNA
Orchid Biocomputer Inc.	SNPstream	Lab
Motorola BioChip Systems	Not named	DNA & Lab
Nanogen Inc.	NanoChip	DNA

In the economic downturn of 2000 and 2001, some 30% of the original players have deicided to leave the field. The most notable is Corning. Just over a year after formally announcing intentions to develop microarray applications, Corning, a fiber optic giant, closed their microarray initiative, Corning Microarray Technologies, as part of a cost-containing restructuring effort.[15]

One possible factor leading to this outcome is Corning's late entry into the market. With companies such as Affymetrix holding hundreds of patents related to microarray technologies, the barriers to entry, even for commercial giants such as Corning, may be too great. But Corning's retreat from the microarray sector appears

---

[15] "Corning axes microarray initiative as part of overall restructuring", *GenomeWeb*, October 18, 2001.

to be mainly a casualty of the drastic downturn in telecommunications business. The company's net income in the third quarter of 2001 was $85 million, down from $317 million in the third quarter of 2000.

Corning's retreat from microarrays and closure of other manufacturing facilities is expected to save $400 million during the next year. Competitors such as Motorola, Affymetrix, and Agilent only stand to benefit, while content providers such as Incyte was negatively affected by Corning's decision.

On October 30, 2001, Incyte Genomics announced closing its microarray facilities,[16] laying off about 450 employees and became the latest gene-delivery company to join the pharmaceutical business.

4   Microchip But Macro Impacts

Using biochips, researchers can in one afternoon confirm work that used to take several years using conventional gene-sequencing processes.

Table 3. With improved technologies, sample size decreases while the number of analyses increases.

Format	Number of Analyses/runs	Reaction Volume (μl)	Year
Microamplification	1	50	1995
Microplate	96	25	1996
Microcard	96	1	1998
Chip	1000	0.5	TBA

The chips are not just about speed. They also allow researchers to do things that were previously impossible, such as uncovering the genetic mechanism behind the complex chemistry of organisms.[17] For example, all the 6,200 genes of the yeast cell can be represented on just four chips. It is then possible to take "snapshots" that reveal which genes are active, which are dormant, and how these patterns change during the organism's life cycle.

Systems biology will be the challenge of the 21st century.

4.1   A Chip off the Old Block

The potential for gene chip technology to revolutionize medical diagnosis and treatment, and drug development in the next century has been likened to the micro-processor revolution of the early 1980s.[18] The similarities between silicon chips and

---

[16] Paul Elias, "Incyte Genomics getting out of crowded genomics tools business", *Washington Post*, November 1, 2001.

[17] Shankara Narayanan, "Magic chip", www.cheresources.com/dnachip.shtml

[18] Optics in gene chip technology, *enLIGHTen*, 3(5), April 1998.

their gene-oriented counterparts begin with the way they are manufactured. Like computer chips, most DNA chips are manufactured by computer-controlled micro printing. In computer chips, the process lays down microscopic circuits and switches. In DNA chips, the process lays down DNA sequences.

This analogy is even more apparent if we see how Affymetrix uses photolithography to produce DNA chips.

In 1971, Intel perfected a way to shrink 2,300 transistors into a single circuit. Through mass production, Intel made microprocessors affordable, launching the personal computer industry years later. A quarter of a century later, Affymetrix adapted the same production technique to fabricate microchips that process DNA rather than electrons. Affymetrix claims its GeneChip systems boost the field of genomics medicine the same way desktop computers helped business by gathering information much more quickly and cheaply than previously possible. Such economies of scales are possible because of Affymetrix's clever adaptation of photolithography, the technique routinely used to make semiconductors. Instead of projecting ultraviolet light through a series of masks to etch multilayered circuits into silicon, Affymetrix's machines uses the masks to build chainlike DNA sequences that rise from a glass wafer. Each mask limits where new links are attached, so adjacent chains can contain completely different combinations of the four DNA building blocks — A, C, G and T. In 32 steps, the automated process can create on a single chip, for example, up to 65,536 unique probes, each eight bases long.

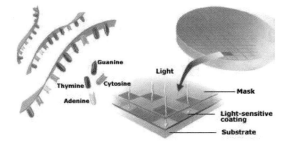

Figure 6. Companies use different micro- and nano-manufacturing techniques to turn glass and plastic wafers into probes. Shown is lithography technique routinely used in manufacturing semiconductors. Instead of projecting ultraviolet light through a series of masks to etch multilayered circuits into silicon, in DNA probe manufacture, machines use the masks to build chainlike DNA sequences that rise from a glass wafer. (Figure adapted from Cliff Henke, DNA-chip technologies).

5   Future Outlook

As DNA chip companies prepare to bring their products to market, major technological and regulatory challenges lie ahead.

5.1   Challenges

Technologically, the trade-off involves finding ways to increase the number of arrays on a single chip, as well as increasing the rate of production to meet expected demand. Currently, few manufacturers are producing chips beyond a pilot-scale production rate.

A second technological challenge involves achieving all these market-oriented parameters at a cost that supports a commercially acceptable price. Currently, DNA chips cost between $500 and $4,000 each. More specifically, a single-use high density DNA GeneChip of Affymetrix costs ~$500–$1,000; a GEM microarray of Incyte Genomics costs ~$2,000–$4,000.* A GenePix4000 scanner of Axon Instruments costs ~$55,000. The latter further adds to the cost of disease diagnosis. In a tight managed-care marketplace, that places a premium on technologies that can either show quick savings or more-efficient results.

As products are readied for the marketplace, they will also encounter regulatory challenges, such as ensuring that manufacturing process meets current quality systems and CLIA (Clinical Laboratory Improvement Act) standards. As companies look to market products outside the research-laboratory and drug development environments, they will face enormous regulatory and market challenges. Chief among the regulatory challenges is developing a set of industry standards for quality assurance. Today, each company has its own standard.

5.2   Future Development

Development is proceeding in pursuit of two basic goals: to increase sensitivity and reliability, and towards systems integration. DNA chip technologies are progressing rapidly in five fronts.

❑   Cost — The bottleneck for DNA (and protein) microarray technology is the cost of making slides that contain thousands of genes or proteins. For instance, to make 34,000 oligonucleotides (a whole genome) using current technology, it would cost about $1 million ($30/oligo). Usually, the company synthesizing the oligo provides them in concentrations of 50 nanomoles per sample. With this amount of sample, only 10,000 slides can be printed using the current deposition technology. Optical Fiber Arrayer technology (OFA) may be able to overcome this limitation. OFA technology

---

* As of November 1, 2001, Incyte Genomics had decided to shut down its microarray facilities in St Louis, Fremont, and Berkeley.

combines optical fibers, microfluidics, microelectronic, deposition technology and stamp technology. Optical fibers are used to deposit samples on to a solid phase substrate that carries positive charges while optical fibers carry negative charges. This technology needs only 30 pL sample for one spot. It overcomes several common problems related to the current arrayer methods such as the need for large samples (about 100 pL per spot), and multiple procedures during the printing. The current technology wastes much of the samples during the washing steps. OFA technology can save a tremendous amount of samples by decreasing the volume of samples and eliminating the need for washing. With this technology, the same sample concentration, 50 nanomoles, can be used to print 1 million slides. Thus, a 100-fold reduction in the cost of printing can be made.[19]

❑ Sensitivity — Target amplification strategies have problems because of the risk of contaminating the sample during amplification steps. There has been increasing movement in product development toward detection of a single molecule or cell in an unamplified sample. This requires increasing sensitivity, which ultimately means either bumping up the signal or getting the background noise down.

❑ Microfabrication — There is evidence that manufacturers will be able to achieve miniaturization on even smaller scales. Researchers are beginning to develop nano-scale devices, or a 1000-fold improvement over micro-scale.

❑ Density — Increasing the density of the arrays packed onto DNA chips is the third area in which manufacturers are making substantial progress. Current technologies enable manufacturers to fabricate single-chip arrays containing roughly 20,000 wells, with each holding about 100 nl. NIST is exploring the possibility of a chip made up of a single molecular layer of DNA bound to a thin film of gold. The surface tethered DNA then binds to a target strand and piezoelectric differences signal the level of hybridization. Problems with hybridization worsen at higher densities and there is a point where the optics of detection reaches its limit of resolution. The theoretical guess limit is 1 million elements on a 1-cm^2 chip.

❑ Integrated chips — The fourth area of development is integration of sample preparation and analysis, the lab-on-a-chip technology. In order for chips to reach its full potential, they should evolve to include sample preparation, analysis, and signal acquisition. Capillary electrophoresis and laser-induced fluorescence have been particularly popular for separation-detection

---

[19] Private communication with Dr. Jack Ye Zhai.

combination. But other techniques have also been explored, including on-chip chemi-luminescence and electrochemi-luminescence assays.[20]

However, not everyone agrees with the lab-on-a-chip approach. These people believe what is critical is that DNA chip technologies are compatible with existing equipment.

A start-up company, GenemiX, Inc.[α], based in Silicon Valley, has come up with novel ways of overcoming the above problems. By using a proprietary method employing optical fiber, and charges, GenemiX promises to cut array cost by about 90%, increase array production throughput by at least 10-fold, and the technology has a greater flexibility for both low and high production.

Table 4.   A comparison of various production technologies of biochip companies in terms of the challenges in the manufacturing process. Struck-out products are no longer in production.

Company/product	Chip Cost	Throughput	Density	Flexibility
Affymetrix, GeneChip	High	High	High	High
Agilent	High	High	High	High
Clontech	Medium	Low	Low	High
GenemiX, Inc. OFD	Low	High	High	High
Incyte Genomics, ~~GEM~~	High	High	High	High
PE Biosystems, NEN	Medium	Low	Low	High
Protogene, ~~Biochip~~	Medium	Low	Low	High

6   Crossing The Chasm From Genomics To Proteomics

Though DNA microarrays have provided useful information from gene expression analysis, mutation detection and disease identification in recent years, it is protein that is most often selected as the target of either a diagnostic product or a pharmaceutical drug. Many of the best-selling drugs either act by targeting proteins or are proteins themselves. It is also becoming increasingly clear that the behavior of proteins is difficult or impossible to predict from gene sequences because of further control by post-translation modification. Owing to the regulation at

---

[20] Guoxiong Xue, Yongbiao Xue, Zhihong Xu, Roger Holmes, Graeme Hammond, and Hwa A. Lim, *Gene Families: Studies of DNA, RNA, Proteins and Enzymes*, (World Scientific Publishing Co., New Jersey, 2001), 305 pages.

[α] President, Alan Lee, M.D.

translation level and varying half-lives of mRNAs and proteins, the mRNA level does not necessarily reflect protein expression amount.[21]

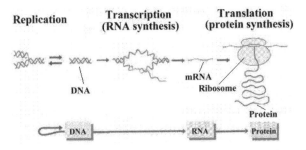

Figure 7. From replication of DNA to transcription (RNA) to translation (proteins). (Figure: Adapted from Burdett Buckeridge Young).[22]

In the June of 2000, the completion of the first draft of the human genome was announced, more than three years ahead of schedule. A natural next step is the Human Proteome Project (HPP). One of the key objectives of the Human Genome Project (HGP) is to provide the basis for understanding the role of genes in disease — leading to drug development — but this was only the first part of the revolution underway in drug discovery processes.

With the genome project came genomics. Analogous to genomics, proteomics is an emerging field conducting study on large set of proteins. Proteomics involves the separation, identification and characterization of proteins in order to understand their functions. The study is complementary to genomics: after the protein is instructed by the gene, proteins interact and undergo change before carrying out their biological function in the body. As most disease processes are manifested at the protein level, global efforts are now turning to proteomics to link genes and the proteins they instruct.

But proteomics is far more complicated than genomics. To fully appreciate the differences, let us take a cursory look at protein synthesis.[23]

6.1   The Cell, DNA and Protein Synthesis

The cell is the basic unit of any living organism. It is a small, watery (90% fluid), compartment filled with chemicals and a complete copy of the organism's genome. The cytoplasm is the gel-like substance in which all cellular components outside the nucleus are immersed. The nucleus is the central cell structure that houses the

---

[21] S. Fields, "Proteomics in genomeland", *Science* 291 (5507), 2001, pp. 1221–1224.
[22] Andrew Goodsall, *Now for the Human Proteome Project*, Burdett Buckeridge Young Limited, ACN 006 707 777, August 2000.
[23] Jack Ye Zhai, "Summary of the human proteome project: Proteome and automated drug discovery to harvest the genome breakthrough", *GenemiX White Paper*, March 13, 2001.

chromosomes which are a packet of compressed and entwined DNA. The organelles are structures found inside the cell cytoplasm which remain in the cell and assist the function of the cell. The lysosome is a spherical membrane-bound vesicle containing digestive enzymes that break down large molecules such as fat. The Golgi apparatus is an organelle in the cytoplasm that packages proteins and carbohydrates into vesicles for export from the cell.

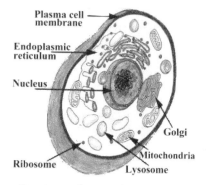

Figure 8. The cell and its constituents: cytoplasm, nucleus, organelles, lysosome and Golgi apparatus. (Figure: Adapted from Burdett Buckeridge Young).

Amino acids are the building blocks of proteins; each amino acid has an amino end, a carboxyl end and a side group. A peptide bond is a chemical bond linking two amino acids during protein synthesis, the peptide bond forms between two amino acids which are held side by side on the ribosome.

The protein is a complex molecule consisting of chains of amino acids joined according to instructions contain in the gene. Proteins are required for the structure, function and regulation of the body's cells, tissues and organs, and each protein has unique functions. Examples of proteins are hormones, enzymes, and antibodies. An enzyme is a protein that can catalyze (increase the rate of) a chemical reaction inside of the cell. Certain proteins cause diseases.

Lipids are hydrophobic molecules which in the case of fats and oils have as building blocks fatty acids and triglycerides. Hydrophobic compounds that do not dissolve in water, such as oil, and are considered "water-hating". Hydrophobic proteins are the proteins in the membrane of the cell, such as receptors and ion channels, which are the targets for drugs. Techniques are needed for identification of hydrophobic proteins.

Just as the genome is the complete gene complement of a cell (all the genes considered together), the proteome is the complete protein complement of a cell. An important distinction is that all cells will have the same genome, but each cell type will have a unique proteome.

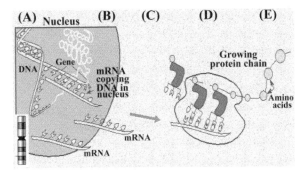

Figure 9. From DNA to disease: (A) Cell nucleus contains chromosomes. Chromosomes are made up of DNA. (B) DNA carries genetic instructions. (C) DNA passes the instructions via mRNA. (D) mRNA construct proteins. (E) Certain proteins cause diseases. (Figure: Adapted from Burdett Buckeridge Young).

Protein production begins in the nucleus at the DNA. A coded message for a protein leaves the nucleus in the form of RNA and may go to a ribosome where synthesis occurs. Proteins made in ribosomes bound to the endoplasmic reticulum migrate to the Golgi and may exit the cell or be incorporated into the cell's surface membrane.

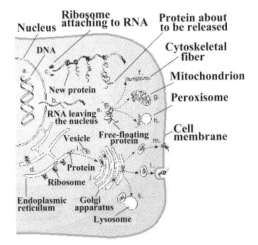

Figure 10. Production of protein. (Figure: Adapted from Burdett Buckeridge Young).

There are several additional layers of complexity in studying the proteome when compared to the genome:

❑ A genome is the complete set of genetic information. A proteome is the protein readout of a single cell, organ, tissue or organism. To date, no complete proteome has been described largely due to inadequacy of technology platform.

❏   DNA and RNA are informational rather than functional molecules. The workhorses of biology are proteins. The difficulty has been that whereas the entire information content (genome) has become available, proteins could only be studied one at a time. Purifying and characterizing a single protein can consume years of a skilled protein chemist's time.

❏   Compared with proteins, DNA molecules are simple. A DNA molecule is a long, spiraling ladder — the famous double helix — composed of just four basic constituents. Proteins are highly complicated structures that fold up into intricate and often unpredictable shapes. They are built from 20 building blocks, called amino acids, each of which has its own unique chemical properties making sequencing proteins a much harder task. We understand completely the 3D structure of DNA, whereas we are at the beginning of understanding all the 3D structures possible with proteins.

❏   In addition, the set of proteins a cell uses is constantly changing. Some proteins are broken down and their components recycled in minutes, while others can persist in the cell for hours or even days. Their surfaces are constantly modified, too — a sugar molecule might be added to one, a phosphate molecule tacked on to another. These additions can activate or deactivate proteins. Scientists estimate that the 30,000–40,000 genes in the human genome can generate perhaps up to a million different proteins.

❏   Though some diseases arise from a single gene defect, i.e., monogenic in nature, monogenic diseases are very rare. Most diseases are multigenic, complex and are results of interactions among many proteins.

Table 5. Human Genome Project versus Human Proteome Project.

Human Genome Project	Human Proteome Project
Lack of support in early days.	Has much greater support from scientific community.
Payoff to do diagnostics, finding susceptibilities in a individual's genome. Automating DNA analysis is relatively simple.	Payoff to do with treatment, finding therapeutic proteins, drugs and vaccines. Automating protein analysis is a much taller order.

Table 6. Key to understanding drug discovery and medical practice.

Understanding	Drug Discovery	Medical Practice
General cell biology	Target selection	Diagnostics
Gene function	Target validation	Markers of response
Genetic regulatory networks	Lead candidate selection	Markers of disease
Signaling pathways	Drug modes of action	Inclusion & exclusion criteria
Metabolic pathways	Toxicology	Post-launch differentiation of competitors

6.2 Proteomics — Direct Impact on Drug Discovery

Given the complexity of protein and the success of genomics, then why bother to attempt to cross the chasm from genomics to proteomics?

Even with the new genomics technologies, it takes 10 or 15 years to get a drug out of the lab, to evaluate it in animal and human trials and get it onto the market, at a cost of between US$300–$500M. Proteomics promises to shorten that by several years by helping researchers identify safer, more effective candidate medicines earlier in the drug-development process and by identifying the populations of proteins affected by the drugs. It is believed that up to 5000 protein targets for drugs exist, but only less than 500 have been identified to date. Proteomics has many significant applications.

6.2.1 Reducing Drug Toxicity

Proteomics will give medical researchers important information about the potential side effects of new and existing drugs, particularly where drugs fail at late stages of clinical development because of unforeseen toxicities. Proteomics can be used to eliminate potentially life-threatening compounds before pharmaceutical companies invest tens of millions of dollars trying to bring them to market. Scientists might, for instance, test how different experimental drugs affect the proteins of a given tissue, say, the liver or the kidney. The drugs that produce the fewest changes are likely to be safest.

6.2.2 Diagnosing Disease

The goal is to find key differences among protein samples and use that knowledge to chart the progression of the disease at the molecular (cellular) level. The U.S. National Cancer Institute has sponsored a program designed to identify protein markers that are specifically linked to early onset of ovarian, prostate, and other cancers. The hope is that these early markers will lead to better diagnostic tests that can identify lesions at the pre-malignant state, when they are just beginning to grow and spread throughout the body. So far, the initiative has uncovered nearly three dozen new protein markers that are consistently seen in pre-malignant cells.

6.2.3 Increasing Throughput of Drug Discovery

Despite the enormous potential of genomics and proteomics for elucidating the nature of human diseases, the mainstay of current and future drug discovery remains drug screening. With the advent of basic technologies, both at the compound synthesis level and at the screening level, the throughput of the screening process has increased by at least two to three orders of magnitude and new technology is expected to increase those multiples. Industry estimates of research expenditure by pharmaceutical companies during 1999 exceed $24 billion, or 20% of sales. This represented a 14.1% increase from 1998. Companies screen hundreds of thousands of potential compounds as part of their drug discovery program, with the expectation that less than 1 in 1,000 will become new drug candidates. Current methods of electrical screening may take a team of eight people several months to measure 50 potential drug compound interactions with targets. Less than one in 5,000 compounds can be expected to complete human clinical trials and receive regulatory approval. Only about 30% of drugs that are commercialized are expected to recover their development costs. Pharmaceutical and biotechnology companies have realized that they need to focus their efforts on improving the quality of drug discoveries in order to improve the success rate through the expensive clinical trials process. As government regulation continues to emphasize precision and understanding of drug compounds entering clinical trials, genomic and proteomic data are likely to become mandatory in the future. Given that 50% of the 20 biggest selling drugs in the world act on electrically active proteins, electrophysiological measures offer a key competitive advantage for drug discovery, yet no high throughput test currently exists. Rewards for the first to market will be significant.

6.3   Turning Discovery Into Cure

Typical high throughput screening methods rely on simple biochemical binding assays which, although rapid, do not generate a great deal of information about the disease process. Newer methods, such as automated cellular and subcellular imaging provide not only high throughput, but also a wealth of detailed information about the precise interaction between the test compounds and the cellular disease targets. This additional information can result in a drastic shortening of the drug discovery process.

❑   Electrophysiology High Throughput Screening (EHTS) — The concept allows simultaneous recording of electrical activity in multiple 'test tubes' to discover ion flows arising when a drug candidate interacts with biological cells. Given that 50% of the 20 biggest selling drugs in the world act on electrically active proteins, electrophysiological measures offer a key competitive advantage for drug discovery, yet no high throughput test currently exists. The manual alternate (the equivalent process can be achieved manually) is said to take up to eight people and a complicated and time consuming process costing around US$400,000 together with high ongoing labor costs.

❑   Image Analysis — Multi-well plates can be thought of as plates of 96 or more miniature test tubes which are used to test the reaction of biological materials when drug candidates are added to each tube. The process begins after the DNA chip scanner has helped identify a potential drug target. It remains the primary method of testing drug candidates on possible drug targets, with analysis required of the reaction and changes over time.

Existing image analysis products capture single exposures, one test tube at a time. New technology in this field combines high-grade images with software to automatically read and interpret the action of potential drugs introduced into multi-well plates containing target compounds.

7 Automation Of Proteomics

Proteomics has reached a critical crossroad — large scale projects are planned to begin proteome mapping, but automation remains underdeveloped. As little as ten years ago, the screening process was carried out manually at a very slow rate, with one scientist examining only a hundred drug candidates per year. Current methodology has accelerated the discovery process to 100 per week, however, equipment which extracts, separates, measures, characterizes and analyses proteins is needed urgently.

The major interests of proteomics are:

❑ Protein expression profiling — expression proteomics,

❑ Protein-protein interaction — cell map proteomics.

Expression proteomics is classic proteomics aiming to study cellular pathways and their perturbation by disease, drug action or other biological stimuli at the whole proteome level, thereby offering potentials to find disease markers and better diagnostic methods.

The technology inadequacy situation is changing rapidly as the approach to proteomics matures. Instead of seeking to purify proteins one at a time, a major advance in thinking has been to approach purification as a parallel process whereby a large number of proteins are purified simultaneously by arraying them using 2D gel technology. With advances in micro instrumentation based on detectors which measure mass (mass spectrometry), it is now possible to identify and characterize proteins that have been arrayed by preparative 2D gel technology, relatively quickly and efficiently.

It is important to understand however that the development of high throughput proteomics is very recent and indeed few laboratories in either industry or the public sector are equipped to operate on an industrial scale. This is due in no small part to the lack of appropriate proteomics instrumentation.

Table 7. Current participants in the proteomics sector. As of this writing, Myriad, Hitachi, and Oracle has formed a proteomic entity with $170 million investment, and Canon is also moving into the sector.

Company Name	Brief Description
AbMetrix (USA)	Has a proprietary digital antibody technology for diagnostics and drug discovery. High resolution and high throughput.
Alpha Gene (USA)	Has gene libraries containing large quantities of full-length cDNA, which include protein-coding sequences.
AxCell Biosciences (USA)	Has technology to map protein-protein interactions.

Table 7. (Continued)

Bioscience Proteomics (USA)	Develops Pro-GEx platform technology which includes high-throughput gel electrophoresis, robotics and computer-based image analysis.
Bio-Rad (U.K.)	Markets equipment and reagents for electrophoresis and analysis software.
Brax Group (USA)	Develops protein sequence tag technology — a high throughput system for protein analysis using small molecules that link to proteins.
Caliper (USA)	Designs and manufactures LabChips, lab-on-a-chip systems that incorporate microfluidics to create miniature devices capable of performing complete analysis on a microchip.
Ciphergen Biosystems (USA)	Designs and manufactures a protein chip that enables protein discovery and characterization.
Curagen (USA)	Provides an integrated and interactive suite of genomic technologies for drug discovery and development.
Genetics Institute (USA)	Develops DiscoverEase, a program that provides access to a library of genes and encoding secreted proteins and the associated expressed proteins.
Geneva Bioinformatics (Sweden)	Provides proprietary databases and software tools and services with emphasis on proteomics.
Genomic Solutions (USA)	Provides research tools for proteomics such as gel electrophoresis, pre-poured gels and an automated gel processor.
Incyte Pharmaceuticals (USA)	Offers database products and services with a focus on both genomics and proteomics.
Invitrogen (USA)	Markets molecular biology research kits and provides research services.
Life Technologies (USA)	Supplies research products, including Gateway Cloning Technology, which provides high-throughput protein expression and access to cDNA libraries.
Myriad Genetics (USA)	Establishes a program for identifying and analyzing protein-protein related interactions for all human proteins.
Oxford GlycoSciences (U.K.)	Has technologies including a protein screening system that uses high-throughput 2D gel systems and fluorescent tags.
Protana (Denmark)	Focuses on research projects applying affinity purification and advanced mass spectrometry for protein characterization.
Proteome Sciences (U.K.)	Focuses on applying proteomics methods to accelerate discovery and development of diagnostic tests and drugs.
Proteome Systems (Australia)	Develops instruments, consumables for arraying (using 2D gel technology) and characterizing proteins and designing informatics tools and databases for high throughput proteomics analysis.
Target Discovery (USA)	Develops and commercializes tissue- and disease-specific proteomics data as well as sequencing instruments.

The barriers to rapid emergence of proteomics as a new intellectual and industrial field are substantially technological. Not only are the picks and shovels yet to be built, but also few are trained to know how to use such instruments. Additionally, because proteomics involves complex systems that require substantially different skills, it is difficult to make progress unless an integrated team is assembled. If the team does not work using a systems approach, discovery is likely to be slow.

8   Picks And Shovels Of Proteomics

The estimates of the market size for the "picks and shovels" of proteomics such as electrophoresis products, mass spectrometry and protein sequencers vary and range from US$1 billion to as high as US$4.5 billion by 2005, according to a U.S. based consultant Decision Resources.

Many of the top selling drugs either act by targeting proteins or are proteins. It is expected that the current understanding of 500 disease proteins will expand to 5000 through proteomics with huge implications for drug discovery. DNA scanner technology and the study of genomics appear simple alongside the challenge of protein analysis.   Currently, proteomics research is performed using gel electrophoresis, mass spectrometry and other protein purification and analysis products. These proteomics tools are labor-intensive and cumbersome. With up to one million proteins in the human proteome, the current range of technology will not allow proteomics and drug discovery to accelerate on an industrial scale, nor be practical as a tool in clinical diagnostics.

However, with the Human Genome Project as our hindsight, it is not unlikely that we may experience similar acceleration in the Human Proteome Project if ancillary technologies make strides forward.

Current methods of protein classification seek to map the position of proteins against a known 'scale'.   These approaches are not easily automated and are essentially biochemical experiments, often plagued by difficulty in reproducing identical results.

8.1   2D Gel Electrophoresis

Two-dimensional gel electrophoresis separates proteins within a gelatin substrate (the gel) according to charge and size.   The first separation consists of moving proteins through a pH gradient with electric current.   Each protein will migrate based on their electric charge (the position in which they stop is called the isoelectric point).   The proteins are then separated in the orthogonal (perpendicular) dimension with electric current again, except this time a charged molecule is added to the proteins which also completely unfolds the proteins.   The electric charge is applied perpendicularly to the first separation and the proteins migrate according to size.   The result is a 2D constellation of proteins, visualized as spots in the gel. Each spot is a unique protein.

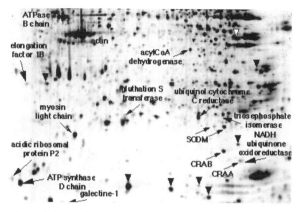

Figure 11. A sample result from a 2D gel electrophoresis. (Figure: Adapted from Burdett Buckeridge Young).

The technique has several significant limitations. For instance, hydrophobic proteins (such as the extremely important receptors and ion channels in the cell membrane) do not separate well, protein maps may be difficult to reproduce between laboratories, and the assay contains many manual and slow steps that cannot easily be automated. Therefore, 2D gel electrophoresis is not the ideal technique for high-throughput analysis.

In combination with mass spectrometry, 2DE-MS is the system that is most widely used in expression proteomics. However, 2D-MS still suffers from the following shortcomings:

❑ Low sensitivity — It is very difficult in a 2DE to visualize proteins expressed at low levels (10–1000 copies per cell) even from 10^8–10^{10} cells.[24]

❑ High cost — Use of mass spectroscopy increases the cost. A mass spectrometer costs about $200,000.

❑ Low throughput — A very well equipped laboratory can only perform 200-400 2DE per week and have to spend months analyzing the data.[25]

8.2   Classification by Absorption

Automatic and rapid chromatographic separations combined with tandem mass spectrometry is another way of classifying proteins. Chromatography is a physical separation method where a mixture of proteins (or protein fragments) can be separated by differential migration through an absorbent material. The materials can vary, but materials that separate proteins according to size can be very precise.

---

[24] V.E. Bichsel, L.A. Liotta, and E.F. Petricoin, 3rd, "Cancer proteomics: from biomarker discovery to signal pathway profiling", *Cancer J.* 7(1), 2001, pp. 69–78.
[25] K.H. Lee, "Proteomics: A technology-driven and technology-limited discovery science", *Trends in Biotechnology*, 19(6), 2001, pp. 217–222.

Once the proteins are separated, the different aliquots can be run through a MS/MS (tandem mass spectrometer) for sequencing and protein identification.

8.3   Emerging Approaches

There are emerging approaches which offer greater potential for precision and automation:

❏   MALDI-TOF (Matrix-Assisted Laser Desorption Ionization-Time of Flight) is a technique that vaporizes proteins. The vaporized proteins are then directed through a magnetic field and the time it takes for each fragment to reach a detector is measured. That measured time of flight is related to size.

❏   ICAT (Isotope coded affinity tags) in which two samples (one control and one diseased) of proteins are labeled with two closely related isotopes. The two samples are mixed together and then proteins are identified using mass spectrometry. Since both samples are labeled with different isotopes, ratios can be calculated to determine which proteins are increased/decreased or did not change in the diseased state.

Other investigations are evaluating the potential of using antibodies in parallel to identify protein from mixtures.

Protein chips, just as DNA chips have massively increased the number of genes that can be identified at one time, will be used to identify tens or hundreds of thousands of proteins at one time. Because proteins are more fragile than DNA molecules, the development of protein chips is not as far advanced as DNA chips. However the technology to fabricate protein chips is rapidly emerging. After the protein chip has been exposed to the unknown protein samples it will be scanned and analyzed using equipment that will be similar to and derived from DNA chip scanners and software.

9   Promise Of Antibody Microarray

Theoretically, any given protein can be specifically detected by one or a few antibodies. Currently, antibody-based immunoassays are the most popular methods for protein detection. A number of high throughput automatic immunoassay systems are available in the market. Immunoassays with various signal amplification systems offer high sensitivity and quantitative detection within a wide dynamic range. Antibody microarray therefore holds great promise in expression proteomics. Several companies are now in the business of developing antibody microarrays or antibody mimicries, but no product is yet available on the market:

There are several obvious obstacles in developing antibody microarrays:

❏   Antibodies cannot be synthesized like index DNA oligos used in DNA microarrays. Development of a highly specific monoclonal antibody may cost more than $10,000 and take several months.

❑   In order to utilize existing signal amplification system to increase detection sensitivity, a sandwich immunoassay has to be considered, which include two specific antibodies recognizing two distinctive epitopes for each target protein. Therefore, 2 million specific antibodies are needed to detect the 1 million proteins in a human body. This is economically not feasible and a Herculean task working with existing technologies. Large Scale Biology claims that it has a technology to develop 1000 antibodies a year, so 1000 years are needed to develop one million antibodies.

❑   To overcome the difficulty, companies are developing antibody-mimicries that can be generated in large scale with various combinatorial library technologies. Unfortunately, none of these technologies have been able to generate enough amount of antibody mimicries with a practically acceptable high affinity.

❑   DNA or mRNA samples can be easily denatured into a hybridizable state, while proteins, in most cases, have to be kept in native forms to be detected. This is technically difficult since a biological specimen often needs to undergo certain chemical treatment prior to testing.

❑   Cross reactivity cannot be avoided when so many antibodies and target proteins are placed in a single reaction solution.

10   Two Useful Twists In The Protein Junkyard

To be creative is to come up with something totally new. To be innovative is to do old things in a new way. Many emerging companies have come up with ways to vault over the above obstacles. Of particular interest is the very innovative approach by the upcoming Silicon Valley-based AbMetrix, Inc.

Table 8. Biotechnology companies that engage in antibody or antibody mimicry microarrays.

Company	Technology	Business Strategy
AbMetrix	Antibody	Digital Antibody Technology, Development stage.
Ciphergen	Antibody	Protein chip, platform only.
GenWay	Antibody	Start up.
Large Scale Biology	Antibody-mimicry	Coop with BioSite, development stage.
Motorola BioChip	Antibody	Development stage.
NeXstar	Antibody-mimicry	SELECT technology, development stage.
Phylos	Antibody-mimicry	HIT^{TM} Chip, development stage.
Zyomyx	Antibody-mimicry	Development stage.

It is clear that the longer a binding epitope is, the higher is the binding avidity and specificity. Empirically, an antibody needs to recognize on average 6 amino acids to ensure its specificity. An antibody binding to a shorter amino acid sequence usually fails to offer enough specificity and is often discarded as junk. However such an antibody is not necessarily a low affinity one. In practice, a lot of such antibodies are of high affinity but are junked immediately without further characterization. Evidences include high affinity antibodies directed against tiny molecule: biotin, BrdU, fluoresceins. Although they are not amino acids, they are smaller than or equivalent to a 3-mer amino acid in size.

The flagship technology of AbMetrix is its Digital Antibody Technology (DAT). DAT is a unique combination of several mature technologies:

❑ A library of digital antibodies (~ 1,000),
❑ A digital antibody database, and
❑ A digital antibody epitope search engine.

A digital antibody is a special type of monoclonal antibody and developing digital antibodies is much easier than developing the same amount of regular monoclonal antibodies. Hence, AbMetrix can achieve its goals very cost effectively in a relatively much shorter time.

DAT will be one of the best tools in the proteomics field because of its two salient features:

❑ High resolution — simultaneously detecting many different proteins in one biological specimen, and
❑ High-throughput — easily handling tens of thousands of specimen at the same time.

The former feature of DAT will allow diagnosis of many diseases with one product, greatly simplifying medical diagnostic procedures and cutting cost. In combination with the second feature, DAT can be used to improve drug screening in the pharmaceutical industry. For example, simultaneous binding of 5 antibodies directed against epitopes QAP, TPG, LTG, VSR and WDQ may be interpreted by a search engine as indication of Hepatitis C virus because HCV NS3 is the only known protein which contains all the five epitopes.

Thus, AbMetrix

❑ Innovalues by creating products for the proteomics market by innovating and adding value to existing products. AbMetrix adds value to the junked 3-mers and 4-mers, and uses computers as a tool to tackle the task of characterization wherever it is humanly impossible.
❑ Cloneverges by a combination of cloning and diverging, i.e., take existing products and rapidly come up with a high-quality, lower cost alternatives with some new wrinkles to the products. AbMetrix utilizes mature technologies, thus avoiding further development cost and marketing cost.

In short, AbMetrix clonverges existing technologies to innovalue junked high affinity antibodies.

11 Paleontology Of Diseases — Beyond Genomics And Proteomics

DNA and proteins are two different ways of looking an organism. One can look at an organism in yet other ways. For example, its metabolic, developmental, regulatory and other functional networks in normal or diseased tissues and organs, at different developmental stages.

Most kids know what a T-Rex or a veloceraptor looks like, thanks to the popularity of the dinosaurs, displays at museums, and the movie Jurassic Park. But how do we know the appearance, physiology and anatomy of a creature extinct tens of millions years ago? All that survived today are a few fossilized bones scattered over the globe, rarely a substantial part of a skeleton.

A paleontologist's approach is that all organ systems in the body are functionally connected. Any single system bears traces of the general work plan of the organism. So, the paleontologist can restore the whole picture of the extinct animal — from looks, morphology, physiology, to behavior — if the paleontologist can recognize and interpret such traces, usually on the skeleton as the best conserved fossilized system. For example, the ratio of certain bone lengths tells about the way and speed the animal moved; the size, the number and the arrangement of the ribs on a bone relates to the size, shape and function of the attached muscles. The holes in the bone are the pathways for nerves and blood vessels; their density and shape allow for reconstructing the shape and the relative load on specific organs. The integration of all these pieces of evidence allows the paleontologist to be amazingly precise in some cases to reconstruct even the skin color and mating behavior of the animal.

Figure 12. Morphological reconstruction of a T-Rex from skeletons by a paleontologist.

Similarly, a functional reconstruction model (FRM) can be used as the skeleton for understanding the whole functionality of a human cell, tissue, disease — and the backbone for the integration of all kinds of biological data scattered in hundreds of private and public databases and lab books at registries. Once assembled, FRM provides very valuable information about the general biochemistry, some regulation and physiology of the cell. The time series of reconstructions (different age, development stages, tissues, and conditions) can help to better elucidate development networks and genetics regulatory cascades.

A unique and powerful approach of functional reconstruction for connecting together expressed sequenced tags (ESTs), known and predicted open reading frames (ORFs), gene expression data, single nucleotide polymorphism (SNP) and other genetic markers into an integrated analytical system is being developed by an upcoming company based in New Buffalo, Michigan. The company, GeneGo, LLC[β], intends to build functional reconstruction models for normal and abnormal tissues, and conditions aiming to correlate gene sequence variability with disease phenotypes. The flagship technology — GeneGo Integrated Systems (GGIS) — has:

❑ a proprietary database (GGDB). By mid Y2001, there are some 2,000 enzymes and 3,000 pathways in GGDB. More than 1,800 probable human-specific pathways have been identified to be used for general metabolic reconstruction.

❑ Supporting software for pathways reconstruction and analysis.

GGIS is directly applicable for condition-based clustering of SNP markers, prioritization and discovery of drug targets for metabolic and developmental diseases, annotation of unknown human genes and alternatively spliced forms, and consumer genetics.

12 Omics This And Omics That

An organism may also be looked at at different scales, may be studied at different levels, and may be viewed from different perspectives. Genes, RNA, proteins, metabolism and methylation are five equally important levels of information in a cell.

Methylation is a modification in cytosine — one of the four bases — either with or without a methyl group. If a methyl group is attached to cytosine in the promoter region of a gene, the gene cannot be activated by transcription factors, thus not producing a protein. A gene can thus be silenced by methylation, leading to the failure to produce the associated proteins and in turn health problems.

Methylation is hereditary through cell division and can be correlated to specific diseases. Subtypes of cancer, diabetes, arteriosclerosis, rheumatoid arthritis have similar symptoms, but subtypes or classes have different genetic and epigenetic origins. Thus, therapeutic molecules used to treat a disease will be more effective if they are custom-tailored to tackle the particular problem of the patient.

With modern technologies, it is possible to design therapeutic molecules specifically for individual gene-products. With the availability of these chemical tools, discovery of the exact molecular problem has become the most crucial step in the development of novel therapies.

The first step in finding the molecular origins of diseases is class-prediction, the correlation of epigenetic and genetic findings with subtle differences in the disease of different patients. Once correlations have been ascertained, the sub-disease has

---

[β] President, Tatiana Nikolskaya.

been identified. Subsequent therapy development is specifically done for those patients with that precise genetic or epigenetic background.

Disease correlation is thus a crucial step each time a patient is diagnosed with a disease. Medication will be administered only after and according to the diagnostic test on which the disease correlation is based. This way, each and every patient may eventually receive the optimal therapy for his or her individual problem.

As a very substantial proportion of SNPs are located at methylated nucleotides, analyzing these SNPs is a part of methylation analysis. Thus current ongoing SNP projects further strengthen methylation research.

Since the Human Genome Project, new emerging subject areas tend to have the post-fix omics. For example, genomics and proteomics. It seems that one is not on the bandwagon if one does not omics this and omics that. Thus we see that these different ways of studying an organism have also assumed more whimsical names like phenomics, ontogenomics.

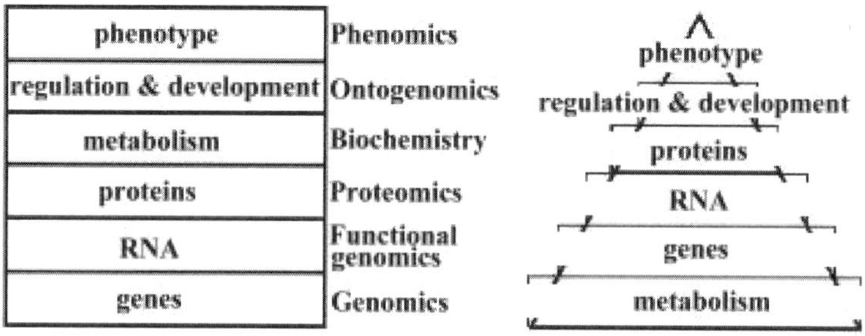

phenotype	Phenomics
regulation & development	Ontogenomics
metabolism	Biochemistry
proteins	Proteomics
RNA	Functional genomics
genes	Genomics

Figure 13. There are different ways of looking at an organism. But because some subject areas have been pursued more extensively than others, how much we know and how much we actually understand are not necessarily the same. We know phenotype, but we understand very little about it. Metabolism has been extensively studied in biochemistry, but we only know moderately about it.

Traditionally, these areas have been pursued independently, with certain areas making more progress than the others. For example, biochemistry is a very well established field. It leads to understandings. of metabolism. However, we only understand metabolism moderately. On the other hand, after the genome project, we know a lot about genes, but that does not translate to we understand genes.

13 Directed Evolution

Efficiency and speed lie at the heart of the new genetic engineering revolution. Nature's production and recycling schedules are deemed inadequate to ensure improved standard of living for a burgeoning human population. To compensate for nature's slower pace, new ways must be found to engineer the genetic blueprints of microbes, plants, and animals in order to accelerate their transformation into useful economic cornucopia.

This brings us to yet another way of looking at an organism — directed evolution. Directed evolution, developed over the past two decades, means evolving and optimizing molecules under laboratory conditions over a short period of time by either recombination or codon-based DNA synthesis for what would take million of years to accomplish by natural selection or Darwinian evolution.[26]

Directed evolution is an alternative to rational protein design — a technique that became popular in the 1980s. In rational protein design, researchers try to craft a new molecule — for example, an antibody or an enzyme — by first studying an existing protein's structure and then modifying it via targeted mutations to the gene that encodes it. But this painstaking methodology can prove to be difficult. Not only must researchers determine the sequence of amino acids — the 20 building blocks that make up all proteins — but they must also understand the complicated pattern of folding that the chain of amino acids undergoes to become a functional, three-dimensional molecule. Even after bringing sophisticated computer tools to bear on the problem, it is hard to unravel the workings of a protein folded up like origami, much less create a new one that has the desired properties.[27]

Frustrated and disappointed with the limitations of rational design, a growing group of researchers are borrowing from nature's tool kit instead, mimicking the basic processes of evolution to generate genetic diversity and select desirable traits to improve proteins, even without understanding their complicated structures. A team led by an MIT chemical engineer, Dane Wittrup, for example, evolved an antibody fragment to bind 10,000 times more tightly to its target in just four rounds of directed evolution. Better-binding antibodies might be used to fight cancer. For example, in attaching a cell-killing agent to an antibody that binds specifically and tightly to molecules found only on cancer cells can wipe out the cancer without damaging healthy tissues.

Pim Stemmer invented his version of directed evolution, called "molecular breeding" in 1983 while working at Affymax, Palo Alto, California. The brainchild of entrepreneur Alejandro Zaffaroni, Affymax was the first company to focus exclusively on discovering new drugs through combinatorial chemistry — a shotgun approach in which huge libraries of unique molecules are randomly generated, and then the useful ones are fished out through clever screening. Extending the combinatorial idea into the realm of protein development, Stemmer struck upon the idea of shuffling DNA. The basic concept is to start with a few different versions of the gene for a protein, cut them up and mix the pieces together to generate a diverse pool of new versions of the gene, fish out the best ones, and iterate. In other words,

---

[26] Vicki Brower, "Directed evolution techniques", *Genetic Engineering News*, 21(14), August 2001, pp. 92.
[27] Kathryn Brown, "Biotech speeds its evolution", *MIT Technology Review*, November/December 2000.

without introducing mutations, genes are cut and paste in a shuffling process, essentially the same process as in sexual reproduction. Rather than starting with a few different versions of a gene and shuffling the DNA, Diversa researchers typically begin with one gene and then introduce a multitude of mutations. This technique generates maximum diversity in the pool of new candidate proteins. Indeed, using a procedure called "gene-site saturation mutagenesis," Diversa researchers can try each of the 20 possible amino acids in each position along the protein chain, like in a number game, in less than two weeks.

Diversa researchers stack the odds in their favor by choosing unusual genes as starting points. The company has hired far-flung scientists to collect microbes from extreme locations — the gut of a bug from the Costa Rican jungle, an industrial dump site or the rotting skin of a submerged whale carcass. By harvesting DNA from bacteria on that dead whale, Diversa scientists collect the raw genes for enzymes that naturally break down polymers or fats in nasty environments. For a high-temperature enzyme to work under an alkaline pH, the company goes looking for places that already have the kinds of conditions. In other words, the company discovers enzymes that are optimal in settings of interest, and then uses directed evolution to push favored traits even further.

Applied Molecular Evolution is focusing on evolving optimized therapeutic proteins solely for human use using codon-based DNA synthesis. For example, researchers at AME has developed a system to alter precisely targeted areas of a protein and avoid other areas where changes could be harmful. The process takes only weeks or months.

Even proteins that evolve beautifully in the laboratory face classic development hurdles. For example, when a detergent enzyme was evolved for Procter and Gamble, the new enzyme passed laboratory tests with flying colors, only to fall apart inside company washing machines. A pressing challenge in applying directed evolution to the development of pharmaceuticals is the immunogenicity of proteins. Human, engineered, and foreign proteins all incite immune responses.

Table 9. Companies in directed evolution business (Table: Adapted from MIT Technology Review).

Company	IPO	Claim to Fame
Applied Molecular Evolution San Diego, California	2000 $88 million	Teaching old medicines new tricks
Diversa San Diego	1999 $174 million	Improving on nature's extremes
Enchira Biotech Woodlands, Texas	1993 $16 million	A new push towards pharmaceuticals
Maxygen Redwood City, California	1999 $96 million	Molecular breeding
Novo Nordisk Biotech Davis, California	-	

The keen competition in directed evolution is clear. Indeed, the $2 billion industrial enzyme market is a logical niche for directed evolution. Fewer than 30 enzymes generate more than 90 percent of industrial enzyme sales. This is not because there is a lack of trying on the part of the chemical companies to make new ones. The problem is that most enzymes fizzle under harsh, real-world conditions. There is also the $300 billion worldwide pharmaceuticals market. Crafting antibiotics and drugs that fight diseases, packing an extra punch into existing drugs by changing their protein structures, and evolving more powerful medicines are all attractive possibilities.

Chapter 9

SUBMICRON TECHNOLOGY AND NANOTECHNOLOGY

"The principles of physics, as far as I can see, do not speak against the possibility of maneuvering things atom by atom. It is not an attempt to violate any laws; it is something, in principle, that can be done; but in practice, it has not been done because we are too big."
Richard Feynman, 1959.

1   Midas! Gold In Potentia

Midas is the legendary king who got his wish that everything he touched would turn into gold. Alchemy is the never-found chemistry of turning base metals into gold. El Dorado is the fabled city of gold that lured the Spaniards to conquer South, Central and parts of North America. The Forty-niners is the discovery of gold that lured tens of thousands of Americans west to California.

The former two: Midas and alchemy are examples of transmutations of one metal to another; the latter two: El Dorado and Forty-niners are examples of extracting inanimate materials from Earth.

Throughout the Industrial Age, we radically altered our society and environment by extracting massive amounts of inanimate materials.   Metals, minerals, fossil fuels are mined, filtered, pumped and dug from Earth. They are burned, forged, soldered, melted, reconstructed and recombined to create machines, structures, and artifacts. The buzzword is manufacturing — an industrial-scale of production.

In the Biotechnology Age, we add living materials to the inanimate matter. But what is the use or profitability of isolating genetic materials unless they
❑   can be copied.
❑   can be reproduced in industrial quantities.

In order to establish an efficient living matter production line process, bioengineers need to invent a process for industrial-scale reproduction of life forms. The buzzword is now, instead of manufacturing, biofacturing — an industrial scale of reproduction.[1]

2   From Alchemy To Algeny

Alchemy, believed to have been derived from an Arabic word meaning "perfection", is said to have originated as a formal philosophy and process in Egypt during the 4th

---

[1] Hwa A. Lim, "Biotechnology as a form of nanotechnology", Talk at Biotechnology and Nanotechnology: Challenges and Opportunities of Drug Targeting in the Post-Genomic Era, Workshop In Honor of Britton Chance, University of Texas Southwestern Medical Center at Dallas, October 18, 2001.

century B.C.. Many historians believe its root dates further back in antiquity. According to alchemists, all metals are in the process of becoming gold.[2,3] They are, in other words, gold *in potentia*. The alchemists were firmly convinced that it was possible to accelerate what they believed to be the natural process of transformation by an elaborate orchestrated set of laboratory transmutations.

The alchemic process began by fusing together several metals into one mass or alloy, which was regarded as a universal base material from which various transmutations could be made. Fire was indispensable to the entire transitional process. It allowed the alchemists to melt, fuse, purify and distill the base material to create new combinations and forms, each one closer to the ideal golden state.[4] Because of the extensive use of fire, we call this pyrotechnology.

As we move from pyrotechnology to biotechnology, a new metaphor is emerging. Algeny, attributed to the Nobel laureate biologist Joshua Lederberg, means to change the essence of a living thing. The algenists view the living world as *in potentia*. The algenic arts are dedicated to the improvement and enhancement of existing organisms and the design of wholly new ones with the intent of perfecting their performance. Their task is to accelerate the natural process by programming new creations that they believe are more efficient than those that exist in the same natural state.

The algenic process begins with slicing out or splicing in genetic material to get genetic chimeras or transgenics from which other chimeras or transgenics may be made. The repertoire of recombinant DNA tools is indispensable to the entire process.

Viewed in the same light as alchemy, algenists do not regard an organism as a discrete entity, but rather as a temporary set of relationships existing in an ephemeral context, on the way to becoming something else. For the algenists, species boundaries are just convenient labels for identifying a familiar biological condition or relationship, but are in no way regarded as impenetrable walls separating various plants and animals.

Table 1. Comparison of pyrotechnology and biotechnology.

Pyrotechnology — Alchemy	Biotechnology — Algeny
All metals are gold *in potentia*	Living world *in potentia*
It is possible to accelerate the natural process transformation by elaborate procedures	To accelerate natural process by programming new creations that are more efficient

[2] Morris Berman, *The Reenchantment of the World*, (Cornell University Press, Ithaca, New York, 1981).
[3] Titus Burckhardt, *Alchemy: Science of the Cosmos, Science of the Soul*, (Stuart & Watkins, London, 1967).
[4] Jeremy Rifkin, *The Biotech Century: Harnessing the Gene and Remaking the World*, (Jeremy P. Tarcher/Putnam, New York, 1999).

Table 1. (Continued)

The process began with fusing several metals into one mass or alloy, which was then regarded as a kind of universal base material from which various transmutations could be made	The process begins with slicing out or splicing in genetic material to get genetic chimeras or transgenics from which other chimeras or transgenics may be made
Fire was indispensable to the entire transitional process	The repertoire of recombinant DNA tools is indispensable to the entire process

3   Now Nanotechnology

On January 21, 2000, Bill Clinton, at California Institute of Technology (Caltech), announced the federal government's commitment to science and technology.[5]   His administration made nanotechnology a top science and technology priority.[6]   This vital federal commitment in long-term, high-risk research will move the U.S. to the forefront of the nanotechnology frontier.

The purpose of nanotechnology is the ability to move matter in molecular or atomic levels.   The ultimate goal is to create functional materials, devices and systems through the control of matters at the nanoscale. The nanoscale is 1~100 nm, or 10^{-9}~10^{-7}m.

However, nanotechnology, as an anticipated manufacturing technology giving thorough, inexpensive control of the structure of matter, has sometimes been used to refer to any technique able to work at a submicron scale. It can also mean what is sometimes called molecular nanotechnology, which means basically "a place for every atom and every atom in its place".   In the latter definition, terms like molecular engineering, molecular manufacturing, etc. have also been used instead.

The entire field of nanotechnology can be viewed as an offshoot of the semiconductor microelectronic revolution, which spawned the development of micro electro mechanical systems.   In analogy with MEMS (micro electro mechanical systems) and BEMS (bio electro mechanical systems), in nanotechnology, we have NEMS (nano electro mechanical systems).

Caltech is no stranger to the idea of nanotechnology — the ability to manipulate matter at the atomic and molecular level.   Over 40 years ago in 1959, Caltech's very own Nobel laureate and quantum physicist, Richard Feynman asked, "The principles of physics, as far as I can see, do not speak against the possibility of maneuvering things atom by atom.   It is not an attempt to violate any laws; it is something, in principle, that can be done; but in practice, it has not been done because we are too

---

[5] President Clinton's address to Caltech on Science and Technology, California Institute of Technology's Audio Visual Services, Electronic Media Publications, January 21, 2000.
[6] *National Nanotechnology Initiative — Leading to the Next Industrial Revolution*, Intragency Working Group on Nanoscience, Engineering and Technology, National Science and Technology Council, 2000.

big."[7]   The neologism "nanotechnology" was first used by Taniguchi to describe ultra fine machining of matter.[8]   Drexler introduced two principal approaches to manufacturing at the molecular scale[9]:

❑   Bottom up approach — self-assembly of machines from basic chemical building blocks
❑   Top down approach — assembly by manipulating components with much larger devices such as manipulators in micron scale.

3.1   Nano Research And Development

Nanotechnology is still in research and very early development phase. Research or deepening knowledge is the activity of making basic breakthroughs into new areas. An example of research is genetic engineering in biotechnology. Development or widening knowledge is the expansion of technological knowledge in already existing areas. An example of development is genetically engineer pharm herds to produce proteins in milk. In between, there is applied research, in which the basic science is in place but some fundamental engineering breakthroughs have to take place to implement empirically what is already known scientifically. A good example of applied research is the Manhattan Project. Great brains like Einstein, Teller and Oppenheimer had shown that it was theoretically possible. The Manhattan Project made atomic bomb a reality.

Because private returns are apt to be much more certain if one is looking for an extension of existing knowledge than if one is looking for a major breakthrough, private firms tend to concentrate their money on the developmental end of R&D process.

Time lags are also shorter in development phase, and in the business sector, speed is a critical success factor. Often inventions become useful because the costs of other factors have come down. The Internet is an excellent precedent. It started in the 1960s as a communications system between military bases using IBM mainframes that cost millions of dollars. Although the technology had obvious commercial applications, the two telecommunications giants of the time, IBM and AT&T, were least interested in participating because computers were not cheap enough to make such a network affordable. Instead, Pentagon's Advanced Research Projects Agency and the civilian National Science Foundation nurtured the Internet for more than twenty-five years until 1994, when sufficient number of companies saw the potential of the Internet.

---

[7] Richard Feynman, "There is plenty of room at the bottom", *Engr. & Sci.*, 1960, pp. 22–36. Reprint: H.D. Gilbert (ed.), *Miniaturization*, (Reinhold, New York, 1961), pp. 282–296.
[8] N. Taniguchi, "On the basic concept of nanotechnology", The International Conference on Production Engineering, Japan, 1974, pp. 18–23.
[9] K.E. Drexler, *Nanosystems: Molecular Machinery, Manufacturing, and Computation*, (John Wiley and Sons, New York, 1992).

Because of the proclivity in the private sector to shy away from research and early phase development, the government has to focus its spending on long-tailed projects for advancing basic knowledge. Nanotechnology is currently an example of these long-tailed projects. Though this is where the private firms will not invest, it is precisely where the breakthroughs that generate a lot of private business opportunities are made. This is why a government that envisions to be a dominant technology leader will support nanotechnology.

Table 2. The investments of six U.S. departments and independent agencies in nanotechnology. (Adapted from National Nanotechnology Initiative).

Department/Agency	FY1997	FY1999	FY2001
NSF	65	85	150
DoD	32	70	110
DoE	7	58	93
NIST	4	16	10
NASA	3	5	20
NIH	5	21	39
Total	**116**	**255**	**422**

Similar to other technologies in science and engineering, nanotechnology can be broadly divided into two major fields: nanoscience and nanoengineering. It is important that nanoscientists first lay the foundations for nanotechnology through new findings so that nanoengineers can develop nano-systems and nano-devices.[10]

In the Nanotechnology Age, we will get essentially every atom in the right place, and make almost any structure consistent with basic laws of physics and chemistry. The process will be futile unless we can

❑ Have manufacturing costs not greatly exceeding the cost of required raw materials and energy.
❑ Positional assembly, which means we need molecular robotics for assemblying.
❑ Self-replication or be able to make copies of self and produce useful products.

The buzzword is, instead of manufacturing or biofacturing, molefacturing (short for molecular manufacturing) — an industrial-scale of production at nanoscales.

---

[10] T.C. Yih, "Nanotechnology: Challenge and opportunity", Biotechnology and Nanotechnology: Challenges and Opportunities of Drug Targeting in the Post-Genomic Era, Workshop in Honor of Britton Chance, University of Texas Southwestern Medical Center, Dallas, October 18, 2001.

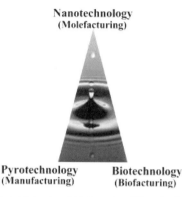

Nanotechnology
(Molefacturing)

Pyrotechnology   Biotechnology
(Manufacturing)   (Biofacturing)

Figure 1. Manufacturing (Industrial Age) and biofacturing (Biotechnology Age) form the base of the current industrial-scale production. Nanotechnology is a taller order and is currently in research phase. It will be some time before this technology has real impact on the economy.

3.2 Self Assembly Versus Positional Assembly

There are two main ways to assemble parts: self-assembly and positional assembly.[11]

In self-assembly, the assemblying parts move randomly under the influence of thermal fluctuations and explore the configuration space of possible orientations. If a particular arrangement is energetically more stable, then it will be preferred. Given sufficient time, this preferred arrangement will be adopted. For example, two complementary strands of DNA in solution will eventually find each other and hybridize together to form a double-helical configuration.

In positional assembly, a restoring force keeps the part positioned at or near a particular location, and two adjoining parts are assembled when they are deliberately moved into close proximity and linked together. While common at the macroscopic world — for example — we hold, position and assemble parts with our hands — this ability is still quite novel at the molecular or microscopic scale. Thermal fluctuations still play a significant role. In this situation, "holding" a molecular part does not provide absolute certainty about its position but rather imposes a bias on the range of positions it can adopt. Using a linear approximation of Hooke's Law[*], the part might be subjected to a restoring spring force F which is proportional to its distance from the desired location, i.e.,

$$F = k_s x$$

---

[11] Ralph C. Merkle, "Biotechnology as a route to nanotechnology", *Trends in Biotechnology*, 17(7), July 1999, pp. 271–274.

[*] This law was first stated by the British scientist Robert Hooke (1635–1703), a contemporary of Isaac Newton, in the form of "Ut tensio, sic vis".

where x is the distance between the part and its desired location, and k_s is the spring constant. Spring constant on the order of 10 N/m (Newtons/meter) or better can be achieved with scanning probe microscopes, which have already demonstrated the ability to move atoms and molecules on a surface in a controlled manner. The fundamental equation relating positional uncertainty, temperature and stiffness (spring constant) is a Gaussian distribution:

$$\sigma^2 = k_b T / k_s$$

where σ is the mean error in position, k_b is Boltzmann's constant (1.380658×10^{-23} Joules K^{-1}), T is the temperature in Kelvins. If k_s is 10 N/m, that of a scanning probe microscope, the positional uncertainty σ at room temperature (293K) is ~0.02 nm. This is accurate enough to permit alignment of molecular parts to within a fraction of an atomic diameter. The actual error, however, can be many times σ. The probability that the actual error is x_{err} is

$$P(x_{err}) = 1 / (\sigma\sqrt{(2\pi)}) \times \exp[-k_s \, x_{err}^2/(2\sigma^2)]$$

Thus errors of a few times σ are common, but errors of many times σ would be extremely unlikely.

Quantitatively, σ can be used to distinguish between self-assembly and positional assembly. The distinction is not binary, but moves continuously along a scale depending on the positional uncertainty σ. When the positional uncertainty σ is large, we are near the self-assembly end of the spectrum. When σ is small, we are at the positional assembly end of the spectrum. Intermediate points along this spectrum are occupied by, for example, a molecule "tethered" to an SPM tip by a polymer; or an object held by optical tweezers (a spring constant of 10^{-4} N/m implies a positional uncertainty s at room temperature of ~6 nm).

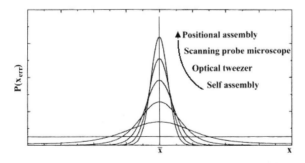

Figure 2. As σ decreases, the Gaussian distribution becomes more peaked. When σ approaches zero, the distribution approaches a delta function. In terms of assembly process, this corresponds to transitioning from self-assembly to positional assembly.

3.3   *Nature's Assembler*

Biomimetics is the art and science of designing and building apparatus by copying nature. The product, biomimetic, thus refers to human-made process, substances, devices, or systems that imitate nature.

While the SPM provides programmable positional control by allowing variations of the Cartesian coordinates x, y and z, a simple form of positional assembly can also be seen in the biological world. Enzymes bind two substrate molecules. The two bound molecules are positioned with respect to each other, thus facilitating their assembly. A limited form of positional assembly is also used in the ribosome, which can position the end of a growing protein adjacent to the next amino acid to be incorporated into that protein.

Ribosomes manufacture all the proteins used in all living things. A typical ribosome is relatively small, of the order of a few thousand cubic nanometers, and is capable of building almost any protein by stringing together amino acids — the building blocks of proteins — in a precise linear sequence. To do this, the ribosome has a means of selectively grasping a specific transfer RNA, which in turn is chemically bonded by a specific enzyme to a specific amino acid, of grasping the growing polypeptide, and of causing the specific amino acid to react with and be added to the end of the polypeptide.

The instructions that the ribosome follows in building a protein are provided by mRNA (messenger RNA). This is a polymer formed from the four bases adenine, cytosine, guanine, and uracil. A sequence of several hundred to a few thousand such bases codes for a specific protein. The ribosome "reads" this "control tape" sequentially, and acts on the directions it provides.

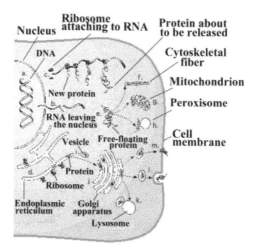

Figure 3. Production of protein. (Figure: Adapted from Burdett Buckeridge Young).

3.4   The Diamond of Living System — Stem Cell

Coal and diamonds, sand and computer chips, cancerous and healthy tissues, throughout history, variations in the arrangement of atoms of these entities have distinguished the well-to-do from the poor, the cheap from the cherished, the diseased from the healthy.[12] Arranging the atoms one way, we have coal; in another way, we have graphite; in yet another way, we have diamond. Similarly, we have, by various arrangements of atoms in H_2O, water vapor, liquid water, snow and ice. Or when helium is cooled down to a few Kelvins, it becomes a superfluid.

In living systems, we have something similar. Stem cells of a human can be coaxed to morphologize into the 200 different cell types in a human body.

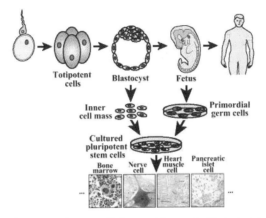

Figure 4. During development, totipotent cells become blastocyst. Pluripotent inner cell mass is used for stem cell research. Cells from aborted fetuses which are committed into becoming testes or ovaries can also be extracted for stem cell research. The stem cells can be coaxed into any of the 200 cell types found in a human body. (Figure adapted from various sources, including Stem Cells: A primer, National Institutes of Health, May 2000).

During the development of an organism, a fertilized egg is totipotent since it has the full potential to develop into a fetus. Approximately four days after fertilization during which several cycles of cell divisions will have taken place, the totipotent cells begins to specialize, forming a hollow sphere of about one hundred cells called the blastocyst. The blastocyst has an outer layer of cells and a cluster of inner cell mass.[13]

The outer layer of cells will develop into placenta and other supporting tissues needed for fetal development. The inner cell mass will develop into virtually all of the tissues of the human body. These inner cell mass are pluripotent rather than totipotent. They cannot form an organism because they cannot develop into the placenta and supporting tissues necessary for the development in the human uterus.

---

[12] K. Eric Drexler, *Engines of Creation*, (Anchor Press, 1986).
[13] *Stem Cells: A primer*, National Institute of Health, May 2000.

The pluripotent stem cells undergo further specialization into stem cells that are capable of developing into cells of particular functions. For example, blood stem cells, skin stem cells, and other human body's 200 different cell types. In laboratory conditions, stem cells have been stimulated to development into the cell type of interest for transplantation.

In this sense, stem cells — just like atoms of coal can be rearranged to get graphite and diamond — can be stimulated into any of the cell types. Thus specializations of stem cells into different cell types are a form of nanotechnology manipulation in nature's tapestry.

3.5   A Caveat in Mimicking Nature

While the power of self-assembly has been amply demonstrated by the wide range of complex molecular and a remarkable range of biological structures, we have barely begun to explore the power of positional assembly at the molecular scale. Despite this, it seems clear that this new capability will play a major role in our future ability to synthesize molecular structures. The power of positional assembly has been amply demonstrated at the macroscopic scale in today's factories and by our own ability to make things with our hands. While its application at the molecular scale will differ in many details, it will provide a new and remarkably powerful tool for extending the range of structures that we can make.

Though living systems have provided us many clues, we cannot blindly mimic nature. Early pioneers of flight took inspiration from watching birds soaring through the air. But "to be inspired by" is not tantamount to "to blindly copy". Airplanes are very different from birds. The artificial self replicating systems that have been envisioned for molecular manufacturing bear about the same degree of similarity to their biological counterparts as a car might bear to a horse.[14]

The functions of horses and cars are to provide transportation. Horses and cars have similar function, yet they are very different.

Horses replenish themselves by consuming potatoes, corn, sugar, hay, straw, grass, and other food. Cars, man-made machines, use only a single artificial and carefully refined source of energy, gasoline.

Man made machines tend to be very inflexible in response to environmental changes. They need roads to travel on. They have to be provided with parts, some of which are difficult to repair. The parts do not self-repair. They cannot cope with complex environment. They work because we design them to work inexpensively in those conditions. If operating them becomes too expensive, it becomes economically not viable to operate them for our purposes.

Living systems like horses are products of millions of years of evolution. They are very flexible and adaptable. Horses can maneuver a narrow trail, and jump over fences. They have a self-repair mechanism built into their genetic system.

---

[14] This example is taken from Zyvex page, http://www.zyvex.com/nanotech/selfRep.html.

In the same way, the artificial self-replicating systems that are being proposed for molecular manufacturing are inflexible. It is difficult to design a system capable of self-replicating in a controlled environment. Designing a system that uses a single source of energy is both much easier to do and produces a much more efficient system.

Horses are more complex at the expense of efficiency. They can produce the many complex proteins and molecules they need from the food they consume. They pay for this flexibility by having a very intricate digestive system capable of breaking down the food into constituent molecules.

Thus, the mechnaical designs proposed for nanotechnology are more like a factory than a living system. Molecular scale robotic arms capable of moving and positioning molecular parts would assemble rigid molecular products using methods more familiar to machine shops than the complex brew of chemicals found within a cell.

Although we are inspired by living systems, the actual designs are likely to owe more to design constraints and our objectives, than to the inspiring living system.

3.6 Drexler's Assemblers

In an analogous fashion, Drexler has proposed an "assembler", a device having a submicroscopic robotic arm under computer control. It will be capable of holding and positioning reactive compounds in order to control the precise location at which chemical reactions take place. This general approach will allow the construction of large atomically precise objects by a sequence of precisely controlled chemical reactions, building objects molecule by molecule. If designed to do so, assemblers will be able to build copies of themselves, that is, to replicate.

Because they will be able to copy themselves, assemblers will be inexpensive. By working in large teams, assemblers and more specialized nanomachines will be able to build objects cheaply. By ensuring that each atom is properly placed, they will manufacture products of high quality and reliability. Leftover molecules would be subject to strict control as well, making the manufacturing process extremely clean.

A la protein synthesis, an assembler will build an arbitrary molecular structure following a sequence of instructions. The assembler, however, will provide three-dimensional positional and full orientation control over the molecular component (analogous to the individual amino acid) being added to a growing complex molecular structure (analogous to the growing polypeptide). In addition, the assembler will be able to form any one of several different kinds of chemical bonds, not just the single kind (the peptide bond) that the ribosome makes.

Calculations indicate that an assembler need not inherently be very large. Enzymes "typically" weigh about 10^5 amu (atomic mass units. 1 amu = 1.66033 × 10^{-27} kg) while the ribosome itself is about 3×10^6 amu. The smallest assembler might be a factor of ten or so larger than a ribosome. Current design ideas for an assembler are somewhat larger than this: cylindrical "arms" about 100 nanometers in length and 30 nanometers in diameter, rotary joints to allow arbitrary positioning of the tip of the arm, and a worst-case positional accuracy at the tip of perhaps 0.1 to 0.2 nanometers, even in the presence of thermal noise. Even a solid block of diamond as large as such an arm weighs only 16×10^6 amu, so we can safely conclude that a hollow arm of such dimensions would weigh less. Six such arms would weigh less than 10^8 amu.

The assembler requires a detailed sequence of control signals, just as the ribosome requires mRNA to control its actions. Such detailed control signals can be provided by a computer. A feasible design for a molecular computer has been presented by Drexler. This design is mechanical in nature, and is based on sliding rods that interact by blocking or unblocking each other at "locks". This design has a size of about 5 cubic nanometers per lock (roughly equivalent to a single logic gate). Quadrupling this size to 20 cubic nanometers (to allow for power, interfaces, and the like) and assuming that we require a minimum of 10^4 locks to provide minimal control results in a volume of 2×10^5 cubic nanometers (.0002 cubic microns) for the computational element. These many gates are sufficient to build a simple 4-bit or 8-bit general-purpose computer.

An assembler might have a kilobyte of high speed (rod-logic based) RAM — similar to the amount of RAM used in a modern one-chip computer — and 100 kilobytes of slower but more dense "tape" storage. The tape storage would have a mass of 10^8 amu or less, or roughly 10 atoms per bit. Some additional mass will be needed for communications to send and receive signals from other computers, and power. In addition, there will probably be a "toolkit" of interchangeable tips that can be placed at the ends of the assembler's arms. When everything is added up a small assembler — with arms, computer, toolkit, and other components — should weigh less than 10^9 amu.

Escherichia coli (a common bacterium) weigh about 10^{12} amu. Thus, an assembler should be much larger than a ribosome, but much smaller than a bacterium.

Table 3. Complexity of self-replicating systems. Estimate of the Internet worm is an approximation of the number of bits in the C source code. For biological systems, the complexity is twice the number of base pairs. The complexity of NASA proposal was taken from Advanced Automation for Space Missions. Note that viruses do not qualify as a self-replicating system as they require a living system to infect. Viruses need additional molecular machinery provided by the host environment.[15]

Complexity of self replicating systems	Kilo Bits
Von Neumann's universal constructor	~500,000
Internet worm	~500,000

[15] Klaus S. Lackner, C.H. Wendt, "Exponential growth of large self reproducing machine systems", *Mathl. Comput. Modelling*, 21(10), 1995, pp. 55–81.

Table 1. (Continued)

Mycoplasma genitalium	~1,160,140
E. coli	~9,278,442
Drexler's assembler	~100,000,000
Human	~6,400,000,000
NASA Lunar Manufacturing Facility	>100,000,000,000

The observation to be inferred from the table is that simple designs for self-replicating systems both exist and are well within current design capabilities. The engineering effort required to design systems of these complexities will be significant. But they should not be greater than the complexity involved in the design of computers or airplanes.

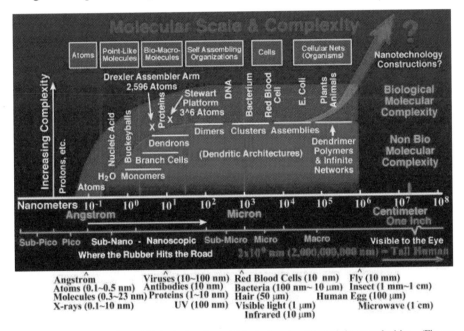

Figure 5. Scales and sizes of biological and non-biological molecules, and the complexities. (Figure: Adapted from NanoTechnology Magazine).

3.7 Manipulations In Nanoscale

Today's scanning probe microscope is still relatively large. To be able to make mole quantities of products, we still need smaller, fast positional devices.[16]

---

[16] Ralph C. Merkle, "A new family of six degree of freedom positional devices", *Nanotechnology* 8, 1997, pp. 47–52.

The Stewart platform is an octahedron. Six of the struts can be lengthened or shortened under programmatic control. The position and orientation of the platform with respect to the base has six degrees of freedom — x, y, z, roll, pitch, and yaw — by adjusting the lengths of the six side struts.[17]

Figure 6. Schematic illustration of a Stewart platform. (Figure: Adapted from Ralph C. Merkle).

The ability to construct an octahedron is not beyond the capabilities of biotechnology. A truncated octahedron has already been demonstrated.[18] A way to shorten or lengthen a strut is to use two overlapping struts. The strut can be made to slide one relative to the other in a controlled fashion.

Figure 7. To translate one strut relative to the other: c-C bonds are first created. Then e-C bonds are created. The c-C bonds are then dissolved, leaving only e-C bonds. Since the e-C bonds are in tension, they will slide the strut one unit relative to each other. The process can be repeated.

Suppose the first strut is composed of three repeats: cehcehcehceh... (denoted by sketches of Britton Chance, Albert Einstein, and Sherlock Holmes), while the second strut comprises CEHCEHCEHCEH.... We first bond the two struts together with c-C bonds. To translate one strut relative to the other, we create e-C bonds. Now we dissolve c-C bonds. The e-C bonds will now be in tension and will move one unit relative to each other. The process can be iterated.

4   From Controlled Environment To Work Environment

If artificial self-replicating systems will only function in carefully controlled environment, how can nanotechnology find applications in complex environments, such as inside the human body?

---

[17] D. Stewart, "A platform with six degree of freedom", *Proceedings of the Institution of Mechanical Engineers*, 180, Part 1, No. 15, 1965–1966, pp. 371–386.
[18] Y. Zhang, and N.C. Seeman, "The construction of a DNA truncated octahedron", *Journal of the American Chemical Society*, 116, 1994, pp. 1661–1669.

Self-replicating systems are key to low cost, but their main domain of operations is the world outside the controlled environment. Medical devices designed to operate in the human body do not have to self-replicate. They can be manufactured in controlled environment and then injected into patients. Similarly, in the controlled environment, self-replicating systems can manufacture simpler and more rugged systems that can be transferred to their final destinations of operations. The resulting devices will be simpler, smaller, more efficient and more precisely designed for the task at hand than the devices designed to perform the same task and self replicate in the work environment.

The essence of nanotechnology is the creation of and utilization of materials and devices at the level of atoms, molecules and supramolecular structures, and the exploitation of unique properties and phenomena of matter at 1 to 100 nanometers. Research is underway to design implants with nanoscale features that will promote the healing of natural tissue. Devices in development promise to meet bacteria and viruses on their own level for the detection and treatment of diseases.[19] Biomedical nanotechnology will make it possible to build nanorobots (nanobots) having cellular dimensions with the ability to eliminate infections, unclog arteries, and a range of other applications. A lab-on-a-chip would be developed that would be able to test for a host of conditions with a single drop of blood and thus replace an entire laboratory.

In principle, the rule of thumb is: optimize device design for the desired function, manufacture the device in an environment optimized for manufacturing, and then transport the device from the manufacturing environment to the environment for which it is designed.

However, we do caution that systems designed in a controlled environment do not always perform satisfactorily when transferred to the environment they are designed for. For example, the oil-digesting bacteria of GE that swung the genetic floodgate wide open. While the bacteria would eat oil in the laboratory, years of test proved that they were too fragile to function in the open seas, where they are most needed. Another example is the detergent enzyme evolved for Procter and Gamble. The new enzyme passed laboratory tests with flying colors, only to fall apart inside company washing machines.

5   From Algeny To Almoly

We can compare biotechnology with nanotechnology in the same way we compared pyrotechnology with biotechnology.

---

[19] Allen J. Menezes, Vik J. Kapoor, Vijay K. Goel, Brent D. Cameron, Jian Yu Lu, "Within a nanometer of your life", *Eye on the Future: Nanotechnology*, The American Society of Mechanical Engineers, 2001.

Table 4. Comparison of biotechnology with nanotechnology.

Biotechnology — Algeny	Nanotechnology — Almoly
Living world *in potentia*	All animate and inanimate materials *in potentia*
To accelerate natural process by programming new creations that are more efficient	To improve and enhance existing materials and design of wholly new ones with intent to perfect their performance
The process begins with slicing out or splicing in genetic material to get genetic chimeras or transgenics from which other chimeras or transgenics may be made	Process begins with moving atoms and molecules
The repertoire of recombinant DNA tools is indispensable to the entire process	Molecular manufacturing will be indispensable to the entire process

The nanotechnology community believes that nanotechnology will begin to have its first impact on the economy in 10–15 years, if not sooner. We stress that even when breakthroughs are made and technological success can be said to have occurred, risk and uncertainties are still high. The pay offs are often not in the expected areas. And when something unexpected pops up, those paying for the R&D may not be in a position to exploit it. Perhaps the developments are not in their line of business, or the company does not have the expertise to take advantage of them.[20] Also, often it takes time to realize the importance and usefulness of what has been discovered. A case in point is the patent attorneys at Bell Labs did not want to incur costs of patenting the laser because at that time, optical waves did not seems to be relevant to telecommunications.[21]

This is how many new companies get started. Researchers from big companies frequently turn up ideas that do not fit in with their employers' business plans, or are too far out in left field, or ahead of its time. When their ideas are turned down, these researchers go off and set up new companies to exploit them.

We thus expect nanotechnology to lead to many offshoot industries.

We also stress that technology leadership is not the same as R&D development. In information technology, Europe is a big spender, but in terms of technology leadership, it is behind. To have pay off, spending has to be followed by activities necessary to embed the newly developed technologies into the economy. For example, the U.S. outclasses Europe not so much in information technology R&D spending, but in investment is information technology hardware and software.

---

[20] Lester C. Thurow, Building Wealth, (HarperCollins Publishers, New York, 1999).
[21] Nathan Rosenbert, "Uncertainty and technological change", *Technology and Growth*, Federal Reserve Bank of Boston, June 1996.

6 Biomimetics — The Confluence Of Biotechnology, And Nanotechnology

The essence of nanotechnology is the ability to work at the atomic, molecular and macromolecular levels in order to create functional materials, devices and systems with fundamentally new properties and functions. The three paths of biotechnology, biomimetics, and atomic positioning are parts of a broad bottom up strategy: working at the molecular level to increase our ability to control matter.[22]

The essence of biomimetics is to design human-made processes, substances, devices, or systems by copying nature to the extent possible. For example, a car is very similar to a horse, yet they are very different in complexity. Possible applications of biomimetics include biochip. A biochip is a substrate wafer that is designed for the purpose of accelerating genetic research. It may also be able to rapidly detect chemical agents used in biological warfare so that defensive measures could be taken. A biochip is used as a kind of "test tube" for real chemical samples. Other possible applications include artificial organs in regenerative medicine, electronic devices, and nanorobots that can seek and destroy disease-causing bacteria in preventive medicine.

Nanotechnology will further improve the design of the biochip and the development of preventive and regenerative medicine. Traditional miniaturization efforts based on microelectronics technology have reached the submicron scale. This can be characterized as the top down strategy. The bottom-up strategy of nanotechnology, however, may provide a viable alternative.

7 Nano Dragon

At the nation's very first Science Activities Week — an event planned to be held annually during the third week of each May — the Shanghai City's Science and Technology Commission held a popular-science exhibition and cross-talk performance on Nanjing Pedestrian Mall's Centruy Plaza. Perhaps more significant was the announcement of Shanghai Stone Nanotechnology Port Co. and Shanghai University to join forces on a 1.8 billion yuan (US$217 million) nanotechnology center. The Nanotechnology Port will be developed near the University of Baoshan District campus over the next five years.[23] The board chairman of Shanghai Stone, Shen Guojun, indicated that Stone would provide the investment capital, and Shanghai University would supply most of the over 100 researchers who would take part in the project. The rest would come from Lanzhou Chemistry and Physics Institute, Chinese Academy of Sciences.

---

[22] "Overview of Nanotechnology", Adapted by J. Storrs Hall from papers by Ralph C. Merkle and K. Eric Drexler.
[23] James Chang, "Big future for tiny-tech", *Shanghai Daily*, Tuesday, 15 May 2001, pp. 1.

The Nano Port will cover an area of 600,000 square meters and comprise a research facility, a product incubator and an information center. There will also be an exhibition center and living quarters for the scientists.

The research will center around the development of environmentally friendly materials, goods made from petrochemicals and bioloical science products. First projects will be to merge nanotechnology with other fields such as textiles and pharmaceuticals. For example, lyocell — a natural fiber that can be made into fire-proof material, and selenium — a medicine that can boost human immune systems.

On November 12, 2001, the day the People's Republic of China and the Republic of China were admitted into WTO, Jilin University of Changchun inaugurated its Alan G. MacDiarmid Institute, the first institute to be named after a non-Chinese. MacDiarmid, together with Alan Heeger of Univeristy of California at Santa Barbara and Hideki Shirakawa of Tsukuba University, won the 2000 Nobel prize in Chemistry for his contribution to conducting polymers. The Institute is committed to be a world class research institute devoted to fundamental and applied research in physical and biological sciences, with emphasis in the scientific and technological cutting edges of nanoscience and nanotechnology.[24]

---

[24] Based on a delegate report by Da Hsuan Feng, Vice President, The University of Texas, Dallas, November 7–17, 2001.

Chapter 10

PATENTABILITY IN BIOTECHNOLOGY

"The effort to decipher the human genome will be the scientific breakthrough of the century — perhaps of all time. We have a profound responsibility to ensure that the life-saving benefits of any cutting-edge research are available to all human beings."
President Bill Clinton, Wednesday, March 14, 2000.

1 Decoding A Genetic Announcement

On March 14, 2000, U.S. President Bill Clinton and British Prime Minister Tony Blair jointly issued a three-paragraph statement.[1] The statement sparked a sell-off of biotechnology stocks, and caused a jitter in the high-tech stock market. The shares of Celera Genomics (Bethesda, DC) fell 20%, the stocks of Incyte Genomics (Palo Alto, CA) plummeted 27%, and the NASDAQ's biotechnology stock index tumbled 13%, the largest one-day loss since the measure was established.

What is less obvious is how the three-paragraph statement sets business, legal and political leaders arguing about the scope and implications of the declaration. This pronouncement could not have come at a more propitious time when a coalition of publicly funded laboratories in the U.S. and U.K. are competing with privately sponsored firms in the U.S. and abroad to decipher the entire sequence of the human genome that controls human development and diseases.[2,3,4]

The cause for the tailspin of biotechnology stocks after the pronouncement can be traced to the issue of patentability. Genes are the "green gold" of the biotechnology century. The economic and political forces that control the genetic resources of the planet will exercise tremendous power over the future world economy, just as the industrial age access to and control over fossil fuels and valuable metals helped determine control over world markets.[5]

To fully understand how and why, let us take a historical *tour de horizon* of patents, for we cannot ignore the more elusive subterranean currents of history as a shaping force.

---

[1] "Call for access to genome data, Clinton, Blair want it open to public", *San Francisco Chronicle*, Wednesday, March 15, 2000, pp. A1.
[2] "Biotech firm falters in bid to control yield from genetic research", *San Francisco Chronicle*, Monday, March 6 2000, pp. A3.
[3] "Putting a price on secret of life", *San Francisco Examiner*, Sunday, March 12, 2000, pp. B1.
[4] "Controversial genome guru defends work", *San Francisco Chronicle*, Monday, March 27, 2000, pp. A3.
[5] Jeremy Rifkin, *The Biotech Century, harnessing the gene and remaking the world*, (Tarcher/Putnam, New York, 1998).

2 Enclosing The Genetic Commons

The worldwide race to patent genes is the culmination of a long odyssey to enclose and privatize all of ecosystems that make up the earth's biosphere. The concept of enclosure is critical to understanding the potential of long-term consequences of current efforts to enclose the world's gene pool.

"Enclosing" means "surrounding a piece of land with hedges, ditches, moat, or other barriers to the free passage of men and animals".[6,7] Enclosure places the land under private control, severing any right the community has to use it. Since the time when humans started to domesticate livestock and cultivate grains, enclosure was already practiced at a very limited and primitive level. Domesticated dogs were used to fend off intruders and trespassers. Modern enclosure movements are more sophisticated. They are carried out by acts of parliament or congress, the agreement of all members of a village commune, and permission from the monarch or order from the appropriate authority such as the court. We shall now turn to this issue by looking at concurrent developments in three continents: Europe, Asia and America.

2.1 Europe

Medieval European agriculture was communally organized. Peasants pooled their individual holdings into open fields that were jointly cultivated. Common pastures were used for grazing of animals. Life was spare, demanding and unpredictable. The village commons existed for more than six hundred years along the base of the feudal hierarchy, under the watchful presence of the landlords, monarch, and pope. Then, in the 1500s, new and powerful political and economic forces were unleashed, first in Tudor England, later on Continental Europe, which eventually destroyed the communitarianism of village life that had bound humans to one another and the land for centuries.

2.1.1 Tudor England

The commodification of the global commons began in Tudor England in the 1500s with the enactment of the great enclosure acts. Increased urban demand for food triggered an inflationary spiral, which in turn increased the cost to landlords whose land rents had been fixed at pre-inflationary rates. These laws were designed to privatize the feudal commons, transforming the land from a shared trust to private real estate that could be bought and sold as individual units of property in the commercial market.

---

[6] Gilbert Slater, *The English Peasantry and the Enclosure of Common Fields*, (A.M. Kelly, New York, 1968).
[7] William Tate, *The English Village Community and the Enclosure Movement*, (Walker, New York, 1967).

The second wave of enclosure occurred in the late 1700s and early 1800s during the reign of King George III. At the time, an expanding textile industry was clamoring for more wool, making sheep grazing an increasingly attractive and lucrative prospect. The forces of gentrification and the need for grazing grounds conspired, creating an irresistible lure to enclose the land. With the financial help of a new and wealthy bourgeois class of merchant and bankers, landlords began to buy up common lands to turn them into pastures for sheep.

2.1.2 Medieval Europe

Much of the economic life of the Medieval Europe was centered on village commons. Although feudal lords owned the commons, they leased them to peasants under various tenancy arrangements. In return for the right to cultivate the land, tenant peasants had to turn over a fraction of their harvest to their landlords or devote a comparable amount of time to tilling the landlords' fields. With the introduction of a moneyed economy in the late Medieval period, peasants were increasingly required to pay rent and taxes in return for the privilege to farm the land.

So between the sixteenth and the nineteenth centuries, a series of political and legal acts were initiated in countries throughout Europe that enclosed publicly held land. In the gentrification process, millions of peasants were dislodged from their ancestral homes and forced to migrate into new towns and cities where, if they were fortunate, they might secure subsistence employment.

2.1.3 Social Impact of Land Enclosure

The subsequent change in people's relationship to the land touched off a series of economic and social reforms that would remake society and reshape humanity's relationship to the natural world of the modern era. Enclosure fundamentally restructured the way people perceived their relationships with each other and the soil. It introduced a new concept of human relationship that changed the basis of economic security and the perception of social life. Land was no longer something people belonged to, but rather a commodity people possessed. Land was reduced to a quantitative status and measured by its exchange value. So too were the individuals caught in between. Neighbors became employees or contractors. Reciprocity was replaced with hourly wages. People sold their time and labor. Human beings began to views each other and everything in financial terms. Virtually everything and anything became negotiable and could be purchased at an appropriate price.

2.1.4 Other Enclosures: Land, Sea, Air and the Invisible

The enclosure of European landmass and the conversion of feudal commons to privately held real estates began a process of privatization around the world. Today, with the exception of the Antarctica, everything single square foot of landmass on Earth is either under private commercial ownership or government control. The Antarctica has been partially preserved as a non-exploitable shared commons by international agreement.

The enclosure of landmass has been followed, in rapid succession, by the commercial enclosure of parts of the oceanic commons, the atmospheric commons, and more recently, the electromagnetic spectrum commons by media stations. The large swath of coastal waters are commercially leased to collect port duties, the air has been converted to commercial air corridors for international travels, and electromagnetic frequencies have been leased to private companies for radio, telephone, television and computer transmission.

2.2   The Middle Kingdom

China has a 5,000-year history. Its curiosity, its instinct for exploration, and its drive to build had created all the technologies necessary for an industrial revolution — something that would not occur for another 350 years![8,9] It had blast furnace and piston bellows for making steel, gunpowder and cannon for military conquest; the compass and rudder for exploration; paper, movable type, and printing press for disseminating knowledge; the wheeled plow, horse collar, rotary threshing machines, mechanical seeders to generate agricultural surpluses; the ability to drill for natural gas; decimal systems, negative numbers, and the concept of zero to analyze what they were doing. Large armadas (28,000 men) were exploring the east coast of Africa at about the same time when the Portuguese and Spaniard were sending out much smaller expeditions down the west coast of Africa. At the beginning of the fifteenth century, about a hundred years before Columbus' voyage, Admiral Cheng Ho — the Muslim eunuch who became the greatest admiral in the Chinese history — commanded a huge navy comprising of 62 junks and 225 other vessels to explore seven times the Indian Ocean with ships four to five times as large as those of Columbus.

2.2.1   West Meets East

During the Dark Ages of Europe, China was under the Mongolian rule (1206–1368), the Yüan Dynasty (also called the Mongol Dynasty). The Mongolian hordes tore across the Siberian plains, through what are modern Russia and Eastern Europe.[10] The Mongolians, a race of nomadic warriors, disdained agricultural people. They

---

[8] David S. Landes, *The Wealth and Poverty of Nations*, (Norton, 1998).
[9] Alain Peyrefitte, *The Immobile Empire*, (Knoft, 1992).
[10] http://www.britannica.com.

probably viewed Western culture as being what it was at that time — a peasant society enduring the yoke of feudalism reinforced by religion.[11]

On the battlefield, European armored men were rather inflexible, each carrying about 20–50 pounds of armor. The Mongolians wearing silk, in contrast, were very adaptable and swift. They caught their European armored men by surprise, outflanked them, and defeated the Europeans at each confrontation. The Mongolians quest reached as far as modern Poland and Ukraine. They were never defeated. The Mongolian assault on Europe and the Middle East stopped short of completion due not to military failure, but to dissension over succession.

When the Mongolian threat subsided, Marco Polo visited Kublai Khan (1215–1294), the grandson of Genghis Khan. By then, the nomadic Mongolians had been absorbed into the refined Chinese culture. Marco Polo was to remain in China for 17 years and upon his return to Europe, Marco Polo brought with him the refinement of culture, which had become an accepted basis of life in China. With the culture, Marco Polo also brought new technologies, including gunpowder. The rocket and multi-staged rockets was reportedly to have been imported from China during this time as well.[12] Around 1150, it had crossed someone's mind to attach a comet-like firework to a four-foot bamboo stick with an arrowhead and a balancing weight behind the feathers. To make the rockets multi-staged, a secondary set of rockets was attached to the shaft, their fuses lighted as the first rockets burned out. Rockets were first mentioned in the West in connection with a battle in Italy in 1380, arriving likely in the wake of Marco Polo.

2.2.2   Maritime Expeditions

During the time when Europe was going through its Dark Ages, for 300 years, the Chinese had been extending their power out to sea. An extensive sea-borne commerce developed to meet the taste of the Chinese for spices and aromatics and the need for raw industrial materials. In the days of no refrigeration, spice trade started to boom in South East Asia in the 1300s AD. The geographically strategic location and the geologically stable region became the focal point of spice nexus. The Indonesian Archipelago and the Malayan Peninsula became the central meeting place for exchange. Tamasek became Singapore, the Malaccan Sultanate of Malaya was at its golden age.

During the Ming Period (1368–1643) of China, the arts of seafaring reached new heights. This led to a rise of maritime expeditions. Admiral Cheng Ho (1371–1435) of China made seven trips to the Southern Seas, and ventured as far as the eastern coast of Africa, at a time when the Europeans were also exploring the

---

[11] Laurence J. Brahm, *China As No. 1 — The New Superpower Takes Centre Stage*, (Butterworth-Heinemann Asia, Singapore, 1996).

[12] Robert Temple, *The Genius of China: 3,000 Years of Science, Discovery, and Invention*, (Simon and Schuster, New York. 1986).

western coast of Africa. The only key difference is that the Europeans colonized the "newly discovered" *Terra Incognito*, while the Chinese did not. In the wake of Cheng Ho's expeditions, he established tributary trade, which lasted to the 19th Century.

2.2.3   East Meets West

The European exploitation of the potential of gunpowder catapulted Europe out of its self-imprisoning armor. Four centuries after the return of Marco Polo to Europe, the Europeans, having perfected the application of gunpowder, would venture back to South East Asia and the gate of China. The military technology would for the first time in history be superior to what the East had to offer. By this time, the Manchurians, NOT the Mongolians, ruled China. The Manchurians of the Qing Dynasty (1644–1911), like their predecessors, absorbed the Chinese culture. The China the West "discovered" was a world of refinement and culture consistent with the marvels which inspired Marco Polo, in sharp contrast to the conquering nomadic Mongolian warriors the West knew more than four centuries ago when the West first met the East.

Coveted by the West, these refinements became targets for possession. With the growth of Western industrialization, the West was looking for new markets into which it could expand and eventually control. As the mindset remained unchanged from their experience with the nomadic Mongolians, the West justified their actions through religious dogma and feelings of superiority. The missionaries saw a vast population of heathens to convert; while politicians saw the opportunity of economic colonization.

The rest is history.[13,14] The Opium War, the Unequal Treaty and the subsequent century of humiliation, the experience from which will still plague the Chinese in doing business with the West in modern days.

2.3   The New World — USA

The New World was "discovered" during the Age of Expeditions, at the time when the Europeans were trying to find a sea-lane to India and South East Asia. The Americans celebrated the quincentenary of its rediscovery in 1992 by Christopher Columbus, and the Brazilians just celebrated its 500th birthday after its discovery by Pedro Alvares Cabral in 1500. After the discovery, Europeans soon colonized the continent.

However, patents did not come for another three hundred years. The U.S. first Patent Act took effect more than two hundred years ago, during the presidency of Thomas Jefferson. Jefferson, an amateur scientist himself, was determined to ensure

---

[13] Immanuel C.Y. Hsü, *The Rise of Modern China*, 5th edition, (Oxford University Press, Oxford, 1995).
[14] Hwa A. Lim, "Smoking gun", Silicon Valley High Tech, May 2000.

that "ingenuity should receive a liberal encouragement." The Act was passed as a law in 1793. It provided for inventors to patent "any new and useful art, machine, manufacture or composition of matter, or any new useful improvement thereof".[15] As established by the law, a patent is a grant issued by the U.S. government giving the patent holder the sole right to make, use, or sell the invention within the U.S. during the term of the patent, generally seventeen years.

The law of 1793 embodies the principle that living things cannot be patented. For the following 140 years, plant breeders tried to get around the ban and get protection for hybridized plants. In the 1930s, they persuaded Congress to bestow limited commercial protection on plant hybrids. The tags hanging from patented roses in nurseries trace their origin to these laws. Over 5 million patents have been issued since 1793.[16]

2.3.1 Opening of the Genetic Floodgate

The incident we are going to present is the precedent-setting case that leads to the patentability of life. Because of its significance, we shall describe it in relative detail.

The case revolved around Ananda Mohan Chakrabarty, a scientist at General Electric (Schenectady, NY) who in the 1970s, genetically altered a bacterium so that it could devour petroleum — a potentially useful tool for cleaning up oil spills or similar disasters. Nature had already produced several strains of bacteria that had the propensity to digest different types of hydrocarbons in oil. These bacteria are all from a family known as *Pseudomonas*. Each of these bacteria has plasmids, auxiliary parcels of genes that break up or "eat" oil. The trick was to somehow put them all together to get a super oil-eating bacterium. Chakrabarty performed the genetic manipulation by fusing together the genetic material from four types of *Pseudomonas*, thereby creating a crossbred version with an enhanced appetite for oil.[17]

In 1971, when GE and Chakrabarty sought to patent his oil-digesting microbe to the U.S. Patent and Trademark Office (PTO), patent officers, following 180 years of precedence, rejected the application after several years of review. The officers cited that if Jefferson or Congress had intended life to be patentable under the 1793 Act, it would have said so in the law. To underscore its contention, the PTO pointed out in the few cases where patents had been extended to life forms (for asexually reproducing plants), it had taken a legislative act of Congress to create a special exception.

---

[15] US Congress, Office of Technology Assessment, *New Developments in Biotechnology: Patenting Life-Special Report*, OTA-BA-370, (US Government Printing Office, Washington DC, April 1989).
[16] "New patents challenge Congress and Courts", *Scientific America*, September 1988, pp. 128.
[17] Giovanna Brel, "An Illinois biochemist wins a crucial patent fight, and a new era of life in a test tube begins", *People*, July 14, 1980, pp. 38.

GE and Chakrabarty appealed to the Court of Customs and Patent Appeals (CCPA). To the shock of many, GE and Chakrabarty won by a 3-2 decision. The majority on the court argued that "the fact that microorganisms... are alive is without legal significance...patented microorganism was more akin to inanimate chemical compositions such as reactants, reagents, and catalysts, than to horses and honeybees or raspberries and roses."

The PTO was not impressed by the CCPA's decision. It held steadfast to its original rejection of Chakrabarty's application. The case then went to the Supreme Court. The Supreme Court did not decide on the case immediately. Instead, it instructed the CCPA to examine *Parker v. Flook,*[18] a recent Supreme Court decision, which admonished courts that "we must proceed cautiously when we are asked to extend patent rights into areas wholly unforeseen by Congress." It seems that the Supreme Court's action indicated it did not favor patenting life. The CCPA was stubborn. It ignored the Supreme Court's advice and reiterated its holding that life was patentable, while the PTO was equally convinced that life was not patentable. The latter re-appealed to the Supreme Court. In October 1979, the Supreme Court decided to end the issue of patentability of life once and for all. It voted 5-4 in favor of granting patent on "anything under the sun that is made by man".[19] The irony of the decision is that by the time the Supreme Court decided to end the debate of the patentability of life, GE no longer had any intention of marketing the bacteria. Apparently, while the bacteria would eat oil in the laboratory, years of test had proven that they were too fragile to function in the open seas, where they are most needed. GE's motive in pursuing its patent quest was simple. It was using the Chakrabarty case as an altruistic test case to establish the ground rules for patenting of life. If the GE patent succeeded, the patent profit floodgates would be swung wide open. GE and other corporations could then maximize their profits in the coming multibillion-dollar biotechnology industry.[20]

2.3.2   Impacts of the Supreme Court's Decision

The Supreme Court's decision laid the all-important legal groundwork for the privatization and commodification of the genetic commons. Corporate America understood the profound implications of the court decision. Chemical, pharmaceutical, agribusiness, and biotechnology start-up companies everywhere sped up their R&D work, mindful that the granting of patent protection meant the possibility of harnessing the genetic commons for vast commercial gain in the years ahead. In the aftermath of the historic decision, bioengineering technology shed its

---

[18] Parke, Acting Commissioner of Patents and Trademarks v. Flook, 437 US 584, 596, 1978.

[19] Sidney A. Diamond, Commissioner of Patents and Trademarks, petitioner, v. Ananda M. Chakrabarty et al., 65L ed 2d 144, June 16, 1980, 144–47.

[20] Andrew Kimbrell, *The Human Body Shop: The Engineering and Marketing of Life*, (Harper, San Francisco, 1993).

pristine academic garb and bounded into the marketplace, where it was heralded by many analysts as godsend, the long-awaited replacement for a dying industrial order. On October 14, 1980, within months after the Supreme Court cleared the way for the commercial exploitation of life, Genentech came on the market, offering one million shares of stocks at a par value of $35. By the time the closing bell rang, the fledgling biotechnology firm had raised $36 million and was valued at $532 million. The astounding thing was that Genentech had yet to introduce a single product into the marketplace.

3   The Final Frontier — Where No Life Scientists Have Gone Before

The international effort to convert the genetic blueprints of millions of years of evolution to privately held intellectual property represents both the completion of almost a millennium of commercial history and the closing of the last remaining frontier of the natural world.

3.1   Biopatentability

While the Supreme Court decision lent an air of legal legitimacy to the emerging biotechnology industry, a PTO decision seven years later in 1987 opened up a floodgate for the wholesale commercial enclosure of the world's gene pool, signaling the beginning of a new economic era in world history.[21]   The PTO reversed its earlier position and issued a ruling that all genetically engineered multi-cellular living organisms, including animals, but excluding human beings, are potentially patentable.   The reason for excluding human beings was that the Thirteenth Amendment to the Constitution forbids human slavery.   On the other hand, genetically altered human embryos and fetuses, as well as human genes, cell lines, tissues, and organs are potentially patentable, leaving open the possibility of patenting all of the separate parts, if not the whole, of a human being.   In a single regulatory stroke, the PTO had placed the global economy on a new course, one that would lead us out of the Industrial Age into the Biotech Century.

At the very heart of the issue of patentability is the question of whether engineered genes, cells, tissues, organs, and whole organisms are truly human inventions or merely discoveries of nature that have been skillfully modified by human beings.   In order to qualify as a patentable invention, the inventor must prove that the object serves some useful purpose.   Against this standard is an equally compelling qualification.   Even if something is novel, non-obvious, and useful, if it is a discovery of nature, it is not an invention and therefore not patentable.   For this

---

[21] Jeremy Rifkin, *The Biotech Century, harnessing the gene and remaking the world*, (Tarcher/Putnam, New York, 1998).

reason, the chemical elements in the periodic table, while unique, non-obvious when first isolated, and very useful, were nonetheless not considered patentable, as they were discoveries of nature.

What makes the Supreme Court decision and subsequent PTO ruling so audacious, and suspect from a legal stand point of view, is that it appears to defy the very logic of previous patent rulings that preclude claiming a discovery of nature as an invention. No molecular biologist has ever created a gene, cell, tissue, organ, or organism *de novo*. In this sense, the analogy between the elements of the periodic table, and genes and living matter is appropriate. No reasonable person would dare suggest that a scientist who isolated, classified, and described the properties of hydrogen, helium, or oxygen, ought to be granted the exclusive rights for 17 years, to claim the substance as a human invention, and charge a royalty for its use. The PTO has, however, said that the isolation and classification of a gene's properties and purposes is sufficient to claim it as an invention.[22]

4   Biopiracy And Biocolonialism

Corporate efforts to enclose and commodify the gene pool are meeting with strong resistance from a growing number of countries and non-government organizations (NGOs). For example, the recent World Trade Organization (WTO) meeting, December 1999, Seattle, and the recently concluded meeting of the World Bank (WB) and International Monetary Fund (IMF), April 2000, Washington DC, met with strong protests from NGOs such as the AFL-CIO, Green Peace, and Ruckus Society. Protestors carried banners reading "Stop All Slavery", "More World, Less Bank", "Human Need, Not Corporate Greed", "IMF-World Bank, Hundreds Rich, Billions Poor"… protesting anything from animal cruelty to world poverty. Some claimed "Free trade lets corporations grow at the poor countries' expense and fosters corruption". Others argue that "The IMF and World Bank are trying to enslave the working people of the world", and that "The IMF and WB are promoting big businesses of the world which will eventually gain control and are going to run the world".[23] The World Bank and IMF countered that the demonstrators are protesting about issues that the World and IMF are currently already addressing. The World Bank and IMF also blamed the demonstrations on the ignorance of most (not all, as emphasized by the World Bank and IMF) of the demonstrators who are not familiar with the roles and functions of the two bodies.

---

[22] US Patent and Trademark Office, *Animals- Patentability*, US Government Printing Office, Washington, DC, April 7, 1987.

[23] "Protests fail to disrupt World Bank and IMF", *San Francisco Chronicle*, Monday April 17, 2000, pp. A1.

4.1 Cash Cows from Crops

While the technological expertise needed to manipulate the new "green gold" reside in scientific laboratories and corporate boardrooms in the Northern Hemisphere, most of the genetic resources that are essential to fuel the new revolution lie in the tropical ecosystems of the Southern Hemisphere. This struggle between the Northern multinational corporations and Southern countries for control over the global genetic commons is likely to be one of the pivotal economic and political struggles of the Biotech Century.

History can be a good lesson here. The history of colonial struggle has been one of continual usurpation and exploitation of native biological riches for the advantage of home markets. The great explorations of the New World were as dedicated to the task of finding new biological sources for food, fiber, dye, and medicine as to discovering gold, silver, and other rare metals. Native knowledge of agriculture and native labor were both exploited to grow new food staples for export onto the world market. Cassava, sweet potatoes, peanuts, maize, beans, and squash were among the new native cash crops.

Explorers, and later Catholic missionaries and embassy personnel, devoted a great deal of time to biological prospecting, with the hope of securing new biological treasures — a form of biocolonialism — that could be transformed into lucrative commercial markets. Many colonial nations, eager to maintain exclusive control over their biological conquests, enacted stiff penalties on plant contraband, including imposition of death penalty for the theft of valuable plants. In some instances, the theft of native resources could affect the future of entire empires. For example, the pirating of rubber plant from Brazil to South East Asia at the turn of the last century gave the British a commercial advantage in the critical world rubber market and led to the collapse of American efforts to control rubber production in the Western Hemisphere.[24] The balance was not tipped until the Second World War, when the Japanese cut off the natural rubber latex supply, and the U.S., out of necessity, invented synthetic rubber.

4.2 Ethnobotany and Bioprospecting

Of the 119 drugs still extracted from higher plants, about 74% were discovered by chemists who were attempting to identify substances responsible for the plants' medical uses in humans.[25] Moreover, there are at least 250,000 species of higher plants, and those 119 drugs come from fewer than 90 species. Add to this the virtually untapped resources of marine flora and fauna, terrestrial animals, and

---

[24] Jack R. Kloppenburg, Jr., *First The Seed: The Political Economy of Plant Biotechnology, 1492–2000*, (Cambridge University Press, Cambridge, 1988).
[25] N.R. Farnsworth, "The role of ethnopharmacology in drug development", Ciba Foundation Symposium, 154, 1990, pp. 2–11.

microorganisms on land and sea, and there one see the immeasurable potential for new medications.

Two types of collection are being conducted — ethnobotany and bioprospecting.

Ethnobotanical methods involves sending botanists and physicians into the field to interview traditional healers or shamans.   Ethnobotany was once a largely taxonomic exercise to study plant use by local populations.   It has now become a multifaceted discipline that includes a role in the discovery and development of drugs based on plant material.   Its emphasis on people also makes it a conservator of traditional knowledge and culture, wellsprings of scientific information that have less economic value.   The challenge is to secure a niche against the competition posed by synthetic methods of drugs discovery, particularly those employing combinatorial chemistry and genetics.

The more conventional method of bioprospecting is a random collection of species with no basis for plant selection.   An organization operating in this mode is the Missouri Botanical Garden of St. Louis, USA.

4.3   Pirating Indigenous Knowledge

Today, plant hunters are giving way to gene prospectors.[26]   Corporate giants are financing expeditions across the Southern Hemisphere in search of unusual and rare genetic traits that might have some commercial value.

The potential stakes are enormous. Consider just the value of new drugs. As it is, more than a quarter of all the prescriptions from pharmacies come from chemicals originally discovered in plants. Another 13% come from microorganisms, and 3% come from animals. Yet these materials are only a minuscule fraction of the ones that are almost certainly available. Up to now, we have named 1.6 million species, but this is far, far short of the total. Scientists estimate that there are between 10 million and 100 million species on the Earth today.

Nearly three quarters of all the plant-based prescription drugs in use today were derived from drugs used in indigenous medicine.[27,28]   For example, curare, an important surgical anesthetic and muscle relaxant, is derived from plant extracts used by Amazonian Indians to stun prey.

Southern countries claim that what Northern companies call "discoveries" are really the pirating of accumulated indigenous knowledge of native people and cultures — a form of biopiracy.   It is true that Northern companies add some value by engineering and modifying the genetic makeup of plants, or by isolating out,

---

[26] Hope Shand, "Patenting the planet", *Multinational Monitor*, June 1994, pp. 13.
[27] Hwa A. Lim, Martin Wang, "Confluence of alternative and conventional medicines and future prospects", *Silicon Valley High Tech*, April 2000.
[28] Robert Nexworth, "Drug wars: Extracting patents and profits from the rainforests' medicinal plants", *Village Voice*, July 21, 1992, pp. 39.

purifying, distilling and mass producing through clonal propagation and other means. Still Southern countries argue that a slight genetic modification of a crop or herb in the laboratory is rather insignificant, especially when measured against the centuries of painstaking stewardship required to nurture and preserve organisms containing those very unique and valuable traits so coveted by scientists in their research.

4.3.1 Taxol for Cancer

There is tremendous impetus to identify the active components from established herbal formulae, after which modification can be made on these functional molecules to maximize the therapeutic effects with current chemical synthesis. For example, Taxol, which had been extracted from the bark of Pacific yew trees (Taxus brevifolia), is a potent form of anticancer natural product. It can now be synthesized in the laboratory from British yew leaf, which is more abundant.[29]

The yew is also known as the "tree of death". The name Taxus comes from the Greek "toxin", which translate to poison or toxin. Its poisonous nature has been mentioned in many cultures. For example,[30]

❑ In ancient Roman literature, Julius Caesar wrote that Catuvolcus, who was king of the Eburones, poisoned himself with the yew because he was old and weak and did not want to endure another war.

❑ Pliny the Elder (Gaius Plinius Secundus) noted that people died after consuming wine stored in barrels made out of the yew tree

❑ In Celtic culture the yew is a sacred tree and is used to carve out religious objects such as the duric staff.

❑ Dioscorides (Greek physician) observed that the yew emitted poisonous fumes (pollen); and

❑ Shakespeare also used the poisonous nature of the yew in writing Macbeth and Hamlet.

The yew tree is very much in demand due to the discovery of Taxol (anticancer agent) in the bark of Taxus brevifolia. The European yew is a source in the production of Taxotere, which is potentially an even better anticancer agent than Taxol. The English yew contains a similar compound to Taxol but is 10 times more concentrated. One of the main delays in the development of Taxol as an anticancer agent was the difficulty in obtaining sufficient quantities of the compound from its natural source. Taxol only constitutes 0.01–0.03% of the dry weight of the inner bark of the Pacific yew tree. This means that vast amounts of the yew tree are needed to isolate even small amounts of the drug.

---

[29] R.A. Holton, et al, "First total synthesis of Taxol", *J. Am. Chem. Soc.* 116, 1994, pp. 1597–1599.

[30] *Discovery*, Department of Chemistry, Imperial College of Science, Technology and Medicine.

Figure 1. Pacific yew trees, tree bark, fruit and leaves. (Photo: Adapted Pioneer Herb).

Taxol, from the bark of the Pacific yew, is a fungicide. It fights water mold, keeping the yew trees safe from that pathogen, killing that fungus the same way it kills breast cancer cells by binding to tubulin.

4.3.2   Rose Periwinkle for Leukemia

The search for plants that contain genes or cell lines that are key to treating diseases is called bio-prospecting. A well-known case is the discovery of *Catharanthus roseus* or rosy periwinkle in Madagascar more than forty years ago. In 1956, two researchers in traditional medicine reported that alkaloids in the periwinkle, a perennial plant known to Madagascar's traditional healers for its use in treating diabetes, also had anti-tumor and anti-leukemia properties. The healing chemicals occurred in minute quantities. A French pharmaceutical company tried to produce the alkaloids synthetically but soon gave up. Eli Lilly stepped in in the 1960s, at a time when Madagascar just attained its independence from the French and had no barriers to exportation of plants. Although Eli Lilly, like the French pharmaceutical company, did not succeed in producing effective synthesized drugs from the plant, it did develop an extraction process to produce drugs from periwinkle grown elsewhere (as they claimed).

Figure 2. Rosy periwinkle. The ancient forests of Madagascar's central plateau are home to the rosy periwinkle, a plant that stands only a foot tall at maturity. Its tiny flowers and shiny, green leaves distinguish it no more than its height, yet the rosy periwinkle has become famous throughout the world, particularly in medical circles. The rosy periwinkle contains over 70 alkaloids with medical applications such as lowering blood sugar levels, slowing bleeding, and tranquilizing. (Photo: Adapted from National Wildlife Federation).

Periwinkle has saved thousands from Hodgkin's disease and childhood leukemia. Chemicals found only in these plants have a unique ability to keep certain kinds of cancerous cells from reproducing. These chemicals have helped increase the survival rate for childhood leukemia from 10% to 95%. The drugs have been highly profitable. Eli Lilly made about US$1.0 billion but not a single dime goes to Madagascar.[31]

4.3.3 Repackaging Indigenous Knowledge for Sale

Unfortunately, patent laws only reward individual innovative efforts in scientific laboratories. Collective efforts, passed down from one generation to another, are rationalized as "prior art" and dismissed altogether. It appears to many in the Third World that biotechnology companies are getting a free ride on the back of thousands of years of indigenous knowledge. Corporates browse the centers of genetic diversity, helping themselves to a rich largess of genetic treasures. They then sell the same entity back to the same people in a slightly engineered and patented form, but at a hefty price, all for the product that have been freely shared and traded among farmers and villagers for all of human history.

Figure 3. Neem trees are attractive broad-leaved evergreens that can grow up to 30m tall and 2.5m in girth. Their spreading branches form rounded crowns as much as 20m across. They remain in leaf except during extreme drought, when the leaves may fall off. The short, usually straight trunk has a moderately thick, strongly furrowed bark. The roots penetrate the soil deeply, at least where the site permits, and when injured, they produce suckers. This suckering tends to be especially prolific in dry localities. The small, white, bisexual flowers are born in axillary clusters. They have a honeylike scent and attract many bees. Neem honey is popular, and reportedly contains no trace of azadirachtin. The fruit is a smooth, ellipsoidal drupe, up to almost 2cm long. When ripe, it is yellow or greenish yellow and comprises a sweet pulp enclosing a seed. The seed is composed of a shell and a kernel (sometimes two or three kernels), each about half of the seed's weight. It is the kernel that is used most in pest control. The leaves also contain pesticidal ingredients, but as a rule they are much less effective than those of the seed. (Photo: Adapted from Dendrology, Virginia Tech University).

The neem tree (botanical name: *Azadirachta indica*) is an Indian symbol and enjoys an almost mystical status in that country. Ancient Indian texts refer to it as "the blessed tree" and the tree that "cures all ailments". The tree has been used for centuries as a source for medicines and fuel and is grown in villages across the Indian subcontinent. New Year begins with ritual eating of tender shoots of the neem tree. Millions of Indians used neem twigs as toothbrushes because of the tree's antibacterial properties. The leaves and bark are used to cure acne and to treat a range of illnesses from infections to diabetes.[32] The tree has proven particularly

---

[31] K. Robinson, "The blessings of biodiversity", *San Francisco Chronicle*, Wednesday, January 19, 2000, pp. A12.

[32] Vandana Shiva, *Biopiracy: The Plunder of Nature and Knowledge*, (South End Press, Boston, 1997.)

valuable as natural pesticides and has been used by villagers to protect against crop pests for hundred of years. In fact, this natural pesticide is more potent (against locusts, nematodes, boll weevils, beetles, and hoppers) than many conventional industrial insecticides and is as effective as Malathion, DDT, and dieldrin, but without the deleterious impacts on the environment.[33,34]

W.R. Grace Company isolated the most potent ingredient in the neem seed, azadirachtin, and then sought and received a number of process patents for the production of neem extract from the Patent and Trademark Office (PTO) by arguing that the processes it used to isolate and stabilize the azadirachtin were unique and non-obvious.

Six years ago the European Patent Office granted a patent to the U.S. Department of Agriculture and the multinational agriculture company W.R. Grace. It covered a method of using the neem tree oil for fungicidal purposes.

Subsequently, the granting of the patent was challenged by Green party politicans in the European Parliament and other pressure groups, including Vandana Shiva. They oppose big business owning the rights to living organisms — what they call biopiracy — because they say the livelihoods of poor farmers in developing countries will be undermined. Indian scientists, in affidavits to the PTO, pointed out that Indian researchers and companies had been treating neem seeds with the same process and solvents as W.R. Grace, years before the company sought a patent on the processes. Moreover, Indian researchers and companies had never sought patent protection because they believe the information about the uses to be the result of centuries of indigenous research and development and something to be shared openly and freely.   The concern was the W.R. Grace patent would grant the company the right to alter the price and availability of neem seeds. Thus depriving local farmers of their ability to produce and use neem-based pesticides.

At a May 2000 hearing in Munich, the manager of an Indian agriculture company proved that he had been using an extract of neem tree oil for the same purpose as described in the patent several years before it was filed. For this reason the EPO said the method could not be patented and revoked the patent.[35]

The decision was hailed by non-governmental organizations as a signal to companies and governments in rich countries that the developing world is going to challenge this type of patent.

4.3.4   Sweet Profit

Skirmishes over usurpation of indigenous knowledge and natural resources are occurring with a greater frequency.   For example, in 1993, Lucky Biotech

---

[33] National Research Council, "Neem: a tree for solving global problems", *National Research Council Report*, (National Academy Press, Washington DC, 1995).
[34] Richard Stone, "A biopesticidal tree begins to blossom", *Science*, February 28, 1992, pp. 1070.
[35] "Neem tree patent revoked", *BBC News*, Thursday, May 11, 2000.

Corporation, a Korean pharmaceutical firm, and the University of California were awarded U.S. and international patents for a genetically engineered sweet protein derived from a plant thaumatin found in West Africa. The thaumatin plant protein is one hundred thousand times sweeter than sugar, making it the sweetest substance known on the Earth. It has been used by local villagers for centuries as a sweetener for food. The thaumatin protein can be engineered into the genetic code of fruits and vegetables, providing low-calorie sweeteners. With the market for low-calorie sweeteners about $1 billion in the U.S. market alone, thaumatin is almost certain to be a cash-cow for the company in years to come, while villagers in West Africa will not share in the good fortune.[36]

4.4 International Governing Bodies

In order to ease the growing tensions between global companies and Southern countries, a number of international institutions and private companies have proposed plans to share a portion of their commercial gain from new patents on biotechnology products with the host countries, local people and other interested parties. The International Plant Genetic Research Institute (IPGRI), headquartered in Rome, has proposed that companies seeking to market agricultural products derived from germplasm stored in international agricultural research centers be required to negotiate royalty arrangements with the source countries.

However, the matter is more complicated.[37] Private companies have entered into agreement with source countries, local organizations, and indigenous people to share portion of the gains from patented products, but on very uneven terms. The most controversial, and highly publicized venture is initiated by Merck & Co. in Costa Rica. The giant pharmaceutical company entered into an agreement with a local research organization, the National Biodiversity Institute, in which Merck agreed to pay a little more than $1 million to the organization in return for securing the company's potentially valuable plants, microorganisms, and insect samples. Critics liken the deal to European settlers giving American Indians trinkets worth of a few dollars for the exclusive ownership of Manhattan. Merck boasts $4 billion in revenue, purchasing bio-prospecting right in a country with one of the richest reservoirs of plant and animal life on the Earth for a trifling $1 million, is nothing more than tokenism. The recipient organization, on the other hand, is granting a right to bio-prospect on land that it has no historic claim to in the first place. Meanwhile the indigenous peoples, who have legitimate claim to negotiate a transfer of germplasm, are locked out of the Merck agreement![38]

---

[36] Hope Shand, "Patenting the planet", *Multinational Monitor*, June 1994, pp. 13.
[37] Hwa A. Lim, and DaHsuan Feng, "A global economy without a global government", *Silicon Valley High Tech*, May 2000.
[38] Vandana Shiva, *Biopiracy: The Plunder of Nature and Knowledge*, (South End Press, Boston, 1997).

5 Just Price

The debate over life patents is one of the most important issues ever to face humanity. Life patents strike at the core of our beliefs about the very nature of life and whether it is to be conceived of as having intrinsic or mere utility value. A great debate of this kind occurred more than half a millenium ago when the Catholic Church and an emerging merchant and banking class squared off over the question of usury. The Church argued for a "just price" between sellers and buyers and claimed that merchants and bankers could not profit off time by charging rapacious rates of interest because time was not theirs to negotiate. God freely dispenses time and it is only His to give and take. In charging interest, the usurer sells nothing to the borrower that belongs to him. He sells only time that belongs to God. He can therefore not make a profit from selling someone else's property.[39] The merchant disagreed and argued that time is money and that charging interest on time was the only way to secure investments in the market. The Church lost the battle over usury and the charging of interest, and the loss set the hurried course towards market capitalism of the modern age.

In the nineteenth century, another great debate ensues over the issue of human slavery, with abolitionists arguing that every human being has intrinsic value and God-given rights, and cannot be made the personal commercial property of another human being. The abolitionists' argument ultimately prevailed, and legally sanctioned human slavery was abolished in every country in the world. Unfortunately, human slavery in the form of involuntary indenture is still a cultural practice in some areas of the world, including the U.S.! The inherent rights of women, minorities, children and animals have become a matter of increasing concern. For example, sex trade is a new form of slavery in the U.S. An estimate indicates that about 700,000 primarily young girls are trafficked into the U.S. from Asia and the Former Eastern Bloc Countries.[40]

Now still another grand battle is unfolding on the eve of the Biotech Century. This time over the question of patenting life, a struggle whose outcome is going to be as important to the next era in history as the debate over usury, slavery, and involuntary indenture have been to the era just passing.

6 The Biotech Century

In the most recent Chase H&Q Life Sciences Conference, Monday, January 10, 2000, San Francisco, the Year 2000 is viewed as the beginning of the genetic revolution. Tools and techniques of biosciences, notably the ability to manipulate

---

[39] Thomas de Chobam, *Summa confessorum*, F. Broomfiled (ed.), Louvain, Paris, 1968.
[40] "Wellstone, Brownback want to curb 'sex slaves' trade", *San Francisco Chronicle*, Tuesday, April 11, 2000.

molecules and create new compounds, are finding applications beyond the realm of medicine, opening markets in agriculture and manufacturing.[41] Given the burst of millennial optimism, executives, scientists, financiers and observers identify trends that will chart the course of biotech 2000 and beyond:

❑ The Genetic Gold Rush — In anticipation of the completion of the human genome project, genomics has become the hottest word in biotechnology. Genomics is the study of genes and their functions. With the completion of the genome project, the next step will be identifying the subtle differences between the genes of healthy and the genes of those whose are predisposed to diseases. These differences are called single nucleotide polymorphism (SNP). However, to develop a drug from SNP databases will take an average of 5–15 years.

❑ Money in Monoclonals — Biotech firms have been trying to produce drugs based on antibodies, the proteins human bodies produce to fight diseases. Monoclonal antibodies (mabs) technique has successfully produced its first product, Herceptin of Genentech (South San Francisco) to treat metastatic breast cancer. Mabs research is expected to intensify, and many more products are expected to come out from the research.

❑ Spare Parts — Organ transplants have saved lives, but they are not very perfect for donors are scarce and rejections are common. Biotechnology companies are already growing simple organs such as skin or to grow liver. Tissue engineering will expand the scope to grow other organs.

❑ Brain Fix — Thanks to new imaging technology, scientists can now watch a brain at work. The new imaging technology gives scientists the equivalent of oscilloscope, capable of detecting brain activities. Seeing how the brain works is only part of the research. Scientists have already started to tweak the brain's electric grid for therapeutic purposes. For example, special electronics can be rigged to allow people with spinal cord injuries to perform certain task.

❑ Love it or Label It — Genetically engineered food will be an area of public debate. The Europeans have successfully forced the governments to enforce labeling of whole or processed food containing bioengineered ingredients. It is likely that the same will happen in the U.S.

❑ Pet Pharmacies — There are 175 million pets (dogs, cats, birds,...) in the US. Consumers spending on animal drugs more than doubled from an estimated $592 million in 1994 to $1.3 billion in 1998 according to a study by the American Health Institute. Today's pet potion go far beyond flea, tick and worm medicine. Novartis, for example, is selling Clomicalm, a drug for separation anxiety to stop dogs from tearing up the house while the master is gone (to work or on vacation). Pfizer is working on Rimadyl, an arthritis medication for dogs.

---

[41] "Biotech 2000", *San Francisco Chronicle*, Monday, January 10, 2000, pp. B1.

❑   Fountain of Controversy — With all the knowledge of genome maps and SNP come ethical, moral and political issues. The issue of privacy of people who has propensity for diseases is at stake. The issue of spiraling healthcare cost will center stage.

❑   Rodney Dangerfield Cured? — In 1997, California had six regional biotechnology clusters, made up of 2,500 companies, hiring 210,000 for $8 billion and registering $4 billion exports. The Bay Area itself boasts 670 biomedical companies, 78,000 employees, about twice the size of the second largest cluster in Orange County. Official recognition of the industry's importance came when the U.S. Senate designated January as "National Biotechnology Month'. The sheer number of aging baby-boomers and the depth of their desire for new medicines will create markets hitherto unseen.

Amidst all the ethical controversies of organ pharming, genetically engineered farming, and pharma race for drugs, patentability of genetic commons will be a subject of debate for some time to come.

7   Who Owns What?

Capitalism began in Great Britain when the enclosure movement converted communal agricultural lands of feudalism into privately owned land. To have an enforceable property right in capitalism, who owns what has to be very clear.[42]

The private ownership of productive assets and the ability to appropriate the output that flows from those assets lies at the core of capitalism.

Consider environmental pollution. What everyone owns, no one owns. As a result, everyone has an incentive to pollute, to use the free disposal system that is available and let someone else downstream or downwind bear the cost of cleaning up. However, on land that is privately owned, pollution market works. Private owners do not let their neighbors dump waste on their property. Someone siezes the need and get into the business of opening up a dump site. The only problem with these private dump sites is that their owners in turn have an incentive to abuse the free pollution rights.

Capitalism cannot deal with pollution because it cannot establish the ownership rights to clean the environment. In certain sense, we were extremely blessed that during the U.S. longest period of prosperity of the Clinton administration, we had an environmental champion Vice President Al Gore.

---

[42] Lester C. Thurow, "Needed: A new system of intellectual property rights", *Harvard Business Review*, Sept/Oct 1997, pp. 95.

Chapter 11

BIOLOGICAL AND BIOLOGY-RELATED INFORMATION AS A COMMODITY AND THE RISE OF BIOINFORMATICS

"If you have $10 dollars and I have $10. You give me your $10 and I give you my $10, we each still have $10. There seems to be a 'conservation of money'. If you have an idea and I have an idea. You give me your idea and I give you my idea. We each now have two ideas. Thus ideas propagate and the number of ideas multiplies."
Hwa A. Lim, Opening Remark, Bioinformatics & Genome Research, Inner Harbor Hotel, 1996.

1   The Dynamic High Tech World

"Ships in harbor are safe, but this is not what ships are built for".[1] This saying applies extremely well to business. Entrepreneurs and venture capitalists go into business using their opportunity cost and capital in hopes of earning a fair return on their investment. But not all businesses are sure bets. In all businesses, there is an intrinsic element of risk. In most cases, the risks are calculated, weighed and then committed in a way to the owners' advantage. The rules may be changed during the operations of the business, either by exploiting the "weak" or by stretching the legal limits. When the stake is high, the return-capital ratio is also expected to be correspondingly higher. Otherwise, it is normally just not worth wagering the risk.

In recent years, we have seen too often that a seemingly healthy company suddenly has to undergo an abrupt metamorphosis, either to stay afloat or to stay profitable in this cutthroat competitive marketplace. The most recent example is Apple Computer during the first half of the 1990s. Apple Computer has been an excellent company in coming out with new ideas, but by the time the ideas get to the market, the effect of products becomes diffused. Of late, with corporate reengineering and the introduction of a new line of "more brain, less brawn" line of products including *i*Macs, Power Macintosh G3, and PowerBook G3, Apple Computer has reemerged out of the woods and regain their market position.

This rapid change in market thrust is true not only in the computer world, but is also true for biotechnology, pharmaceutical and healthcare sectors. In order to compete and advance in this competitive marketplace, the chances of succeeding will be enhanced if a company stays focused in the areas it is good at and in the areas in which it can effectively compete.

---

[1] Peter Burwash, *The Key to Great Leadership*, (Torchlight, New York, 1995).

2   Types Of Focused Companies

A few examples of successful focused companies are in order at this point. There are at least three broad categories of these types of companies: operationally efficient, service-oriented and product-driven companies. Let us look at each in turn.

2.1   Operationally Efficient Companies

There are quite a number of operationally efficient companies, including Wal-Mart, McDonald, Price-Costco, etc.

A classic example of an operationally inefficient company is the failed boutique business of Aamco. In the mid-1970s, Aamco, a leading wholesaler of auto transmission parts, decided to diversify by opening up a chain of gift boutiques called Plum Tree.[2] The plan called for the same system of centralized buying and centralized reselling to the franchise. By centralizing, sales of each item could be monitored and the system kept the inventory low while still maintaining the levels of stock needed to support the sales. The program went off to a great start but soon declined. The system automatically ordered larger quantities of most frequently used transmission parts in order to maintain the flow of stock. The same system also reordered more gift items customers purchased most frequently. But transmission parts and boutiques were not the same: mechanics would usually repeat the same sort of repair day after day; on the other hand, customers would rarely buy the same gifts for their friends time after time. Presented with what they already had, customers stopped coming to the boutique, and the boutique business failed miserably.

In contrast, the success of Wal-Mart is transparent.[3] When Sam Walton opened the first Wal-Mart store in 1962, it was the beginning of an American success story that no one could have predicted. A small-town merchant who had operated variety stores in Arkansas and Missouri, Walton was convinced that consumers would flock to a discount store with a wide array of merchandize and friendly service. At its core, Wal-Mart is a place where prices are low and customer service is high. Because Wal-Mart carefully controls expenses to maintain its low price structure, customers do not have to wait for a sale to realize savings. Backing up the hometown flavor of a Wal-Mart store is the industry's most efficient and sophisticated distribution system. The system allows each store to personalize the merchandise assortment to match the community's needs.

---

[2] Arnold Penzias, *Ideas and Information*, (Touchstone, New York, 1989).
[3] Linda E. Swayne, and Peter M. Ginter, *Selected Cases in Strategic Marketing*, (Simon & Schuster Custom Publishing, Needham Heights, MA, 1995).

2.2   Service-oriented Companies

The service sector now provides almost four-fifths of all jobs in the U.S., making it our largest job provider, our major producer of wealth. According to a study conducted by The Forum Corporation, a global leader in developing workplace learning solutions and in helping companies build lasting brand loyalty, 70% of those customers who change companies do so because of poor service. As we move into the new millennium, it is interesting to see how corporations become increasingly aware of the fact that it is a lot cheaper to keep a customer than to spend more money on advertising to gain a new one.

There are countless successful service-oriented companies, including Nordstrom, Home Depot, and Federal Express. The "Pony Express" will provide a good case study. In the mid twelfth century, the Pony Express of the Khan Dynasty could cover in just a day and a night what would normally have been a ten-day journey. Using the most powerful horses, the riders, wearing wide belts with many bells that announced their approach to the horse post houses. At twenty-five intervals along the route, a fresh horse would be saddled and waiting for the rider, who would press on, towards the Palace of Genghis Khan at Chard.

In more recent times, the short-lived Pony Express in the Wild West United States amounted to little more than a romantic footnote to the history of American frontier. During the brief period April 1860 and October 1861, a small band of brave and resourceful men provided a unique high-speed service. By changing horses every seventy-five miles, riding day and night, rain or shine, and evading hostile highway bandits, historical record shows that they could cover 2,000 miles between St. Joseph (Missouri) and Sacramento (California) in less than ten days. This was very impressive in those days. But they were soon driven out of business by a single strand of copper telegraph wire.

Today, Frederick Wallace Smith, an inventive entrepreneur, created a similar, very successful, counterpart — the Federal Express Company. The company provides fast delivery of documents while Western Union is cutting back. This example shows that speed is not the only factor. The telegraph is a technology that is alphanumeric, i.e., it works only with numbers, letters, and other keyboard characters. Federal Express, on the other hand, can deliver anything at all within legal limits. While efficient use of technology appears to have carried Federal Express to a clear lead over the telegraph, winners cannot rest on their laurels in any high technology world. A new machine, a desktop copier that can send graphical replicas of letter-sized documents, can now offer telegraphic speed document transmission. And we cannot ignore the effect of the Internet and the World Wide Web (WWW) on the industry, and the new regulation recognizing faxed signatures.

This is a good example of why in all areas of business, changes brought about by technology will leave some of today's leader scrambling to catch up to more nimble competitors. Success usually comes to those who apply technology to their advantage.

2.3   Product-driven Companies

The United States is still by far the leading nation in computer technology and biotechnology. Most of these companies have their respective core products. Examples of these computer companies include Microsoft, Intel, Apple, etc. A few examples of product-driven biotechnology companies and their known product lines are: Affymetrix (DNA chip), Advanced Cell Technology (stem cells), Geron (stem cells and cell lines), Chiron (various products), Genentech (several products), Merck (many products). A noteworthy fact from the list is that as we go down the list, the number of products associated with the company increases, i.e., the company is more diversified.

In 1971, Intel introduced the world's first microprocessor, which sparked a personal computer (PC) revolution that would change the world. About 75% of PC in use around the world today are based on Intel-architecture microprocessor. Today, Intel supplies the PC industry with computer chips, boards, systems, software that are components of the most popular computer architecture. They help create advanced and high performance computing systems for PC users. The product line includes: 1) microprocessor chips, especially the ubiquitous Pentium chips; 2) networking and communication products; 3) semiconductor products such as reprogrammable memory for cellular phones, heavy machinery parts. They supply not only to the computer industry, but they also hold market shares in: 1) manufacturers of computers and peripherals; 2) other manufacturers like makers of automobiles and a wide range of industrial and telecommunication equipment; 3) PC users who buy Intel enhancements for business communication, networking, etc; 4) scientists and engineers working on sophisticated computational problems. Through its Research Council, it champions university research to drive innovations and the advancement of technology in the computer industry so that there is a continuous stream of new product line. The timely response and the professionalism with which Intel handled the 6th decimal rounding error of Pentium chips in 1995 is yet another reason why the company is so successful.

3   Biology And Biology-Related Disciplines — Place In Science

The prevailing view is that biology has jostled to the center stage at the expense of the physical sciences. This is a fallacy.

As we close the twentieth century and begin the new millennium, if we look back on the twentieth century, we can conclude that its first half was shaped by the physical sciences but its second by biology.[4] The first half brought the revolution in transportation, communication, mass production technology and the beginning of the computer age. It also, pleasantly or unpleasantly enough, brought in the nuclear

---

[4] G.J.V. Nossal, and Ross L. Coppel, *Reshaping Life: Key issues in genetic engineering*, (Bricknell Nominees Pty. Ltd, Melbourne, Australia, 1990).

weapons and the irreversible change in the nature of warfare and environment. All of these changes and many more rested on physics and chemistry. Biology was also stirring over those decades. The development of vaccines and antibiotics, early harbingers of the green revolution are all proud achievements. Yet the public's preoccupation with the physical sciences and technologies, and the immense upheavals in the human condition which these brought, meant that biology and medicine could only move to the center stage somewhat later. Moreover, the intricacies of living structures are such that their deepest secrets could only be revealed after the physical sciences had produced the tools — electron microscopes, radioisotopes, chemical analyzers, laser, DNA sequencers, and rather importantly, the computer — required for probing studies. Accordingly, it is only now that the fruits of biology have jostled their way to the front pages.

Computer technology, especially computational power, networking and storage capacity, has advanced to a stage that it is capable of handling some of the current challenges posed by biology. This makes it possible to handle the vast amount of biological and biology-related data that have been and are being generated as a result of genome projects, the dissemination of the data on the Internet, and provide the teraflop compute power required for complicated analyses to penetrate the deepest secrets of biology and biology-related disciplines. Consequently, the time was prime for a marriage made in heaven between biology and biology-related disciplines, and computer science in the late 1980s. Thus the birth of the eclectic bioinformatics in the 1990s.

Bioinformatics will continue to be ubiquitous and play a critical role way into the twenty-first century as biotechnology (BT), information technology (IT) and nanotechnology (NT) are converging to form a triumvirate of BIN convergence.

4   Pharmaceutical Companies As Product-Driven Companies

It is not coincidental that the ten largest pharmaceutical colossi are those which are most visionary: Glaxo Wellcome (UK, sales: $11.68M; capitalization: $45.17M), Merck & Co., (US, sales: $10.96M; capitalization: $83.92M), Novartis (Switzerland, sales: $10.94M; capitalization: $78.87M), Hoechst (Germany, sales: $9.42M; capitalization: $18.64M), Roche Holding (Switzerland, sales: $7.82M; capitalization: $67.21M), Bristol-Myers Squibb (US, sales: $7.81M; capitalization: $44.67M), Pfizer (US, sales: $7.07M; capitalization: $43.25M), SmithKline Beecham (UK, sales: $6.6M; capitalization: $29.61M), Johnson & Johnson (US, sales: $6.3M; capitalization: $63.17M), Pharmacia & Upjohn (UK, sales: $6.26M; capitalization: $22.14M).[5] We intentionally post this slightly outdated 1998 chart for comparison with a more recent data. It should be noted how swiftly these companies jostle for relative positions. For example, AstraZeneca, which was

---

[5] *Wall Street Journal*, March 8, 1996, Friday, B7.

number two in the Y2K listing, comes in first in Y2001, having increased its market capitalization from $57 billion to $85 billion. Amgen slipped from number one last year ($75 billion) to second place ($68 billion).   Big pharmaceutical companies grew appreciably as well. Pfizer, which held second place in the Y2K listing with a market capitalization of $125 billion, tops the Y2001 pharmaceutical company chart at $264 billion. Merck & Co. fell from first in Y2K ($140 billion) to third ($166 billion) in Y2001.

Table 1. Top 10 pharmaceutical companies (Source, Telescan Prosearch, No. 5, May 25, 2001).

Company	Market Cap ($billion)
Pfizer Inc	264.3
GlaxoSmithKline plc	167.4
Merck & Co. Inc	166.5
Johnson & Johnson	135.9
Bristol-Myers Squibb Co.	105.3
Eli Lilly & Co.	94.4
American Home Products Corp.	80.2
Abbott Laboratories	79.6
Pharmacia Corp.	63.2
Schering-Plough Corp.	60.9

Table 2. Top 50 biotechnology companies (Source: Telescan Prosearh, No. 5, May 25, 2001).

Position	Company	Position	Company
1	AstraZeneca plc	26	OSI Pharmaceuticals Inc.
2	Amgen Inc.	27	Aviron
3	Genentech Inc.	28	Myriad Genetics Inc.
4	Serono SA	29	Techne Corp
5	Genzyme Corp.	30	QLT Inc.
6	Chiron Corp.	31	Tanox Inc.
7	Celltech Group plc	32	Regeneron Pharmaceuticals
8	Biogen Inc.	33	Cell Therapeutics Inc.
9	IDEC Pharmaceuticals Corp.	34	Albany Molecular Research Inc.
10	Immunex Corp.	35	Titan Pharmaceuticals Inc.
11	MedImmune, Inc.	36	CV Therapeutics Inc.
12	Human Genome Sciences, Inc.	37	Scios Inc.
13	Applera Corp.	38	Genencor International Inc.
14	Gilead Sciences Inc.	39	Xoma Ltd.
15	Abgenix Inc.	40	Trimeris Inc.
16	ImClone Systems Inc	41	Cubist Pharmaceuticals Inc.
17	Icos Corp.	42	Immunomedics Inc.
18	Cephalon Inc	43	Enzo Biochem Inc.
19	Enzon Inc.	44	Ligand Pharmaceuticals Inc.
20	Applera Corp	45	Exelixis Inc.
21	Sepracor Inc.	46	Corixa Corp
22	Vertex Pharmaceuticals Inc.	47	Amylin Pharmaceuticals Inc.
23	Celgene Corp.	48	Arena Pharmaceuticals Inc.
24	Medarex Inc.	49	Biotechnology General Corp.
25	Sicor Inc	50	Transkaryotic Therapies, Inc.

Each of these companies has a well-defined mission and each has various core products. For example, the theme of Bristol-Myers-Squibb is to be in business whose products help to enhance and extend human life. Pfizer has drug of the decade — and Merck has Propecia to treat hair loss. Similarly, Bayer is a diversified, international chemical and pharmaceutical company. Among the best-known names are undoubtedly the drug of the century — Aspirin, and Alka-Seltzer. A good 50% of the company's sales are attributed to products developed in its own research laboratories in the last 15 years.

In the aftermath of the World Trade Center Twin Tower incident of September 11, 2001, we anticipate that companies, such as Eli Lilly, which produce medicines for anthrax and other bio agents to move up the rank.

5   The Economics Of Drug Discovery

Let us digress a little and look at drug discovery as a case example before we return to bioinformatics. Of about 5,000–10,000 compounds studied, only one drug gets onto the market. Even if the compounds make it to clinical trials, only 1 in 5 gets to the market. In the discovery phase, each drug costs about $156 million. The FDA processes I, II & III cost another $75 million. This brings the total to about $231 million (1994 figure. The 2000 figure is about $400 million) for each drug put onto the market for consumers.[6] The time required for approval is equally long, as shown in the figure.[7] These phases constitute parts of the manufacturing, regulatory and cost factors of drug discovery.

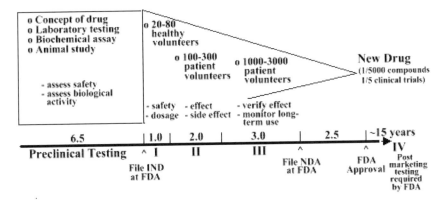

Figure 1. The long and expensive procedure for gaining FDA approval of a pharmaceutical product.

[6] H. Burkholz, *The FDA Follies*, (New York: Basic Books, 1994).
[7] Hwa A. Lim, "Bioinformatics and Cheminformatics in the Drug Discovery Cycle", *Lecture Notes In computer Science, Bioinformatics*, Ed. R. Hofestdaet, T. Lengauer, M. Loffler, and D. Schomburg, Springer, Heidelberg, 1997, pp. 30–43.

By the time a drug gets on to the shelf of a pharmacy, it will have been 15 years since preclinical testing, and about $400–$700 million will have been spent.

6 Future Pharmaceutical Discoveries

Traditionally, large pharmaceutical companies have a cautious, mostly chemistry- and pharmacology-based approach to the discovery and preclinical development program and therefore, do not yet have expertise in-house to generate, evaluate and manage genetic data. Accordingly, many purchase licenses to tap into commercially generated and managed genomic databases. Until large companies build up their expertise and experience in the direction of genomics and functional genomics, there is a target of opportunity for smaller, more adept and savvy companies to fill the gap. Undoubtedly, when larger companies become more sophisticated and comfortable with current technology, the next generation technology-based opportunities will be in the hands of the nimbler, more flexible smaller companies to again cycle and leverage for revenues from their larger counterparts. To constantly stay one step ahead is a winning and survival game plan for small companies.

The accepted consensus is that future pharmaceutical discoveries will stem from biological and biology-related information. Major pharmaceutical companies develop new core products. These companies are either slower in response, do not want to develop full scale sequencing expertise and maintain huge proprietary database in-house (except those pertinent to their product lines), or they do not want to commit the financial resources for such purposes.

But to stay competitive, these companies have to respond quickly and do need access to comprehensive genetic, biological and biology-related information for timely and accurate decision making. Thus, since 1996, we have seen bioinformatics divisions springing up in all major pharmaceutical companies to either partake in this exciting new area, or to partner with smaller, more nimble research institutions or companies. Because of this, smaller companies are constantly being formed to take advantage of the window of opportunities, some of which survive, and many flounder. In general, these small companies try to develop technologies — platform hardware (for example, AbMetrix, Affymetrix and Axon) and software (for example, InforMax and GeneGo), produce a database of some form and then generate revenue from the database by either selling subscriptions to the database, or selling information derived from the database (for example, Celera, deCode and Incyte Genomics).

The current practice to cut down on research and development (R&D) costs favors outsourcing or farming out certain aspects of R&D to smaller companies, universities, and even abroad, for example China and India, where the talent pool is excellent and the labor cost is still very attractive.

Other options include forming strategic alliances and forming consortia such as the Single Nucleotide Polymorphism Consortium.

7   The Rise Of Bioinformatics

On May 3rd, 1996, a BioMASS panel, in conjunction with the BioSCIENCE Career Fair and sponsored by American Association for the Advancement of Science (AAAS), was held at Stanford University Medical Center. The title of the panel discussion was "Ph.D. Promises: The future of the Bioscientist in Academia and Industry".[8]   There were six panelists: P. Gardner, M.D. (VP Research, Alza Corporation & Associate Professor at Stanford University); R. Grymes, Ph.D. (Manager, Outreach Program, NASA); H.A. Lim (HAL), Ph.D. (then Director, Bioinformatics, Hyseq); J. MacQuitty, Ph.D., M.B.A. (CEO, GenPharm International); T. Schall, Ph.D., (Senior Scientist, DNAX Research Institute); J. Shaw, Ph.D. (Founder, The Stable Network); and R. Simoni, Ph.D. (Professor at Stanford and panel discussion moderator). HAL was the bioinformatics panelist. He was rather taken unexpectedly that so many Ph.D. students, post-doctoral fellows and young audience asked him what bioinformatics was.   There were even individuals who commented that the bioinformatics community must be extremely small, so small that they had not heard of it until the panel discussion.

In the following few weeks, the whole matter took a drastic swing, mainly due to AAAS and the Science magazine. Science magazine published a series of articles on the subject and this brought the area a lot of attention and publicity.[9]

Bioinformatics is a rather young discipline, bridging the life and computer sciences.   The need for this interdisciplinary approach to handle biological knowledge is not insignificant. It underscores the radical changes in quantitative as well as qualitative terms that the biosciences have been seeing in the last two decades or so. The need implies:

❑   our knowledge of biology and biology-related disciplines has exploded in such a way that we need powerful tools to organize the knowledge itself;

❑   the questions we are asking of biological and biologically-related systems and processes today are getting more sophisticated and complex so that we cannot hope to find answers within the confines of unaided human brains alone.

The current functional definition of bioinformatics is "the study of information content and information flow in biological and biology-related systems and processes." It has evolved to serve as a bridge between the observations (data) in diverse biologically-related disciplines, the derivations of the understanding (information) about how the systems or processes function, and subsequently the application (knowledge). In more pragmatic terms of a disease, bioinformatics is the creation of a registry, unraveling of the dysfunction, and the subsequent search for a therapy.

---

[8] The meeting was announced in *Science*, May Issue, 1996.
[9] *Science*, July Issue, 1996.

7.1   The Beginning

The interest in using computers to solve challenging biological problems started in the 1970s, primarily at Los Alamos National Laboratory, and pioneered by Charles DeLisi and George Bell.[10] Among the team of scientists were Michael Waterman, Temple Smith, Minoru Kanehisa, Walter Goad, Paul Stein and Gian Carlo Rota.

Photo 1. A group photo of a subset of the 120 participants of the First Bioinformatics Conference. Notables: Alexander Bayev (Chairman, USSR Human Genome Project), George Bell (Acting Director, Los Alamos National Laboratory Genome Project), Charles Cantor (US Department of Energy Human Genome Program Principal Scientist), Anthony Carrano (Director, Lawrence Livermore Genome Project), Charles DeLisi (with Senator Domenici of New Mexico, an early proponent of the Greatest Wellness Project, which would be later called the Human Genome Project), Michel Durand (French Attache), a 6-member delegate from Japan (including RIKEN), and a huge 14-member Soviet delegate, something very unusual during the then Cold War era. (April 1990, Tallahassee, Florida).

In the late 1980s, following the pioneering work of DeLisi and Bell, Hwa A. Lim (HAL) realized the significance of marrying computer science and biology, he tried to come up with a captivating word for the area. In 1988, HAL first coined the word "bio-informatique", with a little bit of French flavor, to denote the subject area. A year later, after making a few surveys, the preparation process for convening the first international conference was begun to bring awareness to the community. The process started by electronic mail since interested participants were most likely computer-savvy. During those days, the smtp mailers on VAX

---

[10] *Computers and DNA*, G.I. Bell and T.G. Marr (eds.), (Addison-Wesley Publishing Co., Redwood City, 1990).

machines (the good hey days of VMS operating system) were not very forgiving. Email frequently bounced because of the hyphen in "bio-informatique". In an attempt to overcome the problem, the word was changed to "bio/informatique", allowing for the fact that the "bio" was just a prefix that could be substituted. The change did not help. Similar email problems persisted, which prompted HAL to drop the "/" completely and the word assumed the form "Bioinformatique". Two conference secretariats commented that the word was a little too French. It was then appropriately modified to "Bioinformatics", in conformance with subjects like "optics", "statistics", "mathematics"... .

Photo 2. Hal, Dr. Charles DeLisi, Professor Charles Cantor, Academician Bayev, and Professor Joseph E. Lannutti at a press conference, Florida Press Club. Hal, who coined the word "bioinformatics" in 1987, was the Chairperson of this very first international conference on bioinformatics. Dr. Charles DeLisi, one of the early proponents of the Human Genome Project, was Dean of College of Engineering, Boston University. Professor Charles Cantor was US DoE Human Genome Project Principal Scientist (present: CSO, Sequanom, San Diego). Late Academician Bayev (1904–1994) was Chairman of the USSR Human Genome Project. Late Professor Lannutti (1926–1998) was initiator of the Supercomputer Computations Research Institute, host of the conference. (April 1990, Tallahassee, Florida).

The very first ever international conference on bioinformatics was chaired and organized by the author (HAL), with help from Professor Charles R. Cantor, then Chairman of the College of Physicians & Surgeons at Columbia University; and late Professor Joseph E. Lannutti, then Director of Supercomputer Computations Research Institute at Florida State University. The first conference was held at the Florida State Conference Center, Tallahassee, from April 10 through 13, 1990. Notable among the participants were: Charles DeLisi (Dean, College of

Engineering, Boston University), Charles Cantor (then Director, Lawrence Berkeley National Laboratory Genome Program), George Bell (then Acting Director, Los Alamos National Laboratory Genome Program), Anthony Carrano (Director, Lawrence Livermore National Laboratory Genome Program), Temple Smith (then Director at Dana Farber Cancer Center of Harvard Medical School), late Alexandar Bayev (then Chairman, USSR Genome Program), Boris Kaloshin (USSR Dept. of Sc. & Tech), M. Durand (French Attaché), N. Shimizu (Head, Department of Molecular Biology, Keio University School of Medicine), I. Endo (RIKEN, Japan), N. Nordén (Sweden), and others (120 participants in total). The conference was funded by U.S. Department of Energy, and Florida Technology Research and Development Authority, Thinking Machines Corp., Digital Equipment Corp., CRAY Research Inc. A proceeding volume was compiled.[11] Note that the sponsors were primarily federal and state agencies, and general-purpose computer companies. Note also the huge 14-member USSR delegate headed by Academician Alexander Bayev, a rather unusual phenomenon during the Cold War era. Also present was a 6-member delegate from RIKEN of Japan.

7.2   Subsequent Years

The conference series continued and The Second International Conference on Bioinformatics, Supercomputing and Complex Genome Analysis took place at the TradeWinds Hotel, St. Petersburg Beach, Florida, from June 4 through 7, 1992. This conference was originally planned for St. Petersburg (Leningrad), USSR. The breakup of the Former Soviet Union forced HAL to come up with an alternative plan in less than seven months. St. Petersburg Beach was chosen partly because of the location, and partly because of its name (just like St. Petersburg of Russia). Participants from more than thirteen countries worldwide took part. A joke that circulated during and after the conference was that some attendees of the conference mistakenly went to St. Petersburg of Russia. The conference was partially funded by Intel Corp., MasPar Computer Corp., World Scientific Publishing Co., Silicon Graphics Corp., Florida Technological Research & Development Authority, U.S. Department of Energy, U.S. National Science Foundation. A second proceeding volume was edited to bring the subject area to the then relatively small community.[12] Notable among the participants was a 12-member delegate from Genethon, France. Note the participation of federal and state agencies, special-purpose computer companies and publishing houses.

---

[11] *Electrophoresis, Supercomputing and The Human Genome*, Charles R. Cantor and Hwa A. Lim (eds.), (World Scientific Publishing Co., New Jersey, 1991).
[12] *Bioinformatics, Supercomputing and Complex Genome Analysis*, Hwa A. Lim, James W. Fickett, Charles R. Cantor and Robert J. Robbins (eds.), (World Scientific Publishing Co., New Jersey, 1993).

Photo 3. A group photo of a subset of the 150 participants of the Second Bioinformatics Conference. Notables: Christian Burks (Los Alamos National Lab., present: VP, Exelixis), Charles Cantor (Boston University, present: CSO, Sequenom Inc.), Rade Drmanac (Argonne National Lab, present: CSO, Hyseq, Inc.), Chris Fields (NCGR, present: VP, Applied Biosystems), Pavel Pevzner (University of Southern California, present: Ronald R. Taylor Chair Professor of Computer Science, University of California at San Diego), Temple Smith (Smith-Waterman Algorithm, Harvard University, present: Professor, Boston University), Robert Robbins (Johns Hopkins University, present: VP, Fred Hutchinson Cancer Center), Chris Sander (EMBL, present: Chief Information Technologist, Millennium), David Searles (University of Pennsylvania, present: VP, SmithKline Beecham), Mark Adams, Phil Green (known for his Phred and Phrap, inducted to the U.S. National Academy of Science, 2001), Edward Uberbacher (Grails), a delegate from Korea Institute of Science and Technology, a delegate from Japan, a delegate from Russia, and a huge 12-member delegate from France (including Genethon). (June 1992, St. Petersburg Beach, Florida).

The third conference, The Third International Conference on Bioinformatics & Genome Research, took place at Augustus Turnbull III Florida State Conference Center, Tallahassee, Florida, from June 1 through 4, 1994. It was partially funded by Compugen Ltd., Eli Lilly and Company, MasPar Computer Corp., World Scientific Publishing Co., Pergamon Press, U.S. Department of Energy, U.S. National Science Foundation, U.S. National Institutes of Health, International Science Foundation. The proceedings were gathered in a volume.[13] A noteworthy point is that the sponsors were federal, state and international agencies, special-purpose computer companies, pharmaceutical companies and publishing houses.

---

[13] *Bioinformatics & Genome Research*, Hwa A. Lim, and Charles R. Cantor (eds.), (World Scientific Publishing Co., New Jersey, 1995).

Photo 4. A group photo of a subset of the 130 participants of the Third Bioinformatics Conference. Notable at the meeting was a panel discussion on sequence databases (Philip Bucher, James Fickett, Murray Smigel, Andrzej Konopka, Douglas Smith, Philippe Rigault, and Tom Slezak), and another panel discussion on technology transfer (Mike Devine, Assoc. VP of Research, Florida State University, Richard Hogg, Provost & VP of Academic Affairs, Florida A&M University, James Ludwig, Eli Lilly & Company, Richard MacDonald, Biosym Technologies). (June 1994, Tallahassee, Florida).

Photo 5. Charles Cantor, Hal, and Robert Robbins. Hal initiated the conference series with the support of these two strong proponents. (June 1994, Augustine Turnbull III Conference Center, Tallahassee, Florida).

7.3   Bioinformatics Conference Going Commercial

Soon after the Third Conference, federal and state agencies went through a period of downsizing and streamlining. Cambridge Healthtech Institute (CHI) and HAL decided that the conference series should go commercial and be self-supporting. Initial negotiations for CHI to take over the biennial conference series started soon after the Conference in 1994. Judging from prior successes of the conference and the rising popularity of the subject area, CHI decided to make the conference series an annual event.

A noteworthy point is that even though the number of participants had been intentionally limited to less than 150 in the first three conferences, the number climbed steadily to 350 in the Fifth Conference, a clear indicator and good measure of the increasing popularity of the subject area. The Opening Ceremony of the Tenth Anniversary of the Conference Series, held at the Fairmont Hotel, San Francisco, on Father's Day Sunday evening (June 17, 2001) and chaired by HAL, was attended by 700 participants. Through the conference, the number of attendance remained the same, and there were 50 corporate sponsors and exhibitors.

8   Genomic Companies As Service-Oriented Companies

Many genomics companies have unique, high-throughput, cost effective technology to do sequencing and to collect biological or biology-related data. But, as shown in the table, data is not commercializable, but information is. This leads naturally to a conceptual flowchart of biological or biology-related data, as depicted in the figure. Or in terms of physical design, the corresponding databases as illustrated.

Table 3. A table to compare and contrast data and information.

Data Are	Information is
Stored fact	Presented fact
Inactive (they exist)	Active (enables doing)
Technology-based	Business-based
Gathered from various sources	Transformed from data

Biodata
⇓
Bioinformation
⇓
Bioknowledge
↓
Next generation genomics

Figure 2. A flowchart to show the paradigm of biodata. The prefix "bio" can very well be substituted for "chem", "health" or any biology-related disciplines.

Database
⇓
Infobase
⇓
Knowledgebase
↓
New order of discovery

Figure 3. The paradigm of biodata presented in a more physical form, i.e., as various databases.

In order to maintain such a scheme, a possible strategic plan is outlined in Figure 4.[14]

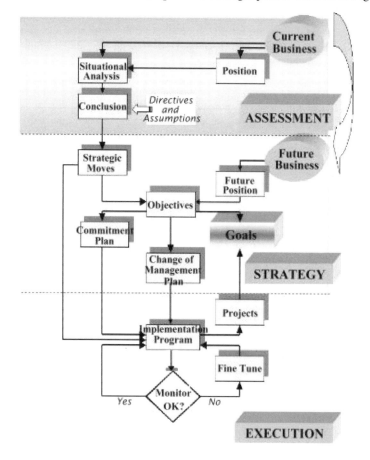

Figure 4. A chart showing the flow and planning of information, in particular, bioinformation. The sequence is: assessment, strategy and execution.

---

[14] B.H. Boar, *The Art of Strategic Planning for Information Technology*, (John Wiley & Sons, Inc., New York, 1993).

8.1   Bioinformatics — Mission and Goals

Typically, some of the goals and missions of a bioinformatics division might include, among many other possibilities and combinations:

- ❑   To perform decision making by centering around intelligent interpretation of existing genetic information, for example, microarray profiles;
- ❑   To identify what information may yet be needed, define what may yet be done, for example, microarray data;
- ❑   To enable corporate partners to accelerate identification of genetic information for gene-based, small molecule-based, pathway-based drug targets;
- ❑   To validate this selection through sequencing-derived drug-genome interaction studies, for example, proteomics;
- ❑   To package this information for efficient decision making throughout a partner's product development cycle.

The goals and mission may vary in accordance with needs, and very much driven by applications and clients.

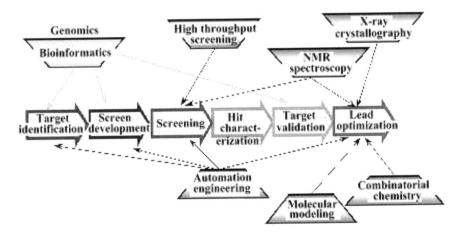

Figure 5. The drug discovery process involves six major steps: target identification, screen development, screening, hit characterization, target validation, and lead optimization. Other supporting technologies can help in accelerating each of the processes. (Figure adapted from Dr. Don Halbert, Abbot Laboratories).

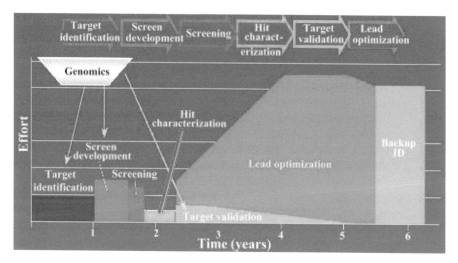

Figure 6. The time involved in each step of the drug discovery process is a very instructive piece of information. From a financial perspective, it is the effort that is involved is more indicative of the risk and cost involved. For example, lead optimization is a big bottleneck in drug discovery. (Figure adapted from Dr. Don Halbert, Abbot Laboratories).

A very seldom cited, yet very important, advantage of using informatics is that information is disciplinary-blind, that is, information can link disciplines that are seemingly unrelated to researchers. For example, it may seem that the pharmaceutical sector and the agriscience sector are not related. But if we compare and contrast drugs and seeds, certain similarities between the two sectors begin to reveal themselves.

Table 4. A comparison and contrast of the pharmaceutical and agriscience sectors.

Pharma	Farmer
Drugs are:	Seeds are:
❑ Chemicals	❑ Organisms
❑ Raw materials for screening	❑ Raw materials for breeding
❑ Sorted by chemistry	❑ Sorted by testing
New drug are:	New seeds:
❑ High margin	❑ Low margin
❑ Low volume	❑ High volume
Genes are	Genes
❑ Linked to diseases	❑ Increase yield

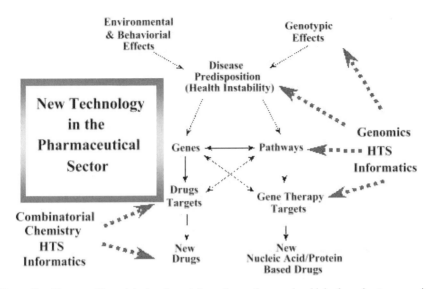

Figure 7. How combinatorial chemistry informatics and genomics high throughput sequencing informatics have shifted the discovery paradigm in the pharmaceutical industry. (Figure adapted from Benjamin Bowen, Pioneer Hi-Bred International).

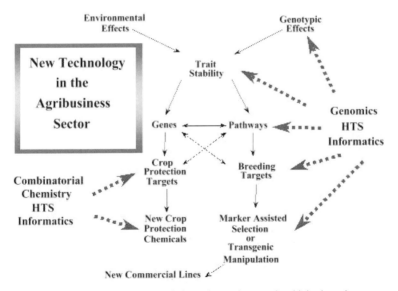

Figure 8. How combinatorial chemistry informatics and genomics high throughput sequencing informatics have shifted the discovery paradigm in the agribusiness industry. Note the similarity between this and the corresponding one for the pharmaceutical sector. (Figure adapted from Benjamin Bowen, Pioneer Hi-Bred International).

8.2   Bioinfobahn

Since bioinformatics is a marriage of computer and biology and biology-related disciplines, it is not surprising that it is well kept abreast with advances in computer technology, in particular, the Internet technology.

The ubiquitous computer network sprang from the mind of a research psychologist — J.C.R. Licklider — working for the Pentagon in the early 1960s. He envisioned what he called an Intergalactic Computer Network.[15] The Pentagon was interested in part because it needed a communications system able to withstand nuclear attacks. A decentralized computer network would satisfy that requirement. Scientists were also intrigued by the idea of making telecommunications systems more efficient by breaking apart messages into "packets" of information that could be routed through different paths around the nation and reassembled by computers at the destination.

Although the technology had obvious commercial applications, the two telecommunications leaders at the time — IBM and AT&T — were not interested in participating because computers were not cheap enough to make such a network affordable.   Instead, Pentagon's Advanced Research Projects Agency and the civilian National Science Foundation nurtured the Internet for more than twenty-five years until 1994, when sufficient number of companies saw the potential of the Internet.

So, the Internet came into being about twenty-five years ago as a successor to ARPANET, a U.S. military network disguised to provide networking capabilities with a high redundancy. The principle behind has remained unchanged and has proven very powerful: to have every computer potentially talk to each other, regardless of what platform, what network path the communication actually takes.

By going cybernized, bioinformation and bioknowledge disseminate at a much timely rate. There are countless electronic publications on the Internet. Genome projects have produced a massive amount of data, resulting in over 400 individual databases at companies and institutions around the world.   These publications appear in all formats — regular ASCII text, postscript, hypertext, Java and other derivations therefrom — supported by the Internet and the World Wide Web.

It is also fair to say that the Human Genome Project would not have progressed as smoothly as it had and completed ahead of the original schedule had it not been because the Internet provided means for disseminating and sharing information in a timely manner.

---

[15] Katie Hafner, and Matthew Lyon, *Where Wizards Stay Up Late: The Origin of the Internet*, (Simon & Schuster, New York, 1996).

9 A New Paradigm Of Discovery

Judging from the current prevailing trends in federal spending, healthcare and social reforms, it is very likely that bioinformation, disease database maintenance, intelligent software for extracting knowledge from these databases, will play a major role in the future of biomedicine. Disease diagnostics, prognostics and therapeutics will rely more on biodata, and bioinformation and bioknowledge derived therefrom, than on guesswork, chemistry or pharmacology.

Up until recently, successful therapeutic approaches target initial causative agents such as infectious microorganisms, or empirically target a single step of a multi-step complex disease process. Current therapeutic intervention, and therefore drug discovery efforts, are aimed at the molecular events of the disease process itself.

Conventional approaches focus on identifying, isolating, purifying targets; determining target sequence and three dimensional structures; applying rational drug design, molecular modeling for docking active sites; synthesizing, screening and evaluating chemical compounds for clinical test and FDA approval.

Traditional methods in molecular biology normally work on a "one gene in one experiment" basis, leading to a very limited throughput, and a global picture of the gene function is hard to obtain. In the past several years, the biochip technology has created a lot of interest and it promises to monitor the entire genome on a single chip so that researchers can have a better picture of the interactions among the thousands of genes in the genome simultaneously.

A biochip technological challenge involves achieving all the market-oriented parameters at a cost that supports a commercially acceptable price. Currently, DNA chips cost between $100 and $450 each. In a tight managed-care marketplace, that places a premium on technologies that can either show quick savings or more-efficient results. When the unit price for biochip reaches about $10, the biochips will be used more widely, and the biochip will be a commodity.

In healthcare, bioinformatics raises a number of future perspectives:

❑ If a target functions in a biological pathway, are there any undesirable effects from interactions of this pathway with associated pathways.
❑ Are there non-active sites that may yield greater specificity and this reduces side effects arising from interactions with structurally and evolutionarily related targets.
❑ The specificity, selectivity and efficacy of small molecules.
❑ Time course of a disease process, i.e., a more dynamical study.
❑ Others.

Table 5. Healthcare in pre-genomic and post-genomic eras. (HAL thanks Dr. David Wang, Motorola BioChip Systems.

Pre-Genomic Era	Post-Genomic Era
❑ Disease description	❑ Disease mechanism
❑ Uniform disease	❑ Disease heterogeneity
❑ Patient homogeneity	Individual variability — SNP
❑ Universal Rx strategy	❑ Patient risk profiling — Pharmacogenomics and targeted care

Though data and information are important for understanding diseases and other healthcare issues, for any data and information to be reliably useful, careful statistical analyses with sufficient samples are necessary. Otherwise, the derived conclusion and inferences may not be a true reflection of the studies. And in each regime of studies, a different approach may be employed. For example, biochips are an appropriate platform for studying pharmacogenomics.

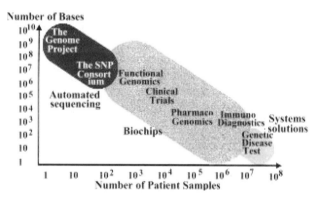

Figure 9. Samples sizes required for different purposes can be different. For example, the Human Genome Project produces 3 billion (3×10^9) bases from the DNA of a few individuals. There is a nucleotide difference in every 1,000 nucleotides. Thus the SNP project has about 10^6 bases involving a few scores of individuals. (Figure adapted from Dr. David Wang, Motorola BioChip Systems).

10   Historical Dé Jàvu?

The crux of hard reality is that if one has no vision and is too inflexible, one is permanently left behind.

The Agrarian Revolution started in Asia, lasted for tens of thousands of years, and ended in Europe. The Industrial Revolution started in Europe, lasted for about two centuries (1760s–1950s), of which 90 years (1860s–1950s) were in United States. History has taught us two important lessons:

❑   With each successive revolution, the time span of the revolution is much shorter than that of its predecessor.

❑   A revolution that initiated in one region of the world does not guarantee that the revolution will end in that part of the world. Indeed, historically, a revolution commenced in one part of the world never ended in that same part of the world.

The twentieth century was very eventful: The First World War, The Great Depression, The Second World War... The first half of the century was shaped by the physical sciences but its second by biology.[16]   The first half brought the revolution in transportation, communication, mass production technology and the beginning of the computer age. In warfare, it brought in the nuclear weapons and the irreversible change in the nature of warfare and environment. The public's preoccupation with the physical sciences and technologies, and the immense upheavals in the human condition meant that biology and medicine could only move to the center stage somewhat later, that is, in the second half.

The Information Revolution began in the United States two and a half decades ago.   So did the biotechnology revolution.   Will these revolutions reach their respective peaks in United States? This only time will tell.*

Now we enter the twenty-first century, biology will likely continue to play a dominant role. First, biotechnology has shed its pristine academic garb and headed into the marketplace for more than two decades. Second, biotechnology has become so data-intensive and information-rich that it is essentially a part of information science. Third, the techniques of biotechnology have now been used for a new kind of war — biowarfare. Similarly, information technology has led to infowarfare. A new kind of war has just begun.   Yet if we weigh the good and the bad, biotechnology and information technology advances have tended to improve life spans and prosperity.

---

[16] G.J.V. Nossal, and Ross L. Coppel, *Reshaping Life: Key issues in genetic engineering*, (Bricknell Nominees Pty. Ltd, Melbourne, Australia, 1990).

* In October of 1999, when HAL presented a lecture on "Turbocharging bioinformation for drug discovery" at an international congress, the remark on the Agrarian, Industrial and Information Revolutions aroused quite a lot of interest.

Chapter 12

TURBO-CHARGING BIOINFORMATION FOR DRUG DISCOVERY

"Unlike material goods, information does not disappear by being consumed, and even more important, the value of information can be amplified indefinitely by constant additions of new information to the existing information. People will thus continue to utilize information which they and others have created, even after it has been used."
Yoneji Masuda.

1 From Gathering To Information Gathering

Since the dawn of modern civilization, we have witnessed at least three watershed revolutions. There are, in chronological order, Agrarian, Industrial and Information Revolutions. The first occurred when modern civilization transitioned from hunting and gathering into a more developed agricultural economy. Hunting and gathering lasted for hundreds of thousands of years. Since then, the duration of each successive economy has gotten shorter and shorter. Agricultural dominance persisted for less than ten thousand years. The industrial economy lasted for slightly less than two centuries (1760s–1950s) globally, with something like ninety years (1860s–1950s) in the United States. We are probably seeing the current information economy halfway through its seven- or eight- decade life span.[1]

The industrial economy produces and emits waste by-products. We call them pollutants or consider them useless, but some of these by-products can be, and have been, put to better use. Natural gas is a classic example. Originally burned off at the wellhead as useless, it became a major industry when infrastructures and markets were developed. Turbo-chargers, another example, recycle engine exhaust fumes to greatly enhance performance and increase power.

2 Information Economy

The information economy also produces information exhaust, and it too can be turbo-charged. Information exhaust can be captured, processed, and recycled to improve business performance. Opportunities exist to provide turbo-charged information services in all businesses and industries. In fact, a new generation of enterprises built around information emitted from older business is already taking shape. The irony is that turbo-charged information service businesses often become worth more than the businesses from which the information is generated in the first place. And examples abound.

---

[1] S. Davis and B. Davidson, *2020 Visions*, (Simon & Schuster, New York, 1991).

For example, Quotron provides information about security prices to brokerage companies. In 1986, Citicorp purchased it for US$628 million. Quotron did not possess proprietary access to securities information. It simply filled a need that brokers had not been attended to by simply capturing securities transaction information and recycling it back to the brokerage industry that generated the information in the first place. In doing so, it created a business with market value greater than the market values of the then leading brokerage firms such as Paine Webber and Smith Barney.

Similarly, Rupert Murdoch acquired TV Guide in 1987 for more than US$2 billion. This publication is essentially a well-packaged listing of TV broadcast schedules, information that is available to anyone from a number of sources. Despite this, the purchase price for TV Guide had a market valuation higher than any one of the major broadcast networks, ABC, CBS, or NBC at that time.

And, Official Airlines Guide (OAG), a listing of monthly flight schedules, sold in 1988 for US$750 million. OAG simply consolidates flight information, yet the basic concept created a business with a greater market value than most airlines, and only slightly less than the market valuation of US Air. More recent examples are Yellow Pages compiled by telecommunications companies, which scoop in more revenue than selling standard telephone services, Hotmail acquired by Microsoft, Infoseek bought out by Go Network (now defunct), and not to mention Yahoo!, eBay, Amazon.com, WebMD and many others.

These examples suffice to show that the proper order of things seems to be standing on its head. Someone, it would seem, has to perform brokerage services so that profits can be made selling financial information services. TV programs are produced and broadcast to create a need for TV Guide. During the downturns of the airline industry, it would seem as if someone has to fly airplanes so that money can be made selling flight information and reservation services. Telephone services are provided to stay in touch so that there is a need for Yellow Pages. Last but not least, the Internet was introduced to transcend socio- and geo-political barriers. Yet in each instance, the turbo-charged information is worth more than the primary business from which the information is derived!

We thus see that information offshoots represent a tremendous business opportunity, and if we take the cue from T.S. Eliot *"The historical sense involves the perception not only of the pastness, but of its presence"*, it would seem every business contains one or more latent info-businesses. And drug discovery is no exception.

3   Drug Discovery

The chronology in drug discovery in the past century is summarized in Figure 1.[2] Drug discovery has come a long way in the past century. In early days, drugs were

---

[2] http://www.phrma.org.

discovered serendipitously. As the science advanced, more systematic and scientific means were used to develop new drugs. Classical drug discovery has been a process of trial and error involving use of natural products and synthetic compounds for the purpose of identifying new chemical entities, many of which fail and a few may reach the marketplace as pharmaceuticals. However, as easier to develop drugs have been "cherry-picked", more and more involved techniques have to be developed to compensate for the complexity.

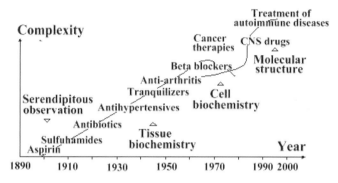

Figure 1. In the past century, drugs have been discovered in various ways. In the early days, drugs were discovered by serendipitous observations. As we progress along the time line, the complexity for discoveries becomes more and more complex and the techniques used for discovery also become correspondingly more advanced, from tissue biochemistry to cell biochemistry to molecular structure. If there is no paradigm shift in the discovery process, it is not difficult to project that in the near future, drug discovery will become very costly and very time-consuming.

Currently, we are at the threshold of this classical way of developing new drugs. The cost of drug discovery and development has spiraled in recent decades because the process has become both very expensive and time-consuming. The cost for developing a new drug is about $400 million and it takes about 12 to 15 years. Figure 2 shows the cost increase in the past quarter of a century.

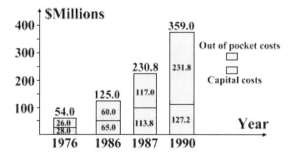

Figure 2. The cost of developing new drugs has been spiraling in the past two and a half decades from $54 million in 1976 to about $400 million in 1990 and to about $700 million in 2001.

Other indicators abound. Table 1 presents the annual prevalence and the cost of the top eight uncured diseases in the U.S. in 1995–1996, while Table 2 lists the pharmaceutical and healthcare expenditures in the world's ten most developed countries in 1994. These indicators suffice to show that if left unchecked, it is conceivable that pharmaceutical and healthcare expenditures will spiral out of reach of the most needy in the very foreseeable future. Thus there is a dire need for a revolution or paradigm shift in the drug discovery and development process.

Table 1. Despite all the news coverage of new discoveries and doctors' claim that they can "cure" diseases, to date, there are only a dozen of curable diseases. The annual prevalence and cost of uncured diseases in the U.S. are rampant (data of 1994/5).

Uncured Diseases	Approx. Annual Prevalence	Approx. Cost ($Billion)	Source
Cardiovascular	56,000,000	$128	Am. Heart Assoc.
Cancer	10,000,000	$104	Am. Cancer. Assoc.
Alzheimer's	4,000,000	$100	Alzheimer's Assoc.
Diabetes	16,000,000	$92	Am. Diabetes Assoc.
Arthritis	40,000,000	$65	Arthritis Assoc.
Depression	17,400,000	$44	Nat. Depression Assoc.
Stroke	3,000,000	$30	Nat. Stroke Assoc.
Osteoporosis	28,000,000	$10	Alliance of Aging Res.

Table 2. Pharmaceutical and healthcare expenditures are not only rampant in the U.S., they are also significant in other ten most developed nations in the world. Note that the expenditure on healthcare is about 5–10 times higher than the corresponding pharmaceutical expenditure. Note also that the expenditure on healthcare in the U.S. has exceeded $1 trillion.

Countries	Pharma Expenses $Millions	Health Expenses $Millions
Australia	N/A	28,657
Canada	7,402	58,438
France	17,922	108,314
Germany	N/A	151,981
Italy	15,503	89,187
Japan	N/A	184,100
The Netherlands	2,692	25,247
Switzerland	N/A	16,048
UK	10,863	70,709
USA	78,577	949,419

4   The Human Genome Project

The paradigm shift has been taking place. The archetypal and exemplar of this revolution is the Human Genome Project (HGP).[3] HGP is often paralleled with the moon-shot program. While this metaphor may be useful in political and funding arenas, it does not convey the true significance of HGP. Moon race affected a negligible number of people substantially and had only negligible effect on a substantial number of people. The impact of HGP will be substantial and it will affect a substantial number of people. The primary purpose of HGP is to understand the molecular basis of human diseases. It will transform the lives of many people through the impact it will have on the pharmaceutical industry.

To the end of understanding human diseases on the molecular basis, HGP has already completed sequencing the first draft of the human genetic materials — the DNA sequences. A detailed sequence is expected in 2003. The human genome contains approximately 30,000–40,000 genes, out of which, by a conservative estimate, 5,000 genes may be associated with diseases.[4,5,6]

5   Enabling Bioinformatics — The Enabler

"Necessity is the mother of invention", and bioinformatics was born out of the need to manage genetic data. Bioinformatics is the study of information content and information flow in biological and biology-related systems. In its most restricted sense, it involves genetic, chemical, and healthcare data collection, management, analysis and dissemination.[7,8,9,10,11,12,13,14,15,16,17,18,19,20,21,22,23]   As such, it plays the

---

[3] *Mapping and Sequencing the Human Genome*, National Research Council, (Washington, D.C.: National Academy Press, 1988).

[4] *Bioinformatics in the Emerging Drug Discovery Marketplace*, (Frost & Sullivan, Mountain View, CA, USA, 1998).

[5] Andrew Lyall, *Bioinformatics: A Revolution in Drug Discovery*, (PJB Publications Ltd., London, UK, 1998).

[6] *Bioinformatics: A Strategic Business Analysis*, (FrontLine Strategic Management Consulting Inc., Foster City, CA, USA, 1998).

[7] *Computers and DNA*, Edited by G.I. Bell and T.G. Marr, (Addison-Wesley Publishing Co., Redwood City, 1990).

[8] *Electrophoresis, Supercomputing and The Human Genome*, Edited by Charles R. Cantor and Hwa A. Lim, (World Scientific Publishing Co. (URL: http://www.wspc.co.uk), New Jersey, 1991).

[9] *Bioinformatics, Supercomputing and Complex Genome Analysis*, Edited by Hwa A. Lim, James W. Fickett, Charles R. Cantor and Robert J. Robbins, (World Scientific Publishing Co., New Jersey, 1993).

[10] *Proceedings of The First International Conference on Intelligent Systems for Molecular Biology*, Edited by Lawrence Hunter, David Searls, and Jude Shavlik, (AAAI Press, Menlo Park, 1993).

[11] *BIOCOMPUTING: Informatics and Genome Projects*, Edited by Douglas W. Smith, (Academic Press, New York, 1994).

[12] *Bioinformatics & Genome Research*, Edited by Hwa A. Lim, and Charles R. Cantor, (World Scientific Publishing Co., New Jersey, 1995).

offshoot role in genetic sequencing, biotechnology and pharmaceutical industries as a turbo-charger of genetic, chemical and healthcare information. The marketplace for bioinformatics has been created by the emergence of new discovery paradigms in large-scale sequencing projects of HGP, which generate an astronomical amount of genetic data. Most of the data are irrelevant ("genetic waste" or "genetic exhaust"). Bioinformatics collects, manages, analyzes, and disseminates, as shown in the figure. In analogy to previous discussion, bioinformatics is enabled by HGP because HGP creates a niche for bioinformatics. However, bioinformatics is also enabling HGP because it creates a paradigm shift from classical drug discovery to a gene-based and chemical-based discovery program, as shown.

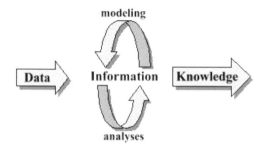

Figure 3. Large-scale sequencing produces a huge amount of genetic data. A large portion of the data is "genetic exhaust". Bioinformatics is the turbo-charger that collects, analyzes and disseminates the data and exhaust as useful information and knowledge.

---

[13] *Bioinformatics: From Nucleic Acids and Proteins to Cell Metabolism*, Edited by D. Schomburg and U. Lessel , (VCH Publishers, Inc., New York, 1995).

[14] *Integrative Approaches to Molecular Biology*, Edited by J. Collado-Vides, Boris Magasnik, and Temple F. Smith, (MIT Press, Cambridge, 1996).

[15] W. Bains, "Company strategies for using bioinformatics", *Trends Biotechnol.*, 14: 312–317 (1996).

[16] D. Benton, "Bioinformatics — principles and potential of a new multidisciplinary tool", *Trends Biotechnol.* 14, 261–272 (1996).

[17] A. Lyall, "Bioinformatics in the pharmaceutical industry", *Trends Biotechnol.*, 14, 308–312 (1996).

[18] Hwa.A. Lim, "Bioinformatics and Cheminformatics in the Drug Discovery Cycle", *Lecture Notes In computer Science, Bioinformatics*, Ed. R. Hofestdät, T. Lengauer, M. Löffler, and D. Schomburg, (Springer, Heidelberg), pp. 30–43 (1997).

[19] *Molecular Bioinformatics — Sequence Analysis and the Genome Project*, R. Hofestädt and Hwa A. Lim (Hrgs.), (Shaker Verlag, Aachen, Germany, 1997), 60 pages.

[20] Hwa A. Lim and T. Butt, "Bioinformatics takes charge", *Trends Biotechnol.* 16 No. 3 (170), 104–107 (1998).

[21] T.V. Venkatesh, B. Bowen, and Hwa A. Lim, "Bioinformatics, pharma and farmers", *Trends Biotechnol.*, 17 No. 3 (182), 85–88 (1999).

[22] Hwa A. Lim, and T.V. Venkatesh, "Bioinformatics: Pre- and post-genomics eras", *Trends Biotechnol.*, 18, April 2000, pp. 133–135.

[23] Hwa A. Lim, *Pathways of Bioinformatics: From data to diseases*, (Henry-Stewart Publishers, London, 2002).

Demand Side

Gene-based informatics
- genome data analysis
- genome datamining

Cheminformatics
- characterization of combinatorial libraries
- optimization of combinatorial libraries

Supply Side
- gene sequencing data generation software
- object-oriented framework for gene sequencing data management
- gene sequence analysis software
- gene sequence data dissemination software

Figure 4. A restricted definition of bioinformatics as applied to the new paradigm of drug discovery. The supply side shows that it is a specialized form of information technology (IT). Gene-based informatics comprises of hardware, database, and software tool components, while cheminformatics encompasses managing and analyzing small-molecule structures. The combination of gene-based and cheminformatics will likely drive the development of molecular diagnostics and therapeutics.

6 Proliferation Of Bioinformatics

A search for the word "bioinformatics" on the Internet using AltaVista^{TM} search engine, for example, produces in excess of a hundred thousand matches, and a search for bioinformatics books using amazon.com produces more two dozen books. These are cyberspace evidences of the popularity of this relatively new field, almost 15 years old since the coinage of the word. In the physical world, the number of biotechnology, pharmaceutical, and information-based companies forming internal bioinformatics divisions is increasing daily. Fledging and emerging bioinformatics companies arrive on the scene daily.[*] These companies produce a plethora of similar but differentiated products and services. They vie for the same pie in the limited marketplace.

7 Market Structures

We alluded to the fact that a market for bioinformatics has been created by HGP. To better understand the bioinformatics market, let us take a cursory look at market structures. In general, there are four types of market structures[24]: perfect competition, monopoly, monopolistic competition, and oligopoly.

---

[*] In the United States, the economic downturn since the beginning of 2001 has put a halt to new companies entering the scene. The same can be said of other parts of the world, when the rippling effect is felt. In October of 2001, more than half of small German companies went under.

[24] W.A. McEachern, *"Economics: a contemporary introduction"*, 4th Ed., (South-Western College Publication, Cincinnati, 1997).

7.1   Perfect Competition

A perfectly competitive market is characterized by many buyers and sellers who exchange a standardized or homogenous product. Buyers and sellers are fully informed of the price and availability of all resources and products. Firms and resources are freely mobile, with no obstacles such as patents, licenses to prevent new firms from entering or existing firms from exiting the market. Two classic examples which come close to this type of market structure are the stock market and the world grain market. The currently very lucrative Internet businesses are probably best described by this market structure. A participant in this market is a price-taker for it has to take or accept the market price.

7.2   Monopoly

"Monopoly" is a Greek word meaning "one seller", just as we understand the word. A feature of this market is the high barrier to entry. Barriers to entry can be any impediments that prevent new firms from competing on an equal basis with existing firms in an industry. Three examples of barriers to entry are legal restrictions (e.g., patents and licenses), economies of scale (e.g., in the cable TV industry where once the cost for laying cable has been sunk, the cost for hooking up additional household declines), and control of essential resources (e.g., the diamond industry).

It is thus clear that in perfect competition, many suppliers offer a homogenous commodity to a market where firms can enter or exit the industry with ease. In contrast, monopoly involves only one seller offering a product with no close substitutes. These polar market structures are logically appealing, but most firms operate in market structures between these two polar structures.

7.3   Monopolistic Competition

As the name implies, inherent in the monopolistic competition market are both elements of monopoly and competition. In this market, many producers offer products with no close substitutes, but the barriers to entry are relatively low so that there are enough sellers to compete. Because of the monopoly, players in this market are price-searchers. The monopoly comes from product differentiation, in sharp contrast to the homogenous products in perfect competition. Product differentiation can be physical appearances (e.g., Burger King's grilled Whopper versus McDonald's fried Big Mac), location (e.g., Seven Eleven convenient stores which are actually selling the convenience of proximity to consumers and opening late hours), services (e.g., Domino Pizza stores who deliver), and product image (this is the image the producer tries to foster in the consumers' mind. For example, Nike shoes are linked with quality and the "Just Do It" is imprinted in consumers' mind).

7.4 Oligopoly

Oligopoly is a market dominated by a few sellers because of the high cost to entry, economies of scale or legal restrictions. In this market, the products can be homogenous, such as in the steel and oil industries, or the products can be differentiated, such as in the automobiles and tobacco industries.

These four market structures are presented pictorially in the figure for easy comparison. Based on these criteria, we shall now argue which of these market structures sequencing, biotechnology, pharmaceutical and bioinformatics companies belong to.

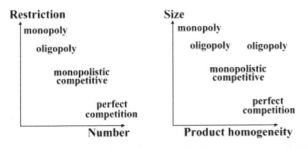

Figure 5. Figures to contrast the number of players, barriers to entry, the "size" and product homogeneity of industries in the four market structures. Note that we say "size", which refers to the relative sizes in that particular market structure, and not size in its absolute sense. Perfect competition is characterized by low entry barrier and product homogeneity. In contrast, monopoly is characterized by high entry barrier and low product homogeneity. Between these polar market structures are oligopoly and monopolistic competition. Note that an oligopolistic company can have homogenous products (for example, steel producers) or differentiated products (for example, automobile manufacturers).

8   Market Structures Of Bioinformatics, Sequencing, Biotechnology And Pharmaceutical Companies

The barrier to entry for high throughput sequencing companies is relatively high because it is costly to set up an enterprise large enough to have a sustainable competitive advantage. Once the system is in place, automation and ultra-high throughput technology provide economies of scale. They each produce raw data (homogenous products). Thus these companies are in the oligopolistic market with homogenous products.

Biotechnology and pharmaceutical companies are in the differentiated product oligopolistic market. For example, GlaxoWellcome has Zantac and SmithKline has Tagamet. Tagamet and Zantac serve more or less the same purpose (ulcer), yet they are not the same. The barrier to entry is extremely high. For example, in addition to patents, it costs about US$500M to discover and develop a drug. These companies are also interdependent, i.e., each firm must consider the effect of its own policies on competitors' behavior. Thus biotechnology and pharmaceutical companies are in the oligopolistic differentiated product market structure.

Bioinformatics companies try to differentiate their products. For instance, the products are

❑ Web-based, graphical user interfaced, enterprise solution, integrated solution, task-specific ("physical appearance") or neural-net based, hidden Markov approach ("approach appearance")

❑ Platform-independent, flexibility, international convention compliant open system ("convenience")

❑ Guaranteed 24×7×365 reliability ("post service")

❑ Use of well-known engines, for example, Oracle for databases, or claim of product use by huge pharmaceutical colossi in press releases or third party validation ("product image").

Many of the products perform similar task. They each claim reliability, accuracy and speed. The barriers to entry are relatively low, as is evident from the number of bioinformatics companies entering the marketplace each day. We thus argue that bioinformatics companies are monopolistically competitors.

An analogy will make the distinction clearer. A monopolistically competitive bioinformatics company is like a professional golf player. The player is striving for a personal best to win the tournament. In contrast, an oligopolistic biotechnology or pharmaceutical company is like a professional tennis player whose action and reaction depend on where the opponent hits the ball to outplay the opponent.

8.1   Goals of Bioinformatics Companies to Monopolize

We shall now investigate the monopoly component of monopolistically competitive bioinformatics industry.

Like any firms in other business sectors, bioinformatics companies would like to maximize profit by capturing a huge share of the market. In an attempt to do so, these companies try to follow the footsteps of successful products and may parallel their products with general-purpose software such MS Office and AutoCAD. For example, they may see that MS Office is not necessarily the best software, yet through good promotion and marketing, Microsoft captures a significant market share. So these bioinformatics companies allocate huge sums for promotion in trade shows and workshops. Huge promotion is not an uncommon feature in a monopolistically competitive market.

Though their products are sometimes greeted with *éclat* in bio-publications, the actual sales may not quite live up to expectations. There are two plausible reasons for this:

❑ Most bio-publications are paid media for promotion. They are not an independent third party for evaluating products.

❑ There are many more players in the bioinformatics market, each vying to putt the hole-in-one.

In addition, there are a number of distinctions between bioinformatics products and general-purpose software such as MS Office:

❑ MS Office's latent client base is PC users in the private, public, and home based sectors. The latent client base of bioinformatics companies is primarily oligopolistic biotechnology and pharmaceutical companies. This client base is not only much more limited, but they are also potential competitors for they may also develop their own bioinformatics tools in-house.

❑ PC users are likely to look for an easy-to-use software (such as MS Office) to get their work done. Expert users of bioinformatics products look for functionalities. It is unlikely that they will accept a product with "force down their throat" features. They likely would want to integrate the bioinformatics products into their in-house enterprise system for proprietary R&D.

❑ As shown in the figure, MS Office is product-focused, i.e. it requires little customization and sells in large volume. Bioinformatics products, in contrast, are highly customized.[25]

Thus, in the current prevailing market, a monopoly of market share is likely an unattainable goal for bioinformatics companies.

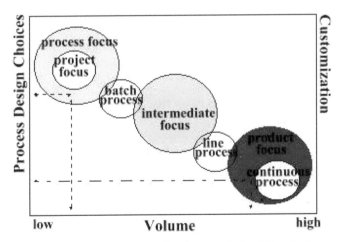

Figure 6. Bioinformatics software and databases for pharmaceutical and biotechnology companies are in the process focus category, i.e., high customization and small latent client base. Healthcare software and databases are in the intermediate focus category. In contrast, commonly used software such as desktop publishing software, MS Office, AutoCAD, is in the product focus category, i.e., low customization and large latent client base.

---

[25] L.J. Krajewski, and L.P. Ritzman, *Operations Management: Strategy and Analysis*, 4th Ed., (Addison-Wesley Publishing Company, Menlo Park, 1996).

8.2   Competition

We now turn to the competition component of the monopolistically competitive bioinformatics industry.

Because of the large number of companies and the keen competitions, each company has to move with celerity to get to the market, sometimes with not fully tested products. Besides competing with each other, bioinformatics companies also have to compete with publicly funded institutions. Because of funding arrangements, these institutions normally provide their R&D results as freeware.

Perhaps the strongest competitions come from oligopolistic biotechnology and pharmaceutical companies integrating backward to form their own bioinformatics divisions. Examples of these include SmithKline, GlaxoWellcome and Wyeth-Ayerst, each of which have a sizable bioinformatics division. Another competition comes from information-based companies concentrically diversifying^α their business into bioinformatics to tap into the short-run profitability.[26] Examples of these include Compaq Computers and the 1997 Reed Elsevier Publishing acquisition of Molecular Design Ltd. (MDL) in San Leandro, California, USA.

With their huge financial resources and established channels of distribution, they can in principle penetrate the market easier than fledging and emerging bioinformatics companies. But size is a double edge sword. A larger size often allows for larger, more specialized machines, hardware and greater specialization of labor. This creates economies of scale. On the other hand, in the thicket of bureaucracy of larger companies, diseconomies of scale may result.

Let us now look at another market, the labor market.

9   Bioinformatics Entrepreneurs And New Entrants

The ranks of entrepreneurs are in constant flux as some emerge from the labor market to form their own enterprises and others return to the labor market after selling successfully or failing miserably. Whatever the reason, the supply of entrepreneurial ability is influenced by a variety of external forces. These forces include the pace of technological changes such as ultra throughput sequencing and computer technologies, government regulations and policies such as Small Business Innovations Research (SBIR) grants, or federal agencies' recommendation to set up federally funded centers and institutions.[27]

Many entrepreneurs prefer the individual freedom that comes from self-employment. Some derive satisfaction from creative processes. Some dream of

---

^{α} Diversification into new product line that has technological and/or marketing synergies with existing product lines, even though the product themselves may appeal to a different group of customers.
[26] P. Kotler, *Marketing Management*, 9th Ed., (Prentice Hall, Upper Saddle River, NJ, 1997).
[27] Report of the NIGMS *Ad Hoc* Bioinformatics Training Committee, March 22, 1999. http://www.nih.gov/nigms/news/reports/bioinformatics.html

founding a corporate empire, and still some may have difficulty finding employment. Essentially, there are three typical roles played by entrepreneurs. They are not mutually exclusive. Entrepreneurs are either:

❑ Broker: These are the so-called "buy low and sell high" entrepreneurs. They contract with resource suppliers and combine the resources to form goods and services. They direct resources to the highest valued use, thus promoting economic efficiency. In such scenario, they attract rivals in the long run, forcing profit down to just a normal level.

❑ Innovator: These are the ones who can make an existing product for less than competitors do, or who introduce successful new products. In this scenario, the possibility of economic profit is a strong incentive for innovations. Whether the profit continues in the long run depends on whether competitors can imitate the cost-saving activities or the new products. Barriers to entry will help economic profit to continue into the long run as well.

❑ Risk bearer: These are those who venture into a world of uncertainty and get paid for their willingness to bear the risk. Examples of this are investment companies and angels.

With the exception of angels and venture capital firms, almost all bioinformatics companies are in categories 1 or 2. In category 1, the bioinformatics company functions like a "hollow corporation" or "network company". It spends most of its time on the telephone or the computer to coordinate suppliers. Examples of this category include bioinformatics companies that license to sell effectively through their distribution channels such as SinoGene.

In category 2, the bioinformatics company normally produces a flagship product, often times by combining products from various producers through licensing. Examples of this category include Compugen, D'Trends, DoubleTwist, Gene Logic, Genomica, NetGenics, and many others. The company can also integrate backward by hiring or buying-out those producers whose products (authors of software tools or databases) are critical to the company's mission. Examples of this include Incyte and those who spend time developing their technologies in incubators, and bioinformatics companies spinning off publicly funded research institutions such as Human Genome Science spinning off The Institute for Genome Research (Bethesda, USA), and eBioinformatics spinning off the Australian National Genome Information Services (University of Sydney, Australia). Yet another category is those companies that farm out their development offshore (such as Russia) to cut down on cost. Examples include GeneGo and InforMax.

9.1 Bioinformatics Labor Supply

Having looked at entrepreneurs, we now turn to the behind-the-scene driving force of bioinformatics companies.

Like any other resources that produce goods and services, the demand for bioinformatics labor resources is a derived demand in the sense that companies looking for bioinformaticists do not value the labor resource itself, but rather the bioinformaticists' ability to produce profitable products.

As depicted in Figure 6, bioinformatics is in the process-focused positioning strategy. This type of positioning requires a flexible work force. Members of a flexible work force are capable of doing many tasks, either at their own workstation or as they move from one workstation to another. This flexibility comes at a cost because it requires greater skill and thus more training and education. Though the cost is high, but the benefits can be large. It is probably the best way to achieve reliable customer service and alleviate capacity bottleneck. It helps absorb the feast-or-famine workloads.

Partially because bioinformatics is a relatively new field, it is still not a regular part of most teaching institutions' curriculum, though the situation is changing rapidly. This, coupled with the rapid growth of organizations doing bioinformatics, creates a dearth of qualified bioinformaticists.[28,29] The shortage is not very elastic in the sense that a qualified bioinformaticist acquires his/her qualification through proper training, which takes time. This short differential should be corrected over time. For now, we must make do and seek more mobile substitutes from closely related disciplines of computer science, mathematics and life sciences.

Bioinformaticians — those who know how to operate bioinformatics packages — on the other hand, are easier to come by. These technicians attend workshops and evening classes, learn to use tools of the trade and graduate with certificates or diplomas in short periods.

It is sometimes said that the private sector "is eating its own seed for bioinformatics" because the private sector is recruiting faculty from the academia to provide a ready source of knowledge and hence spillovers from the academia to the private sector. The practice, where replacement is difficult, impairs the academia's capacity to continue training initiative.[30] There are four interrelated explanations for what appears to be the academia's sluggish response to the dearth of bioinformaticists.

❑ Individual faculty has no incentive to establish bioinformatics programs because traditionally, funding monies are primarily for research and researchers' salaries.

---

[28] "University reforms would help Germany to combat its worrying shortage of bioinformaticists", *Nature* 400, (Macmillan Publishers Ltd., 1999), pp. 93.

[29] Alison Abbot, "German agency to boost bioinformatics", *Nature* 400, (Macmillan Publishers Ltd., 1999) pp. 102–103.

[30] P.E. Stephan and G. Black, "Bioinformatics: does the US system lead to missed opportunities in emerging fields?", Report at National Research Council's Committee on Science, Engineering and Public Policy, Irvine, CA, 1998.

❏   The education system responses differently when the demand is driven by the private sector. Life sciences are arguably not as responsive to demands driven by the private sector as are computer science and engineering.

❏   The interdisciplinary nature of the field creates disincentives to establish the program. The problems of working across departmental lines are difficult because of bureaucracies and turf issues within university administration.

❏   The quick fix of converting life scientists into bioinformaticists is not very easy, given the lack of skills and quantitative abilities of life scientists. And the quick fix of augmenting computer scientists with life science encounters keen competition from the strong job market for computer scientists and engineers in the private sector.

Pressure from the labor market has changed many a risk-aversion university.

It is a noteworthy point that bioinformaticists are professionals. They are typically different from nonprofessionals. They have a strong and long-term commitment to their field of expertise. Their loyalty is more often to their profession than to their employer. To keep current in their field, they need to regularly update their knowledge. This explains the flourishing business of bioinformatics publications (BioInform, BioVenture, BioWorld, Elsevier Trends Series, Genome Technology, etc), bioinformatics reports,[31] conferences (Cambridge Healthtech Institute, Frost & Sullivan, etc), workshops, meetings and online training.

There are other reasons why bioinformatics conferences, workshops, meetings and training are so successful. Though bioinformaticsts and bioinformaticians are *Homo economicus*, lifetime income maximization is not their main interest. In a downsizing environment, job security has come out as more important than maximum wages.[32]

Having described the market structure and the labor market of bioinformatics, let us see how they interact.

10   Technology-Push, Market-Pull Or Interactive Marketing Strategies

In the days of "it is not what you manufacture, rather it is what you know that matters", the distribution channel and marketing strategies are also different.

In the past, product development in the technology industry was largely driven by engineers, who usually designed products based on their own interests and left the marketing department with the task of finding consumers to buy them. This is the so-called bottom-up or technology-push approach:

---

[31] An excellent example is Malorye A. Branca, T.V. Venkatesh, and Nathan Goodman, *Bioinformatics: Getting Results in the High Throughput Genomics*, (Cambridge Healthtech Institute Publication, 2001), 178 pages.

[32] Lawrence Mishel, Jared Bernstein, and John Schmitt, *The State of Working America*, (Cornell University Press, Ithaca, New York, 1998).

Basic research → Applied research → Product development →
Pilot product trial →Manufacturing → Market/sales/services

With the accelerating pace of technology innovation, product cycles are measured in weeks rather than years. Many companies are finding that it makes better business sense to design product based on what customers want rather than what engineers think they want. Increasing globalization of markets also makes the bottom-up approach impractical. For example, a black cellular phone with myriad functions designed for the U.S. market may hold little attention in Latin America, where bright colors and simpler designs are preferred. Thus a top-down approach or market-pull approach is better suited for current technology market environment:

Basic research ← Applied research ← Product development
← Pilot product trial ← Manufacturing ← Market

Ethnography is a branch of anthropology that studies human culture. Social scientists hired by technology companies with the belief that their observations and insights will lead to the development of new products and services. Their task is like a reconnaissance mission — to map out the landscape, constantly foraging for opportunities for new technology. We thus propose an interactive scheme should be considered. The latter is a market-pull scheme couched in advances in the computer and the Internet technology. In this interactive mode, the market is identified through market survey and analysis. Then during development, there is a constant interaction (feedback) between the development team, the market forces, and the technology sector. Companies such as Microsoft and Perkin Elmer also organize workshops to train clients to operate their products. Such a scheme is key in this fast-paced market and rapid advances in technologies. Any entrenched Luddith of new technologies will be permanently left behind.

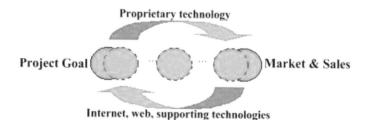

Figure 7. Comparisons of technology-push, market-pull and interactive-pull marketing strategies. In recent product development, the traditional technology-push strategy has been replaced by market-pull strategy whereby the market is first identified and then a product is manufactured to satisfy the need. However, because of concurrent advances of computer and Internet technologies, basic market-pull strategy is quite inadequate to cope with the fast-pace marketplace. To keep abreast with rapid market changes, a more interactive strategy that maintains a constant interaction (feedback) between the development team, the market forces, and the technology sector, has to be adopted.

11   New Paradigm Adoption

We mentioned earlier that we are at the juncture of a new paradigm shift in drug discovery and bioinformatics is the turbo-charger. We now turn our attention to the adoption of this new paradigm.

Virtually all contemporary thinking about high-tech marketing strategy has its roots in the technology adoption curve, a model grew out of social research begun in the late 1950s about how communities respond to discontinuous innovations.[33,34] Along a risk-aversion axis, adopters of the new paradigm self-segregate themselves into roughly a Gaussian distribution, with innovators (2.5%), early adopters (13.5%), early majority (34%), late majority (34%), and laggards (16%). The paradigm shift will be adopted from left to right, with each constituency coming to the fore in sequence. In high tech industries, we can re-label each of the five constituencies as (see figure):

❑ Innovators — technology enthusiasts: These are people who are fundamentally committed to new technology on the ground that sooner or later, it is bound to improve our lives. They derive pleasure from mastering the intricacies, in fiddling with, and love to get their hands on the latest and greatest inventions. They are driven to be the first to explore.

❑ Early adopters — visionaries: These are the true revolutionaries who want to use the discontinuity of any innovation to make a break with the past and start an entire new future. Their expectation is that by being the first to exploit the new capability they can achieve a dramatic and insurmountable competitive advantage over the older order. In contrast to enthusiasts, they are driven to be the first to exploit.

❑ Early majority — pragmatists: These people do not love technology for their own sake. They are neutral about technology and look to adopt innovations only after proven track record. They believe in evolution rather than revolution in the sense that they accept the new technology as a natural extension of the older order (evolution) rather than a paradigm shift (revolution or discontinuity).

❑ Late majority — conservatives: These are pessimists who doubt their ability to gain any value from technology investments and undertake them only under duress — typically the only alternative is to let the rest of the world pass them by.

❑ Laggard — skeptics: These are the gadflies of high tech, the ones who delight in challenging the hype and puffery of high technology.

---

[33] Philip Kotler, *Marketing Management*, 9th Ed., (Prentice Hall, Upper Saddle River, NJ, 1997).
[34] Geoffrey A. Moore, *Inside the Tornado: Marketing Strategies from Silicon Valley's Cutting Edge*, (Harper Perennial, New York, 1999).

Thus, we coin the five "e's" of new paradigm adoption: technology enthusiasts *explore*, visionaries *exploit*, pragmatists *evolve*, conservatives *employ*, and skeptics *elude*.

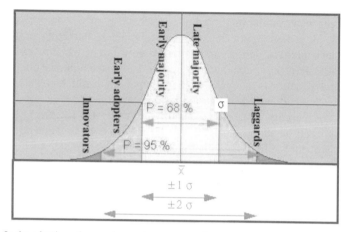

Figure 8. In the adoption of a new innovation or a paradigm shift, adopters self-aggregate themselves into five constituencies. The most risk-immune are the technology enthusiasts, followed by visionaries. Pragmatists are neutral while conservatives are most risk-avert.

12 Who Prevails

With so many external factors — technology, federal policies, competitions, labor resource supply, technology adoption — to worry about, which bioinformatics companies will prevail eventually? To shed light on this important issue, let us look at the microeconomics of supply and demand of bioinformatics products.

Because each monopolistically competitive bioinformatics company offers a product or a gamut of products that differ somewhat from other competitors' products, each company has some control over the price. This means that the demand curve slopes downward, i.e., not completely elastic, as shown in the figure. This is a reasonable assumption since if the quantity produced is decreased, the unit selling price should go up as the firm is a price searcher.

However, monopolistic competitive bioinformatics companies have no guarantee of economic profit, as we shall show presently.

12.1 Short-run Profit/Loss Optimization

If in the short-run, a period too brief to allow a bioinformatics company to enter or leave the market, the company can cover its variable cost[β], it will increase output

---

[β] Any production cost that increases as output increases.

(sales) as long as marginal revenue $(MR)^\chi$ exceeds marginal cost $(MC)^\delta$. The opposite case of MR<MC means the company is operating at a loss and thus there is no reason to expand production. The profit-maximizing level of output occurs where MC intersects MR, and the corresponding profit-maximizing price (p) is the one on the demand curve at that level of output (q), as shown in Figure 9. At an average total cost $(ATC)^\varepsilon$ of ATC', where ATC' is partially below the demand curve D [otherwise (p-c') will never be positive], the company will maximize a short-run profit of q×(p-c'). This is a profit because (p-c') is positive, and it is an extremum as can be seen by varying q. For other q values, (p-c') is smaller.

While at an average total cost of ATC", where ATC" is completely above the demand curve D so that (p-c") is always negative, the company will minimize a short-run loss of q×(c"-p). Again this is an extremum as can be seen by varying q. For other q values, |(p-c")| is larger.

So depending on along which ATC the company is operating, it may be making a profit or incurring a loss. In the latter case, the company may want to shut down temporarily to reevaluate itself. If the loss is expected to persist, the company should consider leaving the industry for good.

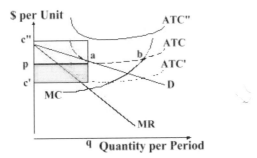

Figure 9. Production level, price, short-run and long-run profit/loss of a monopolistically competitive bioinformatics company. If ATC (average total cost) is partially below the demand curve D so that the difference of unit selling price and cost of production (p-c')>0, the maximum short-run profit is q×(p-c'). On the other hand, if ATC is completely above D, (p-c") is always negative, the minimum short-run loss is q×(c"-p). In the latter case (loss), the company must decide whether to produce or to shut down temporarily. Shutting down temporarily is not the same as going out of business. It allows the company to reevaluate and realign itself. However, if the loss is expected to persist, the company should consider leaving the industry permanently. The monopolistic competitive long-run scenario is depicted by ATC osculating D, in which the company will earn no economic profit.

χ The change in total revenue resulting from a one-unit change in sales. In our examples of perfect competition, marginal revenue equals to market price.
δ The change in total cost resulting from a one-unit change in output. Mathematically, this is ratio of change in total cost to change in output, $(\Delta C)/(\Delta P)$.
ε Total cost divided by output, or the sum of average fixed cost and average variable cost.

12.2 Long-run Economic Profits

Since the barriers to entry are very low, short-run economic profits[φ] of existing bioinformatics companies will attract new entrants in the long-run. Because new entrants offer a product that is quite similar but differentiated to those offered by existing firms, they draw customers from existing firms thereby reducing the demand of each company. Entry will continue in the long-run until economic profit disappears.

On the other hand, if existing bioinformatics companies incur short-run losses, some of them will leave the industry in the long-run, redirecting their resources to activities that are expected to earn at least a normal profit[γ]. As companies leave the industry, their customers will switch to remaining companies, increasing the demand for each remaining company's products. Companies will continue to leave the industry in the long-run until remaining companies have enough customers to earn a normal profit (opportunity cost[η]), but not economic profit (total revenue minus explicit and implicit costs).

In the long-run, entry and exit will alter each company's demand curve until economic profit disappears. This situation is attained when ATC is tangential to D. At the point of osculation, average total cost (ATC) equals the price. At all other level of output, the company's average total cost is above its demand curve, so that the company will lose money if the company reduces or expands its output.

Thus, in the long-run, monopolistically competitive bioinformatics companies will earn no economic profit. The above model assumes that monopolistically competitive bioinformatics companies do not regularly come up with new innovations or that they do not integrate backward, or merge with competitors.

13 Perfect Versus Monopolistic

So what can we infer from the supply-demand of bioinformatics products? Let us compare and contrast bioinformatics business with one of its close relatives, the Internet business, which is probably better described by perfect competition.

In perfect competition, Curve D in Figure 9 is a horizontal line and points a and b collapse to a single concurrent point of intersection of MC, ATC and D at the extremum of ATC resulting in a higher level of output than q shown in Figure 9. Thus perfect competitors in the long-run produce at full capacity at the lowest cost (ATC). Monopolistic competitors have excess capacity. Excess capacity means they can serve more customers to lower the average cost. If the marginal value of production exceeds the marginal costs of production, expanding output would

---

[φ] A firm's total revenue minus its explicit and implicit costs.

[γ] The accounting profit required to induce a firm's owners to employ their resources in the firm. This is also the accounting profit earned when the firm earns their opportunity cost.

[η] The value of the best alternative foregone when an item or activity is chosen.

increase marginal value more than marginal cost, thereby increasing economic welfare. The latter scenario probably describes publicly-funded bioinformatics R&D.

We also point out a key difference between perfect competitors and monopolistic competitors. Monopolistic competitors tend to spend more on advertising and promotion to differentiate their products. These expenses are not reflected in Figure 9. Accounting for these expenses will shift the average cost curve higher. Some people argue that the product differentiation among monopolistic competitors is artificial, while others argue that consumers are willing to pay a little more for a greater selection.

14 Bioinformatics Gorillas, Chimpanzees And Monkeys

A good friend who is also a well-respected authority in bioinformatics once jokingly remarked, "If you want to be rich from bioinformatics, be an attorney". He has a point. Legal matters can be pernicious to any organization. The only winning parties are the attorneys. Let us also not forget in business, one never makes an enemy. A whilom enemy may be a future strategic partner. And strategic partnerships seem to be a way to stay ahead in the competitive bioinformatics market because partnerships help get to the end results or new innovations sooner and more cost-effectively. Else, the god old aphorism of "publish or perish" in the academic sector can be translated into "innovate or perish" in the bioinformatics business. The sole survivors are those who can perform best the *funambulism* of innovation and marketing, those who know how to introduce the future into the present. And ultimately, only a few will survive as oligopolistic bioinformatics companies, be it through mergers or hostile takeovers.

As of now, diseconomies of scale of large biotechnology and pharmaceutical companies are working in the favor of small savvy bioinformatics companies. Because of bureaucracies, large companies have more inertia than smaller bioinformatics to realign to the rapidly changing technologies. Also working in favor of bioinformatics companies are federal agencies' effort to ensure that genetic data do not lie only in the hands of the private sector. Federal agencies do so by investing huge sums into large sequencing projects, two examples of which are the Whitehead Institute of MIT and the Department of Energy Joint Genome Institute in Walnut Creek, California.

As the turbo-charger of biotechnology and pharmaceutical discoveries, bioinformatics will remain popular in the new millennium. However, two to three more years (from 2001) will demarcate an evolutionary epoch of "survival of the operationally fittest" of many small monopolistic bioinformatics competitors. To conclude, we borrow the gorilla-chimpanzee-monkey terminology from Jeff Tarter, a software industry analyst and the editor of *SoftLetter*. By that time, a couple of dominant competitors, the gorillas, will have emerged; a handful strong market contenders, the chimpanzees, will have remained; and a score of opportunists, the monkeys, will seek a small piece of the left-over bioinformatics pie.

Though we like to use September 11, 2001 incident of the World Trade Center Twin Tower as an excuse, the first two quarters of 2001 was a dismaying period of deteriorating economies in the U.S. Poor market performance has ushered in the demise of many of these companies. September 11 incident is only a catalyst, NOT the cause.

15 Early And Late Majority Asian Bioinformatics And Biotechnology Countries

15.1 Singapore — One of the Four Tigers

During November 18–20, 1992, HAL visited the National University of Singapore. After a lot of hard work, Dr. Tan Tin Wee managed to set up the Bioinformatics Centre (BIC). In 2001, the Singaporean government announced another big investment into biotechnology.

Photo 1. Left: HAL visited Hong Kong University of Science and Technology (HKUST), Clearwater Bay, Kowloon, from November 2–7, 1992, during the time when the university was still in phase two of construction. Right: HAL at the Institute of Systems Science, National University of Singapore (NUS) to see the computing facilities. HAL was at NUS from November 18–20, 1992 to present a series of lectures on bioinformatics.

Singapore has been a regular on the international biomedical arena. Of note is the recent Fugu Genome Project. The Fugu Genome Project was initiated in 1989 by Sydney Brenner, Molecular Research Center (MRC, Cambridge, England), along with colleagues Greg Elgar, Samuel Aparicio, and Byrappa Venkatesh. The International Fugu Genome Consortium was formed in November 2000, comprising of U.S. DoE Joint Genome Institute (JGI, Walnut Creek, California, Director: Trevor Hawkins), Singapore Biomedical Research Council's Institute for Molecular and Cell Biology (IMCB, Director: Chris Tan), U.K. Human Genome Mapping Resource Centre (MGMP-RC, Cambridge, U.K.), Cambridge University Department of Oncology, Institute for Systems Biology (Seattle, Washington), Celera Genomics (Rockville, Maryland) and Myriad Genetics, Inc. (Salt Lake City, Utah).

At a recent conference, the Consortium announced the completion of a draft sequence of the genome of the Japanese pufferfish Fugu rubripes.[35] This Fugu draft sequence is the first public assembly of an animal genome by the whole genome shotgun sequencing method. For reassembly, a new computational algorithm, JAZZ, that was developed at JGI was used.

The Fugu genome (365×10^6 bases) is only an eighth the size of the human genome (3×10^9 bases), but it has a similar complement of genes. With far less junk DNA in the Fugu genome to sort through, finding genes and controlling sequences should be a much easier task than doing the same in the human genome.

Sydney Brenner, now a distinguished professor at the Salk Institute for Biological Studies, San Diego, and Board Director of the Singapore Biomedical Research Council, commented, "This represents the culmination of more than a decade of work at Cambridge and Singapore. Without JGI's initiative and Singapore's strong support, the project would have languished. We already know it will illuminate the human genome sequence and help us understand it."

Chris Tan, founding director of IMCB, said, "The draft of the Fugu genome will yield much more accurate estimates of the gene repertoire in humans. We will now be able to refine many of the features of the non-coding regions that may prove to have regulatory control over genes expressed in the human genome." Singapore's IMCB is already leveraging the genome sequence through the creation of a live Fugu genome bioinformatics analysis pipeline producing annotations and comparisons for the genome community.[36]

15.2 Korea — A Tiger

During October 5–23, 1993, HAL visited the Korean Institute of Science and Technology (KIST) to evaluate the cell culture projects and bioinformatics program, and to present a series of lectures on bioinformatics. KIST has now been renamed the Korean Research Institute for Biosciences and Biotechnology.

---

[35] The 13th International Genome Sequencing and Analysis Conference, San Diego, California, October 26, 2001
[36] www.eurekalert.org

Photo 2. Left: HAL preparing lectures in his office at the Korean Institute of Science and Technology (KIST), Daejon, Korea. HAL was at KIST from October 5–15, 1993. Right: Dr. Pal Hajas, CPO, Food and Agricultural Organization (FAO) of the United Nations (UN), and HAL. HAL visited the FAO office in Rome, Italy from October 24–27, 1993 to present findings and recommendations after a United Nations mission.

15.3 The Republic of China — A Tiger

The Republic of China (ROC) has enjoyed phenomenal successes in the computer business in the closing decades of the twentieth century. Former Finance Minister Sun Yin-shien is credited with the successes of the semiconductor industry. Two other principal factors have fueled the successes: one is technology transfer from the United States, mainly in the form of U.S.-trained engineers and computer scientists. The second is the pursuit of a strategy of efficient manufacturing of generic components and computers.

ROC is an island of calm. It weathers the recent Asian financial tempest with business flexibility, low debt, high foreign reserve (US$84 billion), the wits of small- and medium-size businesses, spending heavily on education, and moving rapidly up the value-added chain.[37] ROC's strength has been to focus on specific product categories, build expertise and dominate the world market. The integrated circuit industry was $3.95 billion in Y1996, $4.9 billion in Y1997, $7 billion in Y1998, and an impressive $9.8 billion in Y2000 from wafer and dynamic RAM (DRAM). At the Investment Forum of September 10, 1998, HAL proposed it was high time for ROC to diversify into other technologies, particularly biotechnology because the worry is its increasing reliance on a single sector — electronics — is holding the nation hostage with price cycles.[38]

From November 12 through 19 2000, HAL, as leader of the Biotechnology and Pharmaceutical Delegate at a week-long meeting organized by the Chinese Institute of Engineers (President: Dr. Chintay Shi), visited Taipei and vicinities. Delegate

---

[37] Geoff Hiscock, *Asia's Wealth Club* (Nicholas Brealey Publishing, 1997)
[38] Hwa A. Lim, Lecture at Investment Forum, International Conference Hall, Development Center for Biotechnology, Taipei, September 10, 1998.

members visited and discussed with members of local scientific institutions and business enterprises. Upon completion of the mission, the Delegate met with President Chen Shui-bian, made a report to Premier Chang Chun-xiong and Former Finance Minister Sun Yin-shien. HAL concluded that the biotechnology in Taipei was a few years behind that of the U.S., and recommended the Taipei government investing a sum of $5 billion over a 5-year period. HAL also encouraged Taipei to work with the People's Republic of China, knowing that Taipei had unsuccessfully bid for being a member of the International Human Genome Consortium, and that PRC had been investing a lot of monies into biotechnology. In April 2001, the Taipei government announced plans to invest NT$150 billion (~$5.0 billlion) into the biotechnology and pharmaceutical sectors.

Photo 3. President Chen Shui-bian of Republic of China (fifth from left), HAL (third from left), the Biotech Delegate, and the Local Delegate at the Presidential Mansion, Taipei, November 16, 2000. HAL made a brief report to the President on the findings and recommendations of the Biotech Delegate.

Photo 4. Left: HAL and Premier Chang Chun-xiong of Republic of China shaking hands after reporting findings of the Biotech Delegate at The Grand Hotel, Taipei on November 16, 2000. Right: Former Finance Minister Sun Yun-shien and HAL at a banquet. Mr. Sun is credited for the current semiconductor success in Republic of China (November 18, 2000, The Grand Hotel, Taipei).

15.4   Enter the Dragon — People's Republic of China

As program director of Computational Biophysics and Molecular Biology at the Supercomputer Computations Research Institute, Florida State University, HAL was a delegate member of US-China Joint Conference on Computer and Education, November 9–15, 1992, with a highlight of presentations on November 10th at the Great Hall of the People, Beijing, China. The goal of the conference was to promote the use of computers in specific areas.

From February 12 through March 7, 1999, HAL was at Zhongshan University, Guangzhou, China to help establish the biotechnology and bioinformatics programs in the region. Besides lectures, there were meetings with city officials and trips to Beijing to meet with central authorities to discuss projects and to raise funding. One of the outcomes is the BioIsland in Guangzhou.

Photo 5. Left: HAL explaining to reporters the importance of bioinformatics in drug discovery at a press conference held at the Biopharmaceutical Center, Zhongshan University, Guangzhou, China. HAL was at the Department of Life Sciences of Zhongshan University from February 25 through March 5, 1999. Right: HAL at the entrance to the headquarters of Haizhu Industrial Park. The BioIsland of Guangzhou was at its formative stages then.

Photo 6. HAL and Dr. Xu Anlong at a signing ceremony, witnessed by Lao Bingjian, President and Chair of Haizhu Hi-Tech Park (middle standing), Chui Jianliang, Committee of Haizhu District Economic Development (right standing), and Chen Weilian, Public Relations, Haizhu District (right standing), and others. This was the beginning of BioIsland in Guangzhou.

On March 2, 2000, the North Bioinformatics Association Establishment Conference and Academic Report Meeting was held at the Science Auditorium, Tsinghua University, Beijing. Tsinghua University/Peking University duo is sometimes regarded as the MIT/Harvard duo of the U.S. HAL took part at the Conference as a foreign expert speaker. Also taking part, besides other participants, were some 62 members of the Chinese Academy of Sciences who were very keen to get the bioinformatics efforts in China stepped up, particularly after the Chinese Genome Program had accomplished its goal of sequencing 1% of the human genome as part of the International Human Genome Consortium.

Lest us forget, we must mention the great effort initiated by Dr. Jingchu Luo of Peking University. Luo has made Peking University one of the hubs of bioinformation in China. He is currently working with Beijing Genomics Institute (now renamed Genomics and Bioinformatics Institute, Chinese Academy of Sciences) and has plans to work with Zhongshan University in Guangzhou.

15.5   Rise with the Dragon

The general impression is that these bioinformatics and biotechnology efforts are slightly behind those of the United States and Europe. In some countries along the Pacific Rim, bioinformatics and biotechnology are burgeoning fields. Along the Adoption Curve discussed earlier, most of our Asian colleagues tend to be more pragmatic, and a few are even very conservative. China is leading the pack. Besides Japan, China is the only Pacific Rim country that is a member of the International Human Genome Consortium.

In Y2000 and Y2001, China also has entertained a lot of private funding sources even when fundings are ebbing in other parts of the world. However, the funding levels are uncreatively uniform, always the order of $2 million whatever the projects. These projects are usually projects of returning scholars. The efforts are rather diffused. We anticipate most of these me-to companies, with no proprietary technologies, will not survive beyond first-round financing, or will falter within a year or two of operations. There are yet another mode of business conduct. The government has an invisible hand in most businesses in this vast country. In an effort to encourage a more uniform spread of high-tech pockets throughout China, an unwritten regulation is that in cities where there have been no IPOs, IPOs will be approved more easily. Because of this, new companies — aimed for a quick IPOs — are being formed in remote areas where the infrastructure and distribution logistics are still wanting. Feasibility studies will reveal that most, if not all, of these operations are not viable.

There are commendable exceptions. An example is the Beijing Genomics Institute (BGI). In 2001, BGI changed its name to Genomics and Bioinformatics Institute (GBI), Chinese Academy of Sciences. GBI has numerous very impressive achievements since it started in 1998. It made its name sequencing a part of human chromosome 3 as one of the 16 members of the International Human Genome Consortium. In October 2000, BGI opened a branch in Hangzhou. The two locations, Beijing and Hangzhou, together are churning out 10 megabases or 50,000 reactions a day. Aside from finishing chromosome 3, GBI has embarked on three new projects: finding single nucleotide polymorphisms (SNPs) of the Chinese population, shotgun sequencing the pig with a Danish consortium, and sequencing the *Indica* strain of rice.[39] GBI also has plans for high throughput proteomics project based on traditional Chinese medicinal herbs.

Photo 7. Left: HAL presenting a lecture at HuaDa Genome Sequencing Center, Beijing Genomics Institute (May 7, 2001). Right: HAL at the Hangzhou Branch of Beijing Genomics Institute. To HAL's right is General Manager Tian Wei of Hangzhou Branch (May 18, 2001). The other two individuals are attorneys accompanying HAL on the trip.

The spin-off from Tsinghua University — Capital Biochip Corporation — is a result from the effort of Zhu Rongji — Premier of China, and Cheng Jing, a returning scholar from University of Pennsylvalnia. The corporation, based in Beijing and headed by Cheng Jing, specializes in microfluidics chip (lab-on-a-chip). The chips will be mainly used for laboratory preparation, upstream of microarray hybridization. Capital Biochip Corporation has plans to launch several products in the upcoming year. The company has a window in San Diego. Aviva, formerly Artloon, is doing very well financially.

The only commercial biochip company that currently has products in China is BioWindow Gene Development, Inc. — a holding company of United Gene Holdings, Ltd. Founded in March 1998, they have engaged in large-scale cDNA cloning and sequencing, functional genomics, drug discovery and drug development. Their operation is a copy cat of Incyte Genomics' business model. They entered cDNA microarray field in 1999. The microarray products and services they currently offer are,

❑ BioDoor™ Gene Expression Microarrays
❑ BioDoor™ Classified Gene Expression Microarrays
❑ BioDoor™ Disease Management Microarrays

---

[39] Aaron J. Sender, "Great spiral forward", *Genome Technology*, April 2001, pp. 36–37.

❑   BioDoor™ Diagnostic Microarrays
❑   BioDoor™ Mouse Expression Microarrays
❑   Custom Made Microarrays
❑   Microarray Related Services

When BioWIndow first launched their microarray products in mid 2000, the initial market response was very positive. Their sales of all microarray related products and services since introduction exceeded $1 million. Despite the initial successes, the company just experienced a big layoff.

In Y2000, Peking University and Hong Kong University formed a consortium with Shenzhen Central Hospital. Also, the Tsinghua University Group has established business operations in Shenzhen.

The Zhongguancun Haidan Science Park started as an experimental zone for promoting new technology. It capitalizes on its strategic location in the capital city, and is being fueled by talents from 56 colleges such as Tsinghua University and Peking University, and 138 research institutes such as the Chinese Academy of Sciences. Its One-Four-Eight Project — One new industry policy, four bases of focus areas, and eight aspects of good environment — is on target.[*] Proud spin-offs include Legend and Founder.

The Pudong District of Shanghai is exemplary. For example, recently, Professor Chen Zhu, Director of the Chinese National Human Genome Center in Shanghai and Vice President, Chinese Academy of Sciences, reported a new finding of a collaboration between Sino-German Genome Center and Hoffman-La Roche Genetics.[40] Working as part of the collaboration, Dr. Liu Lisheng of Beijing-based Fuwai Hospital, has identified three to four genetic markers on chromosome 2 that likely cause strokes — the number one killer in China — from 3,000 DNA samples of Chinese stroke patients.

These are commendable business models. But, currently, the Chinese pharmaceutical industry does not have the ability to discover and develop new drugs. For years, the Chinese pharmaceutical companies have been making generic drugs. As a result, the Chinese pharmaceutical companies have never been considered a serious player in an industry that generates multi-billion dollars every year. As the date that China joins WTO approaches, the Chinese pharmaceutical industry is going to face fierce competition and threat from international drug companies even in the domestic market. To survive, Chinese pharmaceutical industry needs help to discover and develop new and effective drugs to meet patients' needs.

---

[*] One policy: promote development of Haidan Science Park. Four bases: software, biology and medicine, high technology, and innovation. Eight aspects of better environments for: innovative activities, open mindedness, legal system, finance resources, intellectual talents, continuing education, industry devlopment, and government management and social services.

[40] Chen Zhu, report in "International Symposium on Environmental Genomics and Pharmacogenetics", May 14–17, 2001, Shanghai.

15.6 Asia and Pacific Rim Rising

Besides these centers along the Pacific Rim, there are also bioinformatics centers in Australia, Indonesia, Japan, Malaysia, Myanmar (formerly Burma), the Philippines, Thailand and Vietnam. The Philippines is part of the International Rice Research Institute.

Currently, HAL is helping establish other bioinformatics and biotechnology programs in the United States, China and India.

A general good sign, nevertheless, is that these Pacific Rim countries are putting in huge efforts and are making great strides. For example, the People's Republic of China, the Republic of China and Singapore have each committed a handsome sum of money into biotechnology. Employing a catch up mode of development by attracting returning U.S.-trained and Europe-trained scholars, these countries should be able to catch up sooner than most people would think.

While we do not encourage outright patent or intellectual property infrigements, we opine copying to catch up is simply the fastest way to advance for these countries. Why duplicate efforts when advances can be made much sooner and cheaper by following trodden research paths? However, this is normally more easily said than done. Most societies resist copying. The copying process has to start by admitting that there is something to learn or that someone else does it better. Believing that if that something was not invented domestically, it cannot be worth copying is a universal human egostic failing. In the 1980s, American firms narrowed the manufacturing quality gap with the Japanese by shamelessly copying them. So why not these countries?

Increasingly, the acquisition of information and knowledge is central for both catch-up countries and stay-ahead countries. Smart developing countries understand the reality facing them. Operating as a monopsonist — a monopoly buyer — these developing countries use their domestic markets as enticements and carrot sticks. For example, China has been saving 30% of its income and has accumulated $100 billion in international exchange reserve. It does not need foreign capital for development. But it can demand the sharing of technology from companies that sell in their market.[41]

Stay-ahead countries deplore the demands. But they should remember how they got to the present position in the first place. For example, U.S. remembers fondly the clever Yankee engineer with a photographic memory who visited British textile mills in the early 1800s and then reconstructed mills in New England.

---

[41] R. Bernstein, and R.H. Munro, *The Coming Conflict with China*, Vintage Book, New York, 1997.

16 Who Owns What?

Capitalism began in Great Britain when the enclosure movement converted communal agricultural lands of feudalism into privately owned land. To have an enforceable property right in capitalism, who owns what has to be clearly defined.[42]

The private ownership of productive assets and the ability to appropriate the output that flows from those assets lies at the core of capitalism. But not all countries are capitalist countries. Different cultures in different parts of the world view and interpret intellectual property rights differently. The idea that people should be paid to be creative is a point of view stemming from the Judeo-Christian-Muslim belief in a God that created humanity in his image.[43] It has no analogue in Hindu, Buddhist or Confucian societies. Knowledge is seen as a free good.

On the other hand, what everyone owns, no one owns. Consider environmental pollution, everyone has an incentive to pollute, to use the free disposal system that is available and let someone else downstream or downwind bear the cost of cleaning up. However, on land that is privately owned, pollution market works. Private owners do not let their neighbors dump waste on their property. Someone siezes the opportunity and gets into the business of opening up a dump site. The only problem with these private dump sites is that their owners in turn have an incentive to abuse the free pollution rights.

Capitalism cannot deal with pollution because it cannot establish the ownership rights to clean the environment. In certain sense, we were extremely blessed that during the U.S. longest period of prosperity of the Clinton administration, we had an environmental champion Vice President Al Gore.

16.1 Loopholes In Intellectual Property System

In this new economy, raw materials can be bought and moved to wherever they are needed. Financial capital is a commodity that can be borrowed, or financed by venture capital. The knowledge that used to be tertiary after raw materials and capital in determining economic success is now primary. The source of future success of a business is apt to be buried in the software and hardware of its electronic information and logistics systems rather than the advertising or the novelty of its products. Intellectual property rights now affect all businesses. It has moved from the periphery to the center of economic success. The most successful companies have some lock on some form of knowledge. So, without a clear and enforceable system, knowledge-based capitalism is not going to work.

Intellectual property is also being used strategically. Patent suits are used to create uncertainties, time delays, and higher start-up costs for competitors. Whether

---

[42] Lester C. Thurow, "Needed: A new system of intellectual property rights", *Harvard Business Review*, Sept/Oct 1997, pp. 95.
[43] Lester C. Thurow, *Building Wealth*, (HarperCollins Publishers, New York, 1999).

used as an incentive for creativity or as a business strategy, historically, efforts to establish and enforce ownership rights to intellectual property have revolved around patents, copyrights, trademarks, and trade secrets. But new technologies are eroding the applicability of this old system, which was designed essentially for the technology of the nineteenth century. For example, this old system allows one to go to a library and browse a book such as this one without paying the publisher or author. But is downloading a book from the Internet equivalent to browsing? If so, how does one sell books when the books can be electronically scanned or downloaded onto a computer free of charge?

The important point is that the current legal system, designed more than a hundred years ago to meet the simpler needs of an economy based on natural resources and mechanical devices, is no longer adequate. The system of intellectual property rights, couched in the language of the current legal system, is an undifferentiated, one-size-fits-all system.

The prevailing wisdom of those earning a living in the current legal system is that some minor tweaking here and there will fix the problem. Opening up the system would be equivalent to opening up Pandora's box. Unfortunately, the time has come for a complete overhaul of the system from ground up. This is unlikely going to happen if left to those making a living in the current system. They have too many vested interests in preserving the system.

Without a reasonable system of protection, companies will defend their economic positions by keeping their knowledge secret. Secrecy is an enemy of knowledge expansion, more so than a monopolistic system of protection of intellectual rights. An investigator who knows what is already known can proceed to the next step, instead of groping in an intellectual wilderness looking for trodden paths. Indeed, a recent research found that 73% of private patents were based on knowledge generated by public sources such as universities and nonprofit or government laboratories.[44]

When it comes to biotechnology, the patent issue can be even more troubling. For example, no society will let anyone have a monopoly on the cure for cancer, nor will biologists be allowed to freely clone or own human beings. This is why the publicly funded consortia such as the Genome Consortium, the Fugu Fish Consortium are commendable worthy efforts. But it is clear that private companies must be allowed to own part of the human beings. Otherwise, no company will invest the funds necessary to find cures. Thus we have Celera Genomics, Human Genome Science and the like.

Biotechnology is a dual-use technologies. The same techniques that can cure human ailments can also be used to enhance humans or make humans more efficient such as smaller but smarter. Thus patents on genetic cures is difficult to differentiate from patents on human enhancements. A possible solution is to draw a distinction

---

[44] William J. Broad, "Study finds public science is pillar of industry", *New York Times*, May 13, 1997, pp. C1.

between fundamental advances in knowledge and logical extensions of existing knowledge. Inventing a new piece of biology that alters the natural characteristics of plants, animals or humans is not the same as discovering how an existing piece of biology works. What a patent means in these instances has to be different.

16.2   Brain Drain Or Gain?

Technology has juxtaposed intellectual property ahead of raw materials and capital. Other factors are also contributing to the need for a better system. From the Second World War to recently, knowledge has flowed freely and cheaply around the world. The U.S. financed for most of the basic research while most other countries invested mainly in development. With the exception of classified technologies, worldwide dissemination was encouraged. During the Cold War, the economic success of other countries was seen as important to the strategic geopolitical position of the U.S. as to its internal economic success.

But the world is now very different. U.S. economic dominance is not as clear as it used to be. As a vivid sign to control the flow of information, members of the Congress call to keep foreign students out of U.S. universities or laboratories. However, we must not forget that most of these people that we get are the cream from these foreign countries. These countries have footed for the most expensive part of their early education. If we retain them after training them, these foreign countries are losing and we are gaining. In fact, this is how the U.S. have come to attain its technologically dominance. We should be thankful that, given the option, these people would prefer to stay and contribute to this nation and call it home.

In summary, what different countries want, need, and should have from an intellectual property system is a function of the level of economic development. The U.S., which plays the role of stay-ahead, would like to have the national system evolve into a *de facto* international system. But this will not happen for countries playing the role of catch-up have the right to a world system that lets them succeed.

On the issue of brain drain or brain gain, it is a matter of perspectives.

Chapter 13

A NEW KIND OF WAR: BIOWARFARE AND INFO WARFARE

"One hundred kilograms of anthrax spores could wipe out an entire city in one go. It is only a matter of time before bioterrorists strike."
Robert Taylor, Bioterrorism Special Report: All Fall Down, New Scientists, September 19, 1998.

1   A New Kind Of War

The full effects of the horrendous September 11, 2001 attacks on the World Trade Center Towers in New York City are still to be felt. During the two fiscal quarters prior to the incident, the United States had been experiencing economic downturn. The September 11 incident exacerbated the gloomy outlook.

Since the end of the Cold War, the Western world has been experiencing an unaccustomed respite from the fears of large-scale violence. No longer do the two superpowers — United States and the Soviet Union — appear ready to bury civilization under a barrage of nuclear missiles. Military analysts warn that we should now be on our guard against a new type of savagery that kills civilians but spares their homes and offices, strikes without warning, and against which there may be no defense. What is more, this threat requires no radically new technology. The laboratories of academia and the biotechnology industry indirectly contribute to its development. The threat is bioterrorism.

Many experts say that it is no longer a question of whether a major bioterrorist attack will occur, but when. When is now. The United States is currently besieged by bioterrorism. For now, the bioagents is anthrax, spread by the U.S. postal system.

Yet hitherto, people understand natural calamities — earthquakes, hurricanes, tornadoes; people understand man-made destructions — explosions, buildings collapsing, viruses and worms on computers; but very few understand chemical and biological weapons. Not everyone lives or works in tall buildings, not everyone travels on airplanes, not everyone has access to the Internet. But everyone receives mail — important mail, much reviled junk mail, unwelcome bills, and now mail laced with deadly agents.

In the terrifying anthrax maelstrom within weeks after September 11, 2001, the mail system has been swept to the center of the vortex. In the aftermath of the September 11 attacks, the U.S. government has sought to drop bombs on and inserting elite ground troops into Afghanistan. So far the pay off has been little. Instead, the U.S. is not only being forced to play defense, but also to play on its home court, whether by the enemy apparent, or by some other unknown parties. The frontline soldiers in the war on terrorism were supposed to be in protective vests, well armed, well trained and well protected. Instead, the battlefield has turned out to be the postal facilities, where workers are girded in nothing more than blue slacks, short sleeve shirts, and shoes.

The U.S. Postal Service handles more than 200 billion pieces of mail per year or 600 million pieces per day. Its efforts to calm the fear of its 800,000 employees, and the 7 million Americans who visit the post office daily have been far from satisfactory. For a few of the employees, it is postal mortem. For a few of the customers, it is death on arrival. For many, it is mass hysteria.

2   United States Of Anxiety

In a fundamental way, the recent events unfolding in the United States of America have transformed the country into the United States of Anxiety. The anxiety is very different from the risks that we are used to facing — the risk of getting cancer, the risk of dying in an automobile accident, the risk of our kids getting hurt playing in the neighborhood — because in all these cases, we as a nation can fundamentally change the odds.[1]

Daniel Creson, a professor of psychiatry at the University of Texas Medical School and a veteran of many disaster relief efforts, describes response to fear as a two-prong phenomenon. The daylight (understanding of what is previously puzzling), rational part of the brain is full of reassurance, but the deeper, instinctual part is not so sure! Even when we are outwardly calm, we are inwardly anxious. Reasons get set aside. Reinforcers of the emotional response surround the public, and rumors run rampant. The psychological impact of chemical or biological weapon is much greater than the physical impact.

Anthrax is not contagious, but fear is. As America learned of new cases of anthrax within weeks after September 11, 2001, an epidemic of vulnerability and panic spread. It was an epidemic with apparent physical symptoms, some of which even wear striking resemblance to early anthrax. In reality, they may portend an outbreak of mass psychogenic or sociogenic illness, more often known as mass hysteria. Mass hysteria emerges from a largely or completely baseless belief that produces ill effect on the mind or the body.[2]

We are warned to watch out for flu-like symptoms of anthrax just as we head into the flu season, a disease that hits 95 million Americans each year and could still kill about 20,000 annually, more than anthrax might. Combine the current crop of mixed messages, the coming flu season — early anthrax symptoms resemble flu, and

---

[1] Katherine Stroup, John Horn, and Adam Rogers, "Facing up to our fears", *Newsweek*, October 22, 2001, pp. 66–69.
[2] M.D. Feldman, J.M. Feldman, *Stranger Than Fiction: When Our Minds Betray Us*, (American Psychiatric Press, Inc., Washington, D.C., 1998).

continuing terrorist threats, the result may be a truly debilitating epidemic of mass hysteria.

3 High Tech, Low Tech And No Tech

A few hundred kilograms of a properly 'weaponized' bacterial preparation, carefully dried and milled to a precise particle size, has the potential to wipe out the inhabitants of an entire city in a single strike. A nuclear bomb in the hands of a deranged person has long been the stuff of nightmares, but the materials needed to make such a device are hard to obtain and exceedingly tricky to assemble. Biological weapons are not nearly so difficult to manufacture, though making them into a form for mass destruction may be quite involved.

Biological weapons have been with us for more than half a century, but military commanders consider them too unpredictable and slow-acting, preferring the touch-of-a-button reliability of explosives. What is more, the international condemnation that the use of biological weapons would provoke gives any rational military strategist pause. Biological weapons were also an unlikely choice for most politically inspired terrorist organizations. Traditionally, political terror groups were trying to get a seat at a negotiation table and to establish the legitimacy of their cause. That goal would not be met by resorting to bioterrorism.

Even so, terrorist experts have feared that the probability of a surprise biological attack on an unprotected city has increased. Many point to a new brand of terrorism — epitomized by Aum Shinrikyo — that lacks the restraints imposed by a political agenda. These are those who do not seem to care about establishing legitimacy, but just want to strike a blow in anger and kill as many people as possible.

3.1 Germ Warfare Agents

In the last century, terrorists used violence to try and get power or approval. Nowadays, those who feel marginalized within the world economy, from religious extremists to the merely unhinged, increasingly just want to kill people or damage industries. So far they have struck mainly with guns and bombs. But the perfect weapon for those who wish only to kill or destroy is germ warfare for which we still have little defense.

Biological warfare (BW) involves the use of living organisms for military purposes. The weapons can be viral, bacterial, fungal, rickettsial, and protozoan. The agents can mutate, reproduce, multiply, and spread over a large geographical terrain by wind, water, and by insect, animal, and human transmission. Once released, biological pathogens are capable of developing viable niches and maintaining themselves in the environment indefinitely. Conventional biological agents include *Yersinia pestis* (plague), tularemia (a plague-like disease), rift valley fever, botulism (caused by a toxin from the common food-poisoning bacterium *Clostridium botulinum*), *Coxiella burnetii* (Q fever), eastern equine encephalitis, anthrax, and smallpox.

Table 1. A biological lethal weapon inventory and their symptoms. (Table adapted from *New Scientist*, 28, February, 1998).

Disease	Agent	Symptoms
Aflatoxin	*Aspergillus Flavus*	Nausea, vomiting, then acute liver failure or cancer
Anthrax	*Bacillus anthracis*	High fever, labored breathing, rapid heartbeat
Botulism	*Clostridium botulinum*	Nausea, fatigue, cramps, headache, respiratory paralysis
Plague	*Yersinia pestis*	Lung infection, pneumonia, haemorrhage
Ricin	*Ricinus communis*	Convulsions, stupor, vomiting, bloody diarrhea

Biological weapons have never been widely used because of the danger and expense involved in processing and stockpiling large volumes of toxic materials and the difficulty in targeting the dissemination of biological agents. Advances in genetic engineering technologies over the past two and a half decades, however, have made biological warfare viable for the first time.

Breakthroughs in genetic engineering technologies provide a versatile form of weaponry that can be used for a wide variety of military purposes, ranging from terrorism and counterinsurgency operations to a large-scale warfare aimed at an entire population. Unlike nuclear technologies, genetic engineered organisms can be cheaply developed and produced. They also require far less scientific expertise, and can be effectively employed in many diverse settings. These factors rekindle military interest in biological weapons. But at the same time, it also generates grave concern that an accidental or deliberate release of harmful genetically engineered microbes can spread genetic pollution around the world, creating deadly pandemics that destroy plant, animal, and human life on a mass scale.

Recombinant DNA designer weapons can be created in various ways. The new technologies can be used to program genes into infectious microorganisms to increase their antibiotic resistance, virulence, and environmental stability. It is also possible to insert lethal genes into otherwise harmless microorganisms, resulting in biological agents that the body recognizes as friendly and does not resist. It is even possible to insert genes into organisms that affect regulatory functions that control mood, behavior, and body temperature. It is also possible to clone selective toxins to eliminate selective racial or ethnic groups whose genotypic makeup predisposes them to certain disease patterns. Genetic engineering can also be used to destroy specific strain or species of agricultural plants or domestic animals, if the intents are to cripple the economy of an adversarial country.

Table 2. Not every bacterium or virus can be made into a weapon, and many would be hard to deploy. This table lists the agents most worry scientists as potential bioweapons. (Table adapted from *U.S. News & World Report*, November 5, 2001).

Weapon	Availability	Means of spread	Counter measures	Symptoms
Botulism (bacterium)	Easily produced. Iraq and Former Soviet Union reportedly have this weapon.	Through air or food.	Neutralized when heated above 85 degrees. Not contagious. Can be treated with antitoxin.	Severe symptoms include paralysis. Can be fatal.
Foot and Mouth disease (virus)	Easily found and cultured. Spread quickly among cattle.	Aerosol. Could be brought into livestock pens.	Cannot affect humans.	Economic disaster.
Plague (bacterium)	Stored in microbe banks around the world. Weaponized by the Former Soviet Union.	Aerosol could lead to outbreak of pneumonic plague. Also spread by personal contact.	Can be destroyed by heat, sun, and disinfectant. Can be treated with antibiotics.	Could affect thousands. Can be fatal.
Smallpox (virus)	Samples stored in U.S. and Former Soviet Union. Iraq, and North Korea are believed to have it.	Aerosol, personal contact.	Vaccine, given soon after exposure, can prevent deadly illness.	Fatality rate higher than 30% if inhaled.
Salmonella (bacterium)	Easy to obtain.	Through food.	Not contagious. Some strains can be treated with antibiotics.	Can be fatal.
Tularemia (bacterium)	The Former Soviet Union produced strains resistant to antibiotics and vaccines in the early 1990s.	By aerosol or through food. Not contagious.	Difficult to stabilize. Killed by heat or disinfectant. Can be cured with antibiotics.	Could affect thousands. Can be fatal.

The widespread use of the basic tools of industrial biology has put the power to create 'traditional' biological weapons in the hands of tens of thousands of people. Advanced biological technologies have spread all over the world. There are many more people who are technically trained, and the methods for culturing large quantities of bacteria are well worked out and commonly employed.

The number of trained biologists has been soaring. Life science Ph.D.s awarded in the U.S. increased by 30 per cent between 1975 and 1991 to more than 5,700 a year. By 1994 England alone had 5,700 biology graduate students. American industry now employs around 60,000 life scientists. There are over 1,300 biotechnology companies in the U.S. and about 580 in Europe. Only 25 years ago there were none. Moreover, many less developed countries, including Iraq, have their own biotechnology industries.

A person who is smart, determined, trained in basic microbiological techniques, and willing to take a few short-cuts on safety and go at a few technical problems in mildly unconventional ways, could conceivably do some horrible things.

Two factors have made the threat of a bioterrorist attack greater than ever before[3]:

❑ First, the unspoken taboo that previously dissuaded terrorists from using chemical or biological weapons against civilians has now been broken. On 20 March 1995, the nihilistic Japanese cult Aum Shinrikyo unleashed nerve gas on the Tokyo subway, killing 12 people and hospitalizing five thousand. Aum was also developing biological weapons.

❑ Second, with the explosive growth of basic biological research and biotechnology, what was once regarded as esoteric knowledge about how to culture and disperse infectious agents has spread among tens of thousands of people.

Many experts say that it is no longer a question of whether a major bioterrorist attack will occur, but when. Indeed, just days before the start of the air war on January 17, 1991, it is believed that Iraq had moved 157 bombs filled with botulinum, anthrax, and aflatoxin to airfields in western Iraq. In addition, 25 warheads missiles filled with the same biological agents were made ready for use at additional sites.[4]

3.2   Aum Shinrikyo — Japanese Cult

John Sopko and his colleagues on the staff of the U.S. Senate Permanent Committee on Investigations found that despite the Japanese cult's ineptitude there was plenty of reason to take notice. In a report presented in 1996 at one of a series of Senate hearings on terrorism, they wrote that the cult, which had more than 40,000 members in Japan and Russia and one billion dollars in assets (higher than the $300 million mentioned above), had recruited hundreds of scientists to assist with its avowed purpose of plunging the United States and Japan into a war of Armageddon from which the cult would arise as the supreme power in Japan.[5]

The cult also had a large biological weapons program, the precise extent of which remains unexplored to this day. There is an Aum laboratory — now sealed — that was devoted to biological agents, which has not yet been fully investigated. As early as 1990 they were trying to aerosolize botulinus toxin. It is believed they had anthrax as well. In 1991, Asahara led an expedition to Zaire to obtain samples

---

[3] Robert Taylor, "Bioterrorism report: all fall down", *New Scientist*, May 11, 1996, pp. 32.

[4] "A new strategic concept for NATO", *Papers on International Security*, May 20, 1997, No. 20.

[5] John Sopko and Alan Edelman, "Global Proliferation of Weapons of Mass Destruction: A Case Study on the Aum Shinrikyo", Staff Statement, in *Global Proliferation of Weapons of Mass Destruction;* in Hearings, Permanent Subcommittee on Investigation, Committee on Governmental Affairs, United States Senate, Part I, 1996, pp. 47–102.

of the Ebola virus. It can only be assumed that they had progressed since then, but how far they got no one knows.

The Japanese cult is now pretty much out of action. The concern is that new groups will look at Aum Shinrikyo's activities, try to copycat and outdo them. Compared with Sarin gas, biological agents might look a lot easier to work with in terms of access to the material, the level of expertise needed, and rather importantly, the higher effectiveness as fatal weapons.

3.3   Anthrax As Bioweapon

Anthrax, a disease of cattle and sheep caused by *Bacillus anthracis*, can also kill humans. *B. anthracis* is a rod-shaped microbe that grows in soil, where it can be ingested by sheep, cows, horses, and goats. That is why anthrax is labeled as a veterinary disease, and why those most likely to contract it work with animals or animal products such as wool. If growing conditions deteriorate, the bacteria form microscopic spores, which can remain dormant but still lethal for decades. In the worst known anthrax outbreak, at least 66 people died when spores were released from a bioweapon plant in Sverdlovsk, Russia, in 1979.

The cutaneous or external form of the disease, which sometimes strikes people who handle infected fleeces, causes unpleasant sores. Animal studies indicate that cutaneous anthrax can be caused by as few as 10 spores. The pneumonic form is far more serious, killing more than 90 per cent of its victims if left untreated. The key to triggering the second form of the disease is to create and disperse spores containing particles of exactly the right size, between 1 and 5 microns, to ensure that they are retained in the lungs. Evidence to date indicates that inhalation anthrax almost never spreads from person to person because infection seems to require thousands of spores. But as few as 8,000 spores per person reliably causes a lethal infection. The spores cross the epithelial lining of the lungs and travel to the lymph nodes, where they germinate, multiply, and then spread to the other tissues, releasing toxins as they go. The first symptoms include vomiting, fever, a choking cough and labored breathing. Antibiotics can cure patients in the earlier stages of the disease. Without antibiotics, death from haemorrhage, respiratory failure or toxic shock follows within a few days.

A U.S. Office of Technology Assessment (OTA) report emphasizes that, for the most part, transforming *B. anthracis* into a weapon is a low-tech procedure. It also notes that on a clear, calm night, a light plane — similar to the one that crashed into the White House in 1994 — flying over Washington, DC, carrying 100 kilograms of anthrax spores and equipped with a crop sprayer, could deliver a fatal dose to up to three million people.[6]

---

[6] Office of Technology Assessment, U.S. Congress. 1993. "Proliferation of Weapons of Mass Destruction", OTA-ISC-559, (Government Printing Office, Washington, DC), pp. 53–55.

Making an anthrax weapon capable of murder on this scale is not a trivial undertaking. But while it may be much more difficult than building a fertilizer bomb, the problems are far from insurmountable. The tricky part is not culturing the agent. Indeed, growing *B. anthracis* is hardly more difficult than growing sourdough starter. But processing the crude colony into a form suitable for dispersal is another matter. Turning bacteria into spores, the only form hardy and stable enough to be spread, requires the tricky step of shocking the bacteria with heat or chemicals without killing them.

A project of this complexity would require months of systematic effort, the practical engineering skills of a clever backyard inventor, and luck. These barriers, however, are not impossibly high. Basic microbiology skills — techniques an undergraduate studying the subject would be taught — should be sufficient to isolate *B. anthracis* from cattle pasture in areas where the disease is endemic, such as small areas of the U.S., and larger tracts of land in Russia and South Africa. Using this as the starter culture, a terrorist with a 100-liter culture vessel — about the size of a home fish tank — could in a few days brew up several kilograms of crude slurry containing billions of spores.

Spores tend to clump. Yet the ideal size of particles must be between one and five microns to enter the lungs and trigger inhalation anthrax. It is not easy to get a sprayer to dispense such a fine mist. The particles have to be dried somehow. Drying the slurry would be tricky, though not impossible. Freeze-drying — a procedure in which material is frozen and put under a vacuum to remove water, and which is used on a small scale throughout the biotechnology industry — could be one option.

The slurry then would have to be adjusted to the right size. Grinding the slurry powder into particles of the desired diameter would provide the greatest challenge, mainly because of the risk of contamination. Indeed, the most likely glitch all round is that the terrorists would become the first victims, or that they infect their neighbors and give the game away. A case in point is the best known (or worst known depending on which perspective you take) anthrax outbreak from a bioweapon plant in Sverdlovsk, Russia, in 1979. At least 66 people died from the incident.

To be used as a weapon, the particles would also have to be loaded into a canister for spraying over targets. To be sure the preparation would work, the isolate should be tested for virulence, the particle size measured and perhaps the sprayer field tested with a non-pathogenic bacterium. All the while the whole operation has to stay clandestine and avoid detection.

The current U.S. postal anthrax cases have not changed one crucial reality: turning pathogens into weapons of mass destruction is still a very difficult task. As far as intelligence agencies can tell, the group that has put the most effort into bioterror was Japan's Aum Shinrikyo cult. They too, went through several futile attempts. In April 1990, members drove an automobile outfitted to disseminate botulinum toxin around Japan's Parliament building. In June, 1993, they tried to disrupt the Royal wedding of Japan's crown prince by spreading botulinum toxin in a similar way. They also tried, for four successive days the same month to spread anthrax from a rooftop in Tokyo. All they had was nine failures in nine attempts. For Aum Shinrikyo who has a war chest of about $300 million, half a dozen laboratories and experienced biologists to fail so miserably can only attest to the difficulty in deploying biological weapons to cause mass casualties.

But the U.S. postal anthrax cases have shown for the first time in history, some person or group seems to have found an effective way to spread highly toxic anthrax spores efficiently enough to kill and sicken people. And a poorly equipped and poorly informed public health system is now struggling to figure out the basics — how spores do their damage, how many it takes, how they are transmitted?

3.4 Smallpox — Eradicated but Not Gone for Good

Smallpox is a highly contagious, deadly, and disfiguring illness that spreads through populations rapidly and for which there is no treatment. It is not easy to use as a weapon, but it is not impossible. Indeed, the U.S., the Former Soviet Union, and other nations have experimented with smallpox as a bioweapon, until these programs came to a halt in 1972. Officially, the U.S. and Russia still have stockpiles of smallpox virus hidden in vaults in the Center for Disease Control in Atlanta and the Institute for Viral Preparations in Moscow. Iraq, North Korea and possibly other nations and terrorist groups may have the virus as well.

The contagious nature of the virus makes it hard for any terrorist to stay safe while creating an aerosol form.

Even when vaccine is available, to vaccinate or not to vaccinate is a complicated issue. Fatal complications with smallpox vaccine is 1 in 500,000 cases. Now we are also concerned about people who are immune deficient such as people with HIV, or transplant, or older people. The vaccine might produce an illness called progressive vaccinia. Progressive vaccinia is difficult to treat and can be fatal. D.A. Anderson, a key Health and Human Services advisor on bioterrorism and the doctor who led the World Health Organization's global fight to eradicate smallpox, advises that it does not make sense to vaccinate everyone when the risk is low. But the balance could change as soon as the first case of smallpox appears.

3.5 Deadly Past in History — Smallpox

Historically, smallpox had proven a particularly vicious killer. It did not, as was typical of most infectious diseases, preferentially attack the most impoverished member of the society.[7] In 45 A.D., it appeared in Asia. A hundred years

---

[7] T. McKeown, *The Origins of Human Diseases*, (Basil Backwell, Oxford, UK, 1988).

afterwards in 165 A.D., the Roman Empire was devastated by an epidemic believed to have been smallpox. The pestilence raged for about 15 years, claiming victims in all social strata in such high numbers that some parts of the Roman Empire lost 25% to 35% of their people.[8]

Over subsequent centuries equally devastating pandemics of the viral disease claimed millions of lives in China, Japan, the Roman Empire, Europe, and the Americas. According to an account, Spanish conquistador Hernando Cortez's capture of Mexico City in 1521 with just a small army of exhausted Spanish irregulars under his command was possible only because the Europeans had unknowingly spread smallpox throughout the land. When Cortez launched his final assault on the capital, few Aztec soldiers were alive and well. Smallpox, together with measles, tuberculosis, and influenza, claimed an estimated 56 millions Amerindian lives during the initial years of the Spanish conquest.[9,10]

Table 3. Smallpox is a relatively recent human disease, seeming to have arisen in India less than 2,000 years ago. In ancient times, medical observers could not clearly discriminate between smallpox and other human-to-human epidemic diseases such as measles, bubonic plague, and typhus. As a result, controversy reigns over modern interpretations of ancient medical records. Nevertheless, according to historians familiar with medical records, several major epidemics that claimed a quarter to a third of the affected populations were likely to have been smallpox. (Table Adapted from A. Patrick).[11]

Epidemic Site	Year, A.D.	To
China	49	
Rome	165	
Cyprus	251	-66
Greece	312	
Japan	552	
Mecca	569	-71
Arabia	683	
Europe, various sites	700	-800

Among the first to deliberately inflict smallpox on an enemy were the British soldiers during the French and Indian Wars of the mid 1700s. They handed out blankets used by smallpox patients to North American Indians to cause epidemics. And it worked. More than 50% of the affected population succumbed.

It is conceivable that in future bioterror attacks, agents other than non-contagious anthrax, may be used.

---

[8] W.H. McNeill, *Plagues and Peoples*, (Doubleday, New York, 1976).
[9] W.M. Denevan, *The Native Population of the Americas in 1492*, (University of Wisconsin Press, Madison, 1992).
[10] Laurie Garrett, *The Coming Plague: Newly emerging diseases in a world out of balance*, (Penguin Books, New York, 1995).
[11] A. Patrick, "Diseases in antiquity: Ancient Greece and Rome", In: D. Brothwell, and A.T. Sandison (eds.), *Diseases in Antiquity*, (Charles C. Thomas, Springfield, Illinois, 1967), pp. 238–246.

3.6 Countermeasures to Combat Bioterrorism

Apart from acting on intelligence, another defense would be to restrict access to the tools of bioterrorism, including starter cultures. In March 1995, Larry Harris, a microbiologist and a member of the Aryan Nations white supremacist group, used a forged letterhead and his professional credentials to order samples of *Yersinia pestis*, the organism that causes bubonic plague, from the American Type Culture Collection, a clearing house for microbiological samples in Rockville, Maryland. The ATCC dutifully mailed the samples, but in the nick of time the staff became suspicious that Harris did not have the expertise to handle plague and the vials were recovered unopened. Harris is being prosecuted for mail fraud–owning plague, it transpires, is not illegal in the U.S. In the U.S., people may keep lethal pathogens at home. But threats to do harm with those pathogens, transporting or storing them improperly, or obtaining them by fraud or theft, are illegal. In Britain, any company that wants to keep lethal pathogens must prove to the government's Health and Safety Executive that it has adequate containment facilities. But the HSE has no jurisdiction over private citizens.

Not that would-be terrorists need obtain their pathogens through official channels. If they know where to look, many can be isolated from the wild.

But perhaps the most neglected area of planning is the medical response to an attack. The scenario is different with the agent used. Philip Russell, former commander of the US Army Medical Research and Development Command in Fort Detrick, Maryland, believes plague is different from smallpox, which is different from anthrax. Russell is now president of the Sabin Foundation, an organization based in New Canaan, Connecticut, which promotes vaccine use against natural diseases. He proposes the need for a group of folks to go through different scenarios and think about what should be in each scenario. For example, plans are needed to ensure that large amounts of antibiotics, and properly trained and equipped people can be rushed to the scene.

In the U.S., these responsibilities fall on the Federal Emergency Management Agency and the Office of Emergency Preparedness of the Department of Health and Human Services, both in Washington, DC. At the moment, although these agencies have adequate plans to cope with floods, earthquakes, and occasional car bombs, OEP head Frank Young told a Senate hearing on 1 November 1995 that there was no coordinated public health infrastructure to deal with the medical consequences of terrorism. This is not to say there are no plans at all. In June 1997, President Clinton told government agencies — including the military — to improve their planning for a massive terrorist strike. But at the Senate terrorism hearing, on 27 March 1998, several key witnesses, among them P. Lamont Ewell, president of the International Association of Fire Chiefs, questioned whether the new plans were adequate and whether they had been sufficiently well rehearsed to cope with a real attack. In Britain, the Home Office takes ultimate responsibility for preventing bioterrorism and for preparing to deal with its aftermath. In the aftermath of the postal anthrax incidents, the Bush administration set up a Home Office in the U.S.

4   Anthrax Vaccines

American forces use a vaccine called MDPH, named after the Michigan Department of Public Health vaccine plant that makes it. The British vaccine is similar. Both immunize against a protein in anthrax toxin called protective antigen. The vaccines, however, may not protect against all natural strains of anthrax.[12]

In experiments on guinea pigs, for example, MDPH gave 100 per cent protection against only one of the five main natural strains of anthrax. In some studies, anthrax killed between 25 per cent and 96 per cent of guinea pigs that had been immunized with MDPH. Primates may be less susceptible.

So far, MDPH has been tested only with natural strains of anthrax. In December of 1998, Andrey Pomerantsev of the State Scientific Centre of Applied Microbiology at Obolensk near Moscow published details of an anthrax strain that he had genetically engineered to produce bacterial toxins called cereolysins. This Russian strain resists six different antibodies including MDPH.

Experts fear that Iraq may have acquired the Russian strain. The UN Special Commission set up to investigate Iraq's biological weapons has found records of Russian sales of biological equipment and materials to Iraq as late as 1995. Pomerantsev's strains were probably developed before 1991, when funding for the Obolensk laboratory dried up.

The search for better anthrax vaccines is hampered by the difficulty of testing them on humans. Hardly anyone in the West is now naturally exposed to anthrax, and it is unacceptable to expose people deliberately. This is why British and American troops are still using the MDPH vaccine that was developed in the 1960s, when workers in wool factories were still exposed to anthrax from sheep.

4.1   Ciprofloxacin

Following the anthrax incidents about a month after the September 11, 2001 horrendous attacks on the World Trade Center Towers, the U.S. State Department ordered all U.S. embassies to buy and store three days' supply of Cipro as a precaution. Bayer, the German pharmaceutical giant, announced that it would increase its production of Cipro by 25% starting November 1 because wholesalers are running low on the drug.

Ciprofloxacin or Cipro is a high potent, high-priced antibiotic used to treat inhaled anthrax.

---

[12] Debora Mackenzie, "Naked into battle", *New Scientist*, 28 February, 1998, pp. 4.

When ingested or inhaled in large quantities, anthrax is far more deadly as it disseminates more widely within the body. Ingesting is normally not a threat for in the U.S. farm animals are vaccinated and meat is routinely inspected. But the lethal case in Florida indicates that inhaled anthrax may pose a growing threat. A victim who breaths in the white powder may feel fine for a week, then flu-like symptoms such as fatigue, fever and muscle aches begin to take hold. Chest pains and labored breathing follow within 24 hours, followed by shock and death in the worst scenario.

Inhaled anthrax is highly treatable if detected within the first few days of exposure, before the organism multiplies to a density detrimental to the host. Drugs are usually rather ineffective once the symptoms begin to show.

Cipro is the first drug used to treat any suspected inhaled case of anthrax, but the vast majority of infections can be managed with penicillin or tetracycline once the spores have been evaluated in the laboratory.

The standard treatment regimen involves 60 days of antibiotic therapy, or 30 days of medication and a series of three-vaccine cocktail shots. Intravenous Cipro, along with clindamycin — an antibiotic that works as an antitoxin are first administered. This combination may be effective because even if the bacteria are killed, the toxins they produce continue to circulate through the body. Finally, rifampin — a drug to treat tuberculosis that can penetrate white blood cells — is added to the treatment regimen.[13]

Anthrax is a deadly germ with real potential as a weapon, but our fear now poses a greater threat than the bacterium itself. Despite events in Florida, New York, Nevada and District of Columbia, a civilian chance of contracting anthrax is still vanishingly small.[14] Taking Cipro may help us feel less vulnerable, but the drug itself has a range of adverse effects. In the face of fear, Valium — a drug to suppress anxiety — may be a healthier choice.

Not that anthrax is a threat to be taken lightly. The infection is rarely fatal when contracted through a skin lesion. In this case, the patient may develop flu-like symptoms and nasty sores. Most recover even without treatment, though antibiotics are almost always curative.

Experts are adamant in urging people not to stockpile medication. Many of the people buying Cipro simply want to keep them on hand for an emergency, but others are ready to take the medication without any exposure. Cipro is rarely prescribed to pregnant women or anyone under 18 except in known cases of anthrax exposure. When prescriptions are not available from doctors, people turn to opportunists online. A 30-day prevention package costs about $299. The package would not fully protect a person who has actually inhaled anthrax because the spores can take six weeks to germinate.

---

[13] See Kaiser Permanente Web site, www.kp.org.
[14] Geoffrey Cowley, Anne Underwood, and Karen Springen, *"A run on antibiotics"*, Newsweek, October 22, 2001, pp. 36–37.

It would not cause a healthy adult much direct harm unless the "patient" is also taking asthma medications such as theophylline. But taking Cipro as a prophylactic precaution is never a good idea. Abusing any antibiotic is a sure way to breed pathogens that can resist it. By taking the drug as a hedge against anxiety, we will exhaust its power against a range of bacterial menaces.

Cipro is still a potent, broad spectrum antibiotic. Because it was introduced only 14 years ago, it can still eliminate bugs that have developed resistance to older medications. It is still our best cure against gonorrhea, infectious diarrhea, typhoid fever, and hospital acquired pneumonia. By stockpiling Cipro and using them casually, we virtually guarantee that all of those diseases will become less treatable, and the effectiveness of the drug against anthrax diminished. For now, our best defense against anthrax is to leave the drug alone. To squander it is to surrender.

Cipro can also cause a range of bizarre effects from psychological problems and seizures to ruptured Achilles tendon or shooting pains to swollen joints. In animal studies, it can disrupt the formation of cartilage. Cipro can be fatal if taken with the asthma medication theophylline. Most antibiotics can cause nausea, but the user cannot take antacids for they reduce the effectiveness of Cipro.

4.2   *Antimicrobial Nanoemulsion*

Bayer is not the only company swept up in America's grim scramble to fend off germ attacks. The $11 million DARPA-funded NanoBio, based in Ann Arbor, Michgan and a spin-off of Michigan University's Center for Biologic Nanotechnology, has created a nontoxic agent that can destroy most virus, bacterium, and fungus around, from influenza to *E. coli* to anthrax.[15]

The agent, a lotion that looks like sunblock, can help prevent people from contracting anthrax but it cannot cure a victim after infection. The microbe-zapping agent is just soybean oil floating in water with nontoxic detergents. It can be rubbed on the skin, used in hot tub, eaten, or put into beverages like orange juice. What makes the stuff potent is how it is made.

The principle is deceivingly simple. When salad dressing is shaken, bubbles of oil are dispersed in the vinegar. These bubbles contain surface tension potential energy. The potential energy is released when the bubbles coalesce. NanoBio's proprietary technology — antimicrobial nanoemulsion — forms these bubbles, as the name stipulates, at the supertiny nano level. A nanometer is about 100,000 times narrower than a human hair. The nanodroplets, stabilized by detergents they float in, are small enough to literally bombard lipids or fats found in bacteria and viruses, blowing them up in the process. NanoBio's formula tricks dormant anthrax spore that ambient surface conditions are ideal for germination into an active bacterium. As the spore germinates, it forms a lipid layer, which the nanoemulsions promptly assault. Within a couple of hours, the anthrax is dead.

---

[15] Julie Creswell, "See this goop? It kills anthrax", *Fortune*, November 12, 2001, pp. 147–148.

Nanobio plans to develop a preventive nasal spray in two years. There are other promising anthrax zappers. A foam developed by New Mexico's Sandia National Laboratories supposedly neutralizes pathogens and chemicals. It was used to decontaminate traces of anthrax found in the NBC New York offices on October 12, 2001.

5 Promises But Not Yet A Sure Cure

Traditional terrorists wanted political concessions but now, some groups have as their main aim mass casualties and mayhem. And their weapon of choice is biological weapons. Terrorists would have little trouble getting their hands on the technology. The apartheid government in South Africa produced terrorist weapons containing anthrax, *Salmonella* and cholera. Former Soviet scientists who have prepared weapons-grade anthrax and smallpox are known to have emigrated, possibly to well-funded terrorist groups.

With bioweapons so readily available, how can governments protect its citizenry from a terrorist armed with anthrax, smallpox or plague? Until now, most biological defense strategies have been geared to protecting soldiers on the battlefield rather than ordinary people in cities. The situations are quite different now, and novel technologies are needed for civilian defense.[16]

5.1 Hypothetical Bioterrorism

The first simulation has taught officials that biological terrorism poses different problems from a chemical attack, and is potentially much more devastating.

Most doctors have never seen a case of plague or anthrax. So it could be days before they realize what they are dealing with. National governments need to stockpile drugs and vaccines, develop and distribute rapid tests for agents used in bioweapons, and come up with effective ways to isolate infected people.[17]

Table 4. Estimates of casualties from a hypothetical biological attack. The numbers are based on a 50-kg airborne agent released over a 2-km radius in a city of 500,000 residents. (Table: Adapted from WHO).

Agent	Casualties	Deaths
Anthrax	125,000	95,000
Tularemia	125,000	30,000
Typhus	85,000	19,000
Tick-borne encephalitis	35,000	9,500
Brucellosis	125,000	500
Rift Valley fever	35,000	400
Q fever	125,000	150

---

[16] Debora MacKenzie, "Bioarmageddon", *New Scientist*, 19 September, 1998.
[17] Nell Boyce, "Nowhere to hide", *New Scientist*, 21 March, 1998.

5.2   Protective Gear

Researchers at Irvin Aerospace in Fort Erie, Ontario, have developed a dome-shaped tent made of ultra tough Mylar that can be filled with a stiff foam — the exact composition of which is a closely guarded proprietary information — that kills germs and also neutralizes chemical weapons. Once covered by the foam-filled tent, a bomb filled with germs can be safely detonated.

But what if germs are already in the air? Geomet Technologies near Washington DC and Irvin Aerospace are about to market civilian bio-suits. In the meantime, other companies are designing protective gear that actually kills pathogens. Molecular Geodesics in Cambridge, Massachusetts, for example, is developing a suit made of a tough, sponge-like polymer that traps bacteria and viruses, which are then destroyed by disinfectants incorporated into the fabric.

5.3   Surveillance and Monitoring

None of this gear will do any good, however, if the emergency services do not know there has been an attack. And an stealthy assault may not be obvious. A terrorist might not use a weapon that goes off with a dramatic bang, or even produces an obvious cloud of germs. The first hint of a biological attack may be a sudden cluster of sick people.

Even that will be missed unless someone is watching. And few are. In the U.S., financial cutbacks have crippled programs to track disease outbreaks, natural or deliberate. Some could be either, such as food poisoning caused by *Escherichia coli* O157 or *Salmonella*. In Europe, disease surveillance is only beginning to be organized on the continent-wide scale needed to track a biological emergency. But in addition to monitoring infected people, Nicholas Staritsyn of the State Research Centre for Applied Microbiology near Moscow says that more effort should be made to find out which bugs live where. For example, a particular variety of anthrax may occur naturally in South Africa, but not in Canada. Having access to such information could help authorities to distinguish between natural outbreaks and deliberate attacks.

Even when infected people start turning up at local hospitals, early diagnosis of their illness might not be easy. The first symptoms of anthrax, plague and many other potential agents of bioterrorism resemble those of flu: headaches, fevers, aching muscles, and coughing. What is more, some of these symptoms might be brought on by panic attacks or hysteria, which are likely to be widespread among people who have just been told that they are the victims of a biological attack.

One solution would be for hospitals to have the type of high-tech detectors being developed to identify airborne pathogens on the battlefield. With a detector at each bedside, doctors could pick out the volatile molecules released by damaged lung membranes at a very early stage of infection and instantly tell whether a patient was a victim of a biological attack.[18]

DARPA, Defense Advanced Research Projects Agency of the U.S. Department of Defense, would like to develop reliable (no false positives), lightweight (<2 kilograms), sensitive (can identify as few as two particles of 20 different biological agents in a sample of air), low cost (<$5000) detectors. Such detectors could be deployed around cities to give early warning of airborne disease.

In the meantime, researchers led by Wayne Bryden at Johns Hopkins University in Baltimore are working on revamping the traditional laboratory workhorse, the mass spectrometer, for use in the field or in hospitals. His group has reduced this unwieldy piece of equipment to a suitcase-sized machine that can distinguish between, say, *Shigella*, which causes dysentery, and *Salmonella*.

Tiny electronic chips that contain living nerve cells may someday warn of the presence of bacterial toxins, many of which are nerve poisons. Like a canary in a coal mine, the neurons on the chip will chatter until something kills them.

While the canary-on-a-chip could detect a broad range of toxins, other devices are designed to identify specific pathogens. One prototype, antibody microarray, consists of a fiber-optic tube lined with antibodies coupled to light-emitting molecules. In the presence of plague or anthrax bacteria, or the toxins botulin or ricin, the molecules light up.

Table 3. Pros and cons of various protocols for detecting bioagents. Dog's nose is solution that turns green on exposure to reagent. (Table: Adapted from Alvin Fox, University of South Carolina, Cepheid, Nomadics, Teracore).

Protocol	Pros	Cons
DNA-based detectors	A prototype machine can identify virulent strains of anthrax in about 30 minutes.	Even the newest device cannot continuously monitor the air. It is another 5 or more years to develop a device of this functionality.
Mass spectrometry	These machines spot anthrax's molecular profile.	To date no machines adapted for anthrax are available. Development is years away.
Antibody-based tests	Antibodies interact with spores and change color. It is cheap, fast, and is available now.	This device is rather insensitive. It cannot tell a virulent from a harmless strain.
Dog's nose project	This device is portable. Synthetic compounds instantly glow when they detect distinctive particles in the air.	This device has been used to detect traces of TNT. It may take another 5 years to adapt it for anthrax.

[18] A Strengthened Biological and Toxin Weapons Convention (BTWC): Potential Implications for Biotechnology — An International information and discussion forum on the potential implications for biotechnology R&D and production of the legally binding protocol being negotiated to strengthen the BTWC 28–29. May 1998, Institute of Applied Microbiology, Vienna, Austria.

Devices based on antibodies are far from foolproof. First, the correct antibodies have to be identified, not easy when one considers the vast number of pathogens that need to be included, and their ever-changing repertoire of surface proteins. Even the right antibodies can identify only what is on the outside of a particle. Bugs can be encapsulated in gels or biological polymers to foil antibodies, or normally harmless bacteria engineered to carry nasty genes.

To overcome this, researchers are developing identification techniques based on RNA analysis. Unlike DNA, which is now used to identify unknown organisms, RNA is plentiful inside cells and need not be amplified before identification begins. And messenger RNA molecules reveal not only what a microorganism is, but what toxins it is making.

Once the biological agent has been identified, what measures should be taken to combat it? Vaccinating people before they are exposed is one answer. This is the strategy the military is betting on. In 1997, the U.S. military launched a program to develop vaccines against potential biological weapons. It will create jabs for diseases for which none exist, such as Ebola, and improve existing vaccines, including the 30-year-old MDPH anthrax vaccine being given to 2.4 million American soldiers.

5.4   Quick Counter Jabs

But vaccines are no panacea. An attacker needs only generate a germ that sports different antigens to those used in a vaccine to render that vaccine ineffective. In addition, as bioterrorists get more sophisticated, they will develop novel, possibly artificial, pathogens against which conventional vaccines will be useless. To get around these problems, the U.S. military is looking at ways of developing vaccines quickly enough for them to be created, mass-produced and distributed after an attack. The first step, which many researchers including those in the fast-paced field of genomics are now working on, involves speeding up DNA sequencing so that an unknown pathogen's genes could be detailed within a day. The resulting sequences could then be the basis for developing an instant DNA vaccine.

Making the vaccine is only half the problem, however. Soldiers can be ordered to take shots, but immunizing the rest of the population is another matter. Civilians are unlikely to volunteer for the dozens of vaccinations that would be necessary to protect them against every conceivable biological threat during peace time. An attack would make many change their minds, but in such circumstances there might not be enough to go around.

Kanatjian Alibekov, now reincarnated as the American resident Ken Alibek and author of *Biohazard*,[19] was a former second-in-command of the Soviet germ warfare

---

[19] Ken W. Alibek, and Stephen Handelman, *Biohazard: The Chilling True Story of the Largest Covert Biological Weapons Program in the World-Told from the Inside by the Man Who Ran It*, (Dell Publishing Co., New York, 2000), 336 pages.

program. Alibek, who revealed in 1997 that the Soviets had weaponized tons of smallpox, argues that it is short-sighted to put too much effort into developing vaccines. Instead, Alibek, who is now at the Battelle Institute in Virginia, argues that researchers should concentrate on ways to treat victims of biological weapons. Today's antibiotics may be useless because germs could be equipped with genes resistant to all of them. For example, Russian scientist Andrey Pomerantsev is believed to have already created such a strain of anthrax.

For any treatment to be effective amid the potential chaos of a bioterrorist attack, speed will be of the essence. Researchers are developing drugs that work against a wide variety of infections and so can be used even before definitive diagnosis. Some are trying to develop broad-spectrum drugs by taking advantage of recently identified similarities in the way many pathogens produce disease. For example, Ebola, anthrax and plague all kill their victims by inducing a widespread inflammatory reaction similar to toxic shock syndrome. A team in Cincinnati is testing an anti-inflammatory drug that could stop all of them. Another gang of bacteria, including plague, *Salmonella*, *Shigella* and *Pseudomonas aeruginosa* (one of the bacteria that can cause pneumonia and meningitis), relies on very similar proteins to latch onto human cells and inject toxins. Drugs that block this system might save people from all these germs.

6 BTWC Treaty — Firm But Unfair

One hundred and forty countries, including Iraq, have ratified the 1972 Biological and Toxin Weapons Convention (BTWC), which prohibits the manufacture and acquisition of organisms or their toxins for military use.

However, most governments agree that as it stands the convention is ineffective. Unlike the treaties that ban nuclear and chemical weapons, the BTWC provides no legal means to check if countries are complying. Treaty members are now trying to strengthen the convention to include a system of verification. Europe and many developing countries want UN inspectors to make random visits, at short notice, to any factory or laboratory in any country capable of producing lethal organisms.

But the U.S. government, under pressure from its drug and biotechnology industries, rejects this idea. The companies fear that such visits would expose trade secrets. Supporters of random inspections point out that while the U.S. is willing to go to war to back the UN's right to inspect any sites it chooses in Iraq, it will not grant the UN the same right to inspect itself or the other members of the BTWC. This is the double standard most countries complain about.

Once Iraq had joined the BTWC after its defeat in 1991, the only legal way to find and destroy the biological weapons that its own generals had claimed it had was for the UN Security Council to set up a special commission, UNSCOM. Through its inspections over the past decade, UNSCOM has tested many of the ways in which a verification regime would work for the BTWC. These include the compulsory declaration of all research and development involving biological weapons, and of any facilities that could be used to make them. An inspection team then compares this declaration with other evidence, such as government documents, trade records, interviews with scientists and visits to laboratories and factories.[20]

All signatories to the BTWC accept this approach in principle. The sticking point is how extensive the inspections should be. Everyone, including the biotechnology industry, agrees to what are known as "challenge" inspections. If there is "substantial and convincing evidence" of a breach, such as an unexplained outbreak of anthrax, a majority of treaty members can demand an inspection. But challenge visits will never be frequent enough to be a sufficient deterrent. They require too much evidence and political risk for the country making the charge.

Europe and most developing countries want random, "non-challenge" inspections. Officials would be able to visit any biological facility at short notice merely to check that everything was in order although this would still not solve the problem of secret facilities. Negotiations on the BTWC have been stuck in a deadlock for four years because the U.S. and the world's biotechnology and drugs industries will not agree to this. In late January of 1998, as the Iraqi crisis deepened, American President Bill Clinton announced he would support limited non-challenge inspections to clarify unclear declarations. But he explicitly rejected random visits.

The Pharmaceutical Research and Manufacturers of America (PhRMA), which represents American drugs companies, maintains random inspections would expose industry to the loss of its legitimate competitive trade secrets. It is also worried that an inspection for biological weapons would be disastrous for a company's public relations.

Some wonder if the U.S. is trying to hide more bioweapons research than it cares to admit by refusing random inspections. The only way to prove it is not, in Los Alamos as in Al Hakam (home of Iraq's anthrax bomb), is to let the inspectors in. The U.S. is ready to go to war to impose inspections on Iraq. It must set a good example and allow the UN to impose them on everyone, including U.S. industry.

Inspection techniques exist that could protect legitimate secrets without hindering verification. DNA probes that screen for specific DNA sequences, possibly coupled with polymerase chain reaction, as well as immunoassays, which use antibodies to reveal specific molecules, are the leading candidates for use in a compliance regime.

These techniques would need to be developed further before the BTWC could use them. But once they were ready, factory managers could supervise the tests at every step, protecting legitimate secrets without hindering the inspectors. For example, instead of taking live microorganisms out of the plant, a company would kill sampled organisms in front of inspectors and scramble the DNA enough to protect proprietary genes without disguising the species. The inspector could then run either PCR or immunoassay tests on the dead organisms with portable kits.

---

[20] Debora MacKenzie, "Deadly Secrets", *New Scientist*, 28, February, 1998.

The Chemical Weapons Treaty, which came into force in 1997, already allows random inspections, with "managed access" guidelines to protect the industry. These guidelines could be adapted for biological plants.

7   Forcing Genie Back Into The Bottle?

The example of Iraq has shown how even a relatively undeveloped country can produce an impressive biological arsenal in secret. And it has shown how hard it is to force that genie back into its culture flask. Former President Clinton admitted there is no obvious way to destroy a country's biological weapons capability with bombs. And it is terrifyingly easy to develop anthrax strains that resist both antibiotics and the West's only anthrax vaccine, MDPH.

So the way forward must be deterrence, plus inspections that can catch cheats before they get too far.[21] It was when UN inspectors in Iraq made routine monitoring visits on short notices to apparently innocent plants that they started noticing things were amiss. That is what the UN should do in every country under the verification regime now being negotiated.

As the nuclear arms race escalated in the early 1950s, the U.S. launched Atoms for Peace, a drive to promote the good things the atom might do. Whatever one thinks of nuclear power, it is hard to deny that nuclear physics and radioisotopes for medicine and research have brought benefits.

Recent alarm about biological weapons may now give us a similar opportunity. We could call it Germs for Peace.[22] Bioweapons have yet to produce a Hiroshima, or Nagasaki. But it is real. It is happening in the U.S. postal services. It may as well be U.S. postal mortem.

Some demented people have tried bioweapons. So countries are now negotiating a long overdue verification agreement to go with the 1972 treaty banning bioweapons. If they succeed, member states will have to declare what they are doing with microbes, and allow investigators in if anyone raises serious suspicions.

But the original treaty was also about the peaceful uses of biology. It called for rich countries to help poor ones combat diseases. This was not merely altruistic. Rich countries wanted poor ones to join the treaty, but most developing countries have more pressing concerns than bioweapons. So the rich nations promised biomedical investment for those that signed.

Little of this ever materialized, but now the poor countries want action as part of the planned verification agreement. This time it may happen, in the form of a plan to help poor countries to monitor diseases.

In Germs for Peace, rich countries stand to benefit as much as poor ones, as emerging infections caused by novel pathogens are one of the 21st century's more

---

[21] "Firm but fair", *New Scientist*, 28 February, 1998.
[22] "Germ for peace", *New Scientist*, 8 April, 2000.

egalitarian menaces. The world is now a global village and physical distances have shrunk thanks to modern transportation system. To face emerging diseases, we need to keep a close watch on infections worldwide. This requires top-quality medical laboratories for diagnosis and epidemiological analysis. These are rare. Even India could not reliably diagnose a suspected pneumonic plague outbreak in 1994. It got foreign help, but fast local action would have been more effective. Africa is especially short of laboratories, yet nearly half the potentially world-threatening, novel infections investigated by the WHO are in Africa or originate from Africa. The peaceful uses part of the bioweapons treaty may yet bring rich and poor together.

The issue is politically sensitive. To assess whether outbreaks are natural or illicit, the epidemiological background has to be ascertained, and this will require international monitors. Developing countries quite rightly want any investment in disease monitoring to be just that, and not a way in for foreign military snoopers or reconnaissance.

The investment, if it happens, will have to be purely civilian. Members of the treaty are discussing collaborating on regional epidemiological laboratories, quite separate from any formal effort to watch for biological attack. Ironically, if this comes to pass, the disease monitoring collaboration may become the biological weapons treaty's greatest achievement. It may or may not stop biowarriors. But natural diseases are a much deadlier enemy. Any effort to even monitor its onslaught deserves support.

8   Cyber Warfare — A Keyboard Is Truly Mightier Than A Gun

The end of the Cold War has not really brought about world peace. We have seen the end of one conflict between two superpowers — the U.S.A. and the U.S.S.R — and the beginning of a new one. The new conflict is much more spread, with potentially many more players. This new conflict is a global economic war in which espionages and new technologies will again play an important role in determining the final victors.

Beginning in World War II and continuing throughout the Cold War, the world's major intelligence agencies — the CIA, KGB's First Chief Directorate, MI6, etc. — employed the most state-of-the-art technologies available to assemble, communicate and analyze information from friendly and hostile countries. At the same time, counterintelligence agencies — the FBI, KGB's Second Chief Directorate, MI5, etc. — employed other technologies in efforts to identify and eliminate foreign espionages domestically. The new global economic warfare will see these basic roles continue, but with important changes in four major areas[23]:

---

[23] H. Keith Melton, "Spies in the digital age", *CNN Cold War Experience*, Espionage Page, www.cnn.com/SPECIALS/cold.war/experience/spies/melton.essay/.

❏   The primary targets of spies for all intelligence services have shifted.
❏   The traditional roles of "friends and foes" continue to blur.
❏   New technologies are changing the traditional methods and techniques, the tradecraft, by which spies operate.
❏   The traditional tradecraft of spies, if still in use, are applied in new ways.

In other words, the fictional James Bond is obsolete. A computer mouse is truly mightier than a gun.

8.1   The Mighty Electron

In the final days of the Cold War, the crumbling Soviet Union possessed the nuclear weapons to destroy the world but lacked the economic and informational infrastructure to compete as a world power. While the preeminent weapon for most of the latter half of the twentieth century was the hydrogen bomb, it has never been used. It has been displaced and replaced by the awesome capability of a single electron — the electron that surges in a computer to perform all the functions of this mighty device! This is not tantamount to saying hydrogen bombs will never be used. It only says that the electron is now the weapon of choice. Future superpowers will be those nations with the greatest capability to harness the power of the electron for both economic warfare and cyber warfare, or digital warfare, or info warfare, or invisible warfare, however we call the latter.

The desire of foreign spies to uncover and obtain military secrets will continue, but with critical variations. We are witnessing the migration of a national defensive infrastructure that has historically been based on "bullets" for physical destruction into one based upon "information" for economic sabotage. Success by spies targeting an opponent's information will ultimately prove more valuable, or detrimental, whichever view you choose to take.

8.2   WWW — World Wide Weapon

During World War II, the U.S. Office of Strategic Services (OSS) and British Special Operations Executive (SOE) coordinated resistance activities in German occupied Europe to disrupt communications, transportation and manufacturing. Those daring individuals risked their lives to sabotage telephone poles, derail trains and delay the shipment of raw material to factories producing war materials.

In the new world of digital spies, these same activities can be accomplished from a computer keyboard thousands of miles away. By electronically sabotaging enemy computer networks, cyber spies can accomplish the same result as their OSS and SOE predecessors.

Computer viruses and other computer agents have been developed and deployed that will be activated in time of war. Imagine the consequence of embedding a Trojan horse in the operating system software that runs critical components of computer systems of both friends and foes. A Trojan horse, once activated, can selectively disable the computer infrastructure of a hostile opponent and cripple its economy, communications and defense. It is checkmate even before the chess game has begun.

8.2.1   Computer Plagues

Dark Avenger and a handful of other viruses — Michelangelo, Jerusalem, Pakistani Brain, Frodo, and other newer ones such Love Bug and Sir Camelot — have transformed the way people experience computers. These cyber plagues launch a new lucrative antivirus trade, and leave in the minds of PC users a palpable fear that any file, no matter how innocuous, might carry with it a rapacious, information-destroying agent.[24]

Though we have experienced many computer viruses, worms, trojans and logic bombs in recent times,[25] much of the speculation about cyber terrorism has been dire, and a sober look suggests the nation probably is not at risk for a sweeping cyber calamity. But cunning, targeted efforts launched by terrorists at home or abroad could have devastating effects. The cyberspace, a technological frontier where the outlaws are sophisticated computer renegades and keyboard criminals — hackers, phreakers and virus writers — is borderless.[26] The U.S. National Academy of Engineering (NAE) is helping the U.S. government to pull together experts to consider countermeasures if there were a concerted attack by a state-sponsored terrorist group or by someone who had really thought deeply about how to attack the U.S. computer systems. According to William Wulf, President of NAE, the U.S. is absolutely, totally unprepared for such an as yet unprecedented cyber attack.

Viruses and break-ins have become a way of life for Internet sites. According to the Cert Coordination Center of Carnegie Mellon University, in 1999, the number of assaults was under 10,000. In 2000, the corresponding number was 22,000.

The key to cyber terror is the same feature that makes the Internet so resilient — the decentralized design. If some trouble develops in one location, the net traffic can quickly be rerouted. This is why we hardly felt any communications problem on and right after September 11, 2001, when communications lines were knocked out in lower Manhattan. That same freeform feature allows anyone from anywhere with a computer connected to a telephone line to get onto the network, without any need to identify oneself, to infiltrate the protected and secured computers at the Pentagon, NATO, or NASA at will if the perpetrator is sophisticated enough.

---

[24] David S. Bennuhum, "Heart of darkness", *Wired Magazine*, November 5, 1997.
[25] Paul Mungo, and Bryan Clough, *Approaching Zero: The extraordinary underworld of hackers, phreakers, virus writers, and keyboard criminals*, (Random House, New York, 1992).
[26] Christopher H. Schmitt, and Joellen Perry, "World Wide Weapon", *US News & World Report*, November 5, 2001, pp. 60–61.

8.2.2 Cyber Weapons

Among the possible offensive weapons are:[27,28,29]

- ❑ Computer viruses — a code fragment that copies itself into a larger program, modifying that program in the process. A virus executes only when its host program begins to run. The virus then replicates itself, infecting other programs as it reproduces. It could be fed into an enemy's computers either remotely or by mercenary technicians.

- ❑ Worms — an independent program that reproduces by copying itself in full-blown fashion from one computer to another, usually over a network. Unlike a virus, it usually does not modify other programs. Its purpose is to self-replicate ad infinitum, thus eating up a system's resources. An example is the infamous worm that crashed the entire Internet network in 1994.

- ❑ Trojan horses — a malevolent code fragment that hides inside a program and performs a disguised function. It is a popular mechanism for disguising a virus or a worm. A well written Trojan horse does not leave traces of its presence and because it does not cause detectable damage, it is hard to detect.

- ❑ Logic bombs — a bomb is a type of Trojan horse, used to release a virus, a worm or some other system attack. It is either an independent program or a piece of code that has been planted by a system developer or programmer. It can lie dormant for years. Upon receiving a particular signal, it would wake up and begin to attack the host system.

- ❑ Back doors and trap doors — a trap door, or a back door, is a mechanism that is built into a system by its designer. The function of a trap door is to give the designer a way to sneak back into the system, circumventing normal system access privileges.

- ❑ Chipping — just as software can contain unexpected functions, it is also possible to implement similar functions in hardware. Chipping is a plan to slip booby-trapped computer chips into critical systems sold by foreign contractors to potentially hostile third parties or recalcitrant allies. According to some sources, this was originally proposed by the CIA.

- ❑ Nano machines and microbes — a nano machine provides the possibility to cause serious harm to a system. Unlike viruses, it attacks not the software but the hardware of a computer system. A nano machine is a tiny robot that could be spread at an information center of the enemy. It crawls through the halls and offices until it finds a computer, enters the computer through slots and shut down the electronic circuits.

---

[27] Yael Shahar, "Information warfare — the perfect terrorist weapon", *ICT*, February 26, 1997.

[28] Deborah Russel and G.T. Gangemi, *Computer Security Basics*, (O'Reilly & Associates, 1994).

[29] Reto Haeni, "Introduction to information warfare", August 23, 1996. www.guest.seas.gwu.edu/~reto/infowar.

A special breed of microbes, genetically engineered to eat silicon would destroy all integrated circuits in a computer, thus causing the computer inoperational.

A few other weapons in the arsenal of information warfare are devices for disrupting data flow or damaging entire systems, hardware and all. Among these — High Energy Radio Frequency (HERF) guns, which focus a high power radio signal on target equipment to put the target out of action; and Electromagnetic Pulse (EMP) devices, which can be detonated in the vicinity of a target system. Such devices can destroy electronics and communications equipment over a wide area.

8.2.3   The Bulgarian Virus Factory

In 1989, the first Bulgarian viruses appeared. By the end of that year, one — Dark Avenger — had spread with enough velocity to attract media attention. Dark Avenger secretly attached itself to MS-DOS .com and .exe files, adding 1800 bytes of code. Every sixteenth time the infected program was run, it would randomly overwrite part of the hard disk. The phrase "Eddie Lives... somewhere in time" would appear, followed by garbage characters. Embedded in the code was another message: "This program was written in the city of Sofia © 1988-89 Dark Avenger". The computer, now infected and self-destructing, would eventually crash, with some precious part of its operating system missing, smothered under Dark Avenger's relentless output.

Viruses spread, most of the time even the affected do not know about them. Programs passed along in schools, offices, and homes — from one disk to the next they carried the infection along, and by 1991, an international epidemic was evident. One-hundred and sixty documented Bulgarian viruses existed in the wild, and an estimated 10 percent of all infections in the United States came from Bulgaria, most commonly from the Dark Avenger. Dataquest polled 600 large North American companies and Federal agencies early in 1991 and reported that 9 percent had experienced computer virus outbreaks. Nine months later, the number had risen to 63 percent. Anecdotal stories of companies losing millions in sales and productivity due to virus attacks became commonplace. The press seized upon the threat and beat the war drums of fear, first in Europe, which was closer to the epicenter. Newspapers carried lurid pieces describing the havoc the Dark Avenger had wreaked.

The origins of the Bulgarian virus factory go back to the 1980s. In the early 1980s, Todor Zhivkov, then president of Bulgaria, decided his country was to become a high-tech power, with computers managing the economy while industry concentrating on hardware manufacturing to match that of the West. Zhivkov envisioned Bulgaria functioning as the hardware manufacturing nerve for Comecon, the now defunct Eastern Europe's Council for Mutual Economic Assistance. Bulgaria would then trade its computers for cheap raw materials from the Soviet Union and basic imports from the other Eastern Bloc socialist countries.

The prevailing environment in Bulgaria was very promising. Bulgaria had many well-educated young electronics engineers. However, its archaic infrastructure and ill-managed economy were a recipe for failure. Neither were there particularly useful applications for the hardware.

In the second half of the 1980s, clones of IBM and Apple appeared. While factories continued to manufacture PCs, the country did not have any software to make the machines function. In pirating Western programs and operating system, the Bulgarians had to crack copy-protection schemes that stood in the way, and in so doing, they became better and better at hacking.

On record, the first Bulgarian virus arrived in the West in 1989. It started as harmless as Yankee Doodle to the more destructive Eddie to the deadly Nomenklatura, which attacked the House of Commons library, rendering valuable information irrecoverable.

By 1993, Bulgaria was no longer a significant source of new viruses. But the damage was done. At its peak, 1990–1991, both the alarm and the reality of the Bulgarian blight had spread exponentially, from computer to computer, and mind to mind. Today, Bulgaria exists as a kind of cybernetic bogeyman, the birthplace of viruses.

8.2.4   The Computer Plague Threat

As the world population of computer plagues grows exponentially, so does the potential for a real disaster. Computer plagues will affect computer users first, but then, many other innocent people who have never even touched a computer will be affected. For example, a virus let loose in a hospital computer could harm vital patient records and might result in patient receiving the wrong treatment regimen; workers could suffer job losses in virus-ravaged businesses; dangerous radiations could be released from nuclear power plants if the computers were compromised.

On record, there has not been a loss of life or jobs due to a virus. The only loss to date has been financial. But hospitals have already found viruses lurking in their computer systems, the military has been affected, and a Russian nuclear power plant's central computer has been shut down by a virus.

It is only a matter of time before there is a real catastrophe. Consider[30]

❑   During the Gulf War of 1991, mercenary Dutch hackers stole information on U.S. troop movements from the U.S. Department of Defense computers and tried to sell it for $1 million to the Iraqis, who thought it was a hoax and spurned the offer.

❑   During the 1991 Gulf War, Allied forces had to contend with at least two separate virus attacks affecting over 7,000 computers. One of the incidents was caused by the Jerusalem bug, and the other by a "fun" virus, Stoned, from New Zealand, which displayed a message "YOUR PC IS NOW

---

[30] John Christensen, "Bracing for guerrilla warfare in cyberspace", *CNN Interactive*, April 6, 1999.

STONED" on the screen. The two assaults caused computer shut downs and loss of data.

❑   In March of 1997, a 15-year-old Croatian youth infiltrated computers at a U.S. Air Force base in Guam.

❑   In 1997 and 1998, an Israeli youth calling himself The Analyzer allegedly hacked into Pentagon computers with the help of California teenagers. Ehud Tenebaum, 20, was charged in Jerusalem in February 1999 with conspiracy and harming computer systems.

❑   In February 1999, unidentified hackers seized control of a British military communication satellite and demanded ransom in return for control of the satellite. The report was vehemently denied by the British military, which said all satellites were "where they should be and doing what they should be doing".

In a normal day, the U.S. Department of Defense experiences 40 to 60 unauthorized intrusions, or once every 20 minutes or so. Of these, about 60 every week are considered serious attacks.[31] With 2 million computers, 100,000 local area networks, and more than 100 long distance networks, securing information is a formidable task for the agency.

8.2.5   Democratization of Hacking

There are about 30,000 hacker-oriented sites on the Internet. This brings hacking and terrorism within the reach of even the technically challenged, creating a form of democratization of hacking. The tools and programs can be downloaded, and with a click on the keyboard, the virus or bomb can be sent to a network to wreak havoc. According to an estimate, the Internet connects over 110,000,000 computers in 2001. This number is growing at a rapid pace. Consequences of any nefarious attempts to wreak havoc can be considerable.

Another threat is posed not by countries, terrorists, keyboard criminals, nor any odd balls with weird agendas, but by gophers, squirrels and farmers. In 1995, a New Jersey farmer yanked up a cable with his backhoe, knocking out 60 percent of the regional and long distance phone service in New York City and air traffic control functions in Boston, New York and Washington. In 1996, a rodent chewed through a cable in Palo Alto, California, and knocked Silicon Valley off the Internet for hours.

8.2.6   The Nimda Worm — A Case Study

The recent attacks by the Nimda or W32/Nimda worm demonstrate the Internet and Web vulnerability. The first public report of Nimda infections occurred on Tuesday, September 18, 2001, between 8:30 and 9:00 a.m. The worm modified

---

[31] Katherine McIntire Peters, "Information insecurity", *Government Executive*, 31(4), April 1999, (National Journal Group, Inc., Washington), pp. 18–22.

Web documents — files ending with .htm, .html, and .asp — and certain executable files found on the systems it infects. It then created numerous copies of itself under various file names, scanned the network for vulnerable computers and propagates through email, thereby causing some sites to experience denial of service or degraded performance. Computers that had been compromised were at high risk for being used for attacks on other Internet sites.[32] One of Nimda's features was to attack computers that had been compromised by the Code Red worm and left in a vulnerable state. It also targeted home users' computers, which were among the most vulnerable. Because of the network traffic generated, Internet Service Providers (ISPs) for home users suffered a negative impact from the worm.

The September 2001 Nimda had several means to infect computers. For example, the worm not only propagated through email attachments and through compromises of vulnerable Internet Information Servers (IIS), but it also spread through shared files on a file server and through Web pages containing JavaScript that had been altered on a compromised server.

The algorithm used to spread the worm concentrated for the most part on local networks. The primary adverse effect of the worm occurred at the "edges" of the Internet. Operators of the backbone of the Internet, though not significantly affected, did experience an increase in customer service calls. Victims could not reach the Internet because of the local scanning and email traffic caused by the worm. They thought that the Internet was down. In other words, they were denied service by the worm!

Nimda is the first significant worm that attacks both computers that act as servers and those that are desktop computers. A server provides services such as a Web site. Code Red exploited the Internet Information Server (IIS), which is a Web server. The Melissa virus spread by means of users' email on desktop computers. Nimda merges the damaging features of both Code Red and Melissa, and more.

The Nimda worm spread so fast that system administrators, users, and vendors did not have time to prepare. Quick response was a challenge because there was no lead time for advance analysis. In contrast, with Code Red, analysts had a small amount of lead time to examine an early version of the worm before a more aggressive version began causing serious damage. These new no-lead-time attack technologies are causing damage more quickly than those created in the past. The Code Red worm of 2000 spread around the world faster than the Morris worm moved through U.S. computers in 1988, and faster than the Melissa virus in 1999. With the Code Red worm, there were days between first identification and widespread damage. The Nimda worm caused serious damage within an hour of the first report of infection.

---

[32] Richard D. Pethia, "Information technology — Essential but vulnerable: How prepared are we for attacks?" Testimony before the House Committee on Government Reform, Subcommittee on Government Efficiency, Financial Management, and Intergovernmental Relations, September 26, 2001.

Analysis of Nimda was hampered by the lack of the source code for Nimda. The source code is the original form of the program, basic code that reveals how the worm works. Thus, it was not possible to determine quickly what the worm did and what it could potentially do. Analysts quickly obtained the binary code, but it was time consuming to decompile this code and analyze the inner workings of the worm. Analysis through decompiling can take hours, days, or even weeks, depending on the complexity of the program.

8.3   The New Villain

Assassination was once considered as a tool of warfare and tactically applied or attempted by some intelligence services during World War II. During the Cold War, the Soviet Bloc utilized assassination to silence exiles taking refuge abroad. The KGB assassinations of Ukrainian exiles Lev Rebet (1957) and Stepan Bandera (1959) in West Germany, as well as the infamous Bulgarian "umbrella assassination" of Georgy Markov (1978) in London, are all cases in point.

8.3.1   CV Please

In the digital world, potential targets of assassination have shifted. Even with the emphasis of advanced computer developments, all nations depend on imbedded computer chips of varying age, some decades old. For example, NASA is still using the VMS operating system, and most state agencies are still using COBOL. Neither VMS nor COBOL are any longer parts of normal computer curriculum. These critically important components control the switching systems in power grids, telephone systems and transportation networks, and commands of space flights. The devastating effect of losing an antiquated but functioning system becomes a reality when the key and indispensable person charged with its upkeep is eliminated. The result of assassinating a political leader pales when compared with the effect in future wars of eliminating key computer programmers and network specialists.

For professional intelligence services, their primary goal is, and will remain, the acquisition of information, not murder. Oleg Tsarev, a retired officer of the KGB's First Chief Directorate and author, accurately stated that "intelligence stops when you pick up a gun".

Instead of eliminating key specialists, it can be equally advantageous to lure away the specialists of a nation. Former Soviet scientists are known to have emigrated, possibly to well-funded terrorist groups. In the dilapidated economy after the collapse of the U.S.S.R., government funding fell sharply, and impoverished researchers fled overseas in a massive brain drain. The U.S., being in an economically advantageous position, has been the most fortunate beneficiary, not only from the U.S.S.R., but also from many other countries such as Germany after World War II, Hungary in the 1960s, China after the Tien Anmen incident of 1989.

8.3.2   911, Help Please

A Swedish teenager disabled South Florida's 911 system in 1997. It is conceivable that the nation's 911 system could be under attack again. For example, by flooding the service with calls.

Imagine, if 911 is in trouble, who are they going to call for help? 911?

The telephone system is far more complicated than it used to be. It has a lot of nodes that are programmable and databases that can be hacked. Also, the deregulation of the telephone and power industries has created another Achilles heel. To stay competitive and cut costs, companies have reduced spare capacity, leaving them more vulnerable to outages and disruptions in service. Still another flaw is the domination of the telecommunications system by phone companies and Internet service providers (ISPs) that compete fiercely and do not trust each other. As a result, the systems do not mesh seamlessly and are vulnerable to failures and disruptions. There is almost no way to organize systems built on mutual suspicion. Subtly changing the underpinnings of the system and not changing the way these systems are built will keep creating cracks for hacking.

8.3.3   Power Grid

The U.S. has so many different and complex systems of power grids. This would in principle impede a coordinated raid. It is unlikely that an attack on the power grid would trigger an across-the-board collapse. But a concerted assault could be very disruptive. To maintain the vital balance of supply and demand, generators, distributors and traders are constantly in contact, mostly over the Internet.

Though remote, an intruder could use the Internet to leapfrog into the computers that control switches, relays, and breakers. This could lead to slow or freeze operations, destabilizing the grid and causing outages.

There is another concern. With deregulation, there is an increasing interest in energy futures trades at the commodities exchange on Wall Street. Hackers might use social engineering techniques to obtain passwords to computers with access to the networks containing sensitive information from these sources.[33] Social engineering is a technique used to obtain key information, such as passwords, just by talking to employees.

8.3.4   Psychological Warfare

The scenarios described above, other similar tactics and combinations thereof belong to a subset of information war, commonly called "hacker warfare". However, the term "infowar" includes other ways of manipulating information, among them "psychological warfare". A psychological warfare is an attempt to warp the opponent's view of reality, to project a false view of things, or to influence

---

[33] Gene Koprowski, "Hacking the power grid", *Wired News*, June 4, 1998.

its will to engage in hostile activities. Psychological warfare includes a variety of actions that can be divided up into categories according to their targets. Strategic analyst Martin Libicki proposes four categories:

❑   operations against troops,
❑   operations against opposing commanders,
❑   operations against the national will, and
❑   operations designed to impose a particular culture upon another nation.

This is usually called "netwar".[34] Netwar refers to information-related conflict at a grand level between nations or societies. Its intention is to disrupt or damage what a target population knows or thinks it knows about itself and the world around it. A netwar may focus on public or elite opinion, or both. It may involve diplomacy, propaganda and psychological campaigns, political and cultural subversion, deception of or interference with local media, infiltration of computer networks and databases, and efforts to promote dissident or opposition movements across computer networks.

Using the media as a weapon of information warfare is nothing new. The attempt to influence the human element in a conflict is an old tactic. Armies have always tried to make their forces seem stronger or weaker than they are, or to convince enemy soldiers that they have no escape but to surrender peacefully. The only difference is the means have changed. Recently, to this component in the military arsenal has been added a relatively new technique of mass information transfer. Now psychological warfare includes the endeavor to manipulate the populace of an enemy country to oppose the war effort, or to depose the reigning government. The means to this end reside in the mass media, and more recently in the Internet. Examples abound.

For example, in the 2001 U.S.-Afghanistan confrontation in the wake of the September 11 incident, the U.S. used the media to sway public opinion and dropped leaflets in Afghanistan to persuade the local population that it was a friendly force to try to depose off an unpopular regime. Under the guise of humanitarian aids, the U.S. delivered food with messages in packages, clearly labeled "The United States of America". There were even attempts to drop radios so that the local people could tune in. Concurrently, the U.S. counterintelligence services also scan the Internet to rid off any undesirable messages.

Similarly, Osama bin Laden's videotaped address, aired shortly after the U.S. strikes, aims to incite Muslims in a holy war. As a countermeasure, the U.S. authorities urged and banned the media, very successfully, not to broadcast the videotaped messages.

Psychological warfare through the media has also been used with success by the U.S. in the Gulf War of 1990–91. The Iraqis were led by media reports to believe that the air war was to be a short-term strike, followed by an immediate ground war, in which they felt themselves to have the advantage of numbers and territorial

---

[34] John Arquilla and David Ronfeldt, "Cyberwar is Coming!", article for RAND, 1997.

dominance. They were also kept busy along the Kuwaiti coasts, by means of disinformation pointing to an imminent American coastal offensive.

Another example of psychological warfare was the American propaganda war in Haiti. The Pentagon launched a sophisticated psychological operation campaign against Haiti's military regime to restore depose President Jean-Bertrand Aristide. Using market-research surveys, the Army's 4th Psychological Operations Group divided Haiti's population into 20 target groups and bombarded them with hundreds of thousands of pro-Aristide leaflets appealing to their particular affinities. Before U.S. intervention, the CIA made anonymous phone calls to Haitian soldiers, urging them to surrender, and sent ominous email messages to some members of Haiti's oligarchy who had personal computers.

With CNN and BBC beamed into almost all countries in the world, the U.S. has a great advantage over any other nations in the netwar. But America has not always been on the winning side in psychological warfare. Democracies, by their very nature, are acutely sensitive to public opinion, making them vulnerable to manipulation through the media. American troops left Somalia after the loss of just nineteen American Rangers in a conflict with the forces of Somali leader Mohammed Aideed. That conflict reportedly cost Aideed about fifteen times that number, roughly a third of his forces. And yet it was the Americans who conceded defeat. Why? Photographs of jeering Somalis dragging corpses of U.S. soldiers through the streets of Mogadishu transmitted by CNN to the United States led to souring of TV audiences at home on staying in Somalia. U.S. forces left, and Aideed, in essence, won the information war.[35]

It is thus not surprising that the U.S. authorities carefully monitored news media coverage of the recent U.S. air strikes on Afghanistan. There were a number of misfires and casualties. Most of the general populace were not aware of these mishaps because of a lack of news coverage.

There are many other examples. These examples suffice to show mass psychology can be manipulated to one's end in a war.

8.4   He Who Lives in Glass Houses Should Throw No Stones

While all this seems to point to an increasing advantage of technologically advanced nations over those less advanced, there is a certain catch to this war game. American strategists are very leery of the prospects of using the more malicious forms of information warfare, for the same reason that American policy forbids the assassination of foreign leaders. We can assassinate a foreign leader, they can easily do what we do upon them. Similarly, the more technologically advanced a nation is, the more vulnerable it is itself to the techniques of information warfare. No nation is more dependent upon the information infrastructure as the U.S. So it is not

---

[35] Martin Libicki, "What is Information Warfare?", article for the Institute for National Strategic Studies, 1997.

surprising that American policy makers are quick to point out that infowar scenarios are being studied at present mostly with an eye toward defense rather than offense. The U.S. is living in a glass house of information. It should avoid throwing any stone.

Compounding to the wariness of the American strategists is the fact that it is the civilian sectors that are most vulnerable, with consequences in both the military and the political sphere. Military infrastructure relies for the most part on civilian infrastructure. Nearly every aspect of the military industry, from basic research and development to paying personnel depends on civilian information networks. Indeed, over 95 percent of military communications use the civilian network. Military bases depend on the national electric power grid. Soldiers travel by means of the national bus cooperative. There is no way that the military can protect all of these networks from a focused infowar attack.

Government sites are not better protected. They are riddled with weaknesses, ranging from failure to rotate computer passwords to unauthorized software installation by IT managers. The Department of Defense (DoD) is more secure. Still it has weaknesses. For example, its computers most likely run a standard operating system such as Microsoft operating system. The network has an electronic mail link to the outside. While an emailed virus is not likely to bring hijack to the DoD systems, it could bring down the network or corrupt the data.

8.5   Friend or Foe?

The traditional Cold War alignment of the East versus West is gone forever. At the height of Cold War solidarity, the slogan was "the enemy of my enemy is my friend". Superpowers collected intelligence and attacked the ciphers or codes of friends as well as enemies.

The national interests of former friends and foes are now being redefined in terms of competing economic interests. Cultural and historic friendships between nations will continue to fade as they are replaced by trading partnerships and other interdependent economic relationships. For example, Premier Zhu Rongji of China, together with his counterparts in the Association of Southeast Asia Nations, agreed during the week prior to China and Taipei joining the WTO on November 12, 2001 to establish, within a decade, a 10+1 free trade zone. With 1.7 billion people, this will be the largest free trade zone in the world. It also has sparked the discussion of building a bullet train from Singapore to Kunmin, China.

The new slogan is "the friend of my enemy may also be my friend", if the price is right. For example, in the 1991 Gulf War, George H. Bush forfeited the $7 billion loan to Egypt so that the U.S. could use Egypt as an air base to strike Iraq. A decade later, the son, George W. Bush when he first moved into the White House in early 2001, accused the Russians of exercising atrocities against the Chechens. After the September 11 incident and in an effort to court the Russians into supporting the U.S. attack on Afghanistan, the same president, in the same year, and speaking from the same White House, applauded the Russian's attack on Chechnya for the latter supposedly harbored Al Qaeda operations.

The U.S. benefited greatly from Jordan in its efforts against Al Qaeda operations. Close intelligence cooperation between the United States and Jordan dates back to 1990 when the late King Hussein warned U.S. leaders about the emergence of a network in Afghanistan headed by bin Laden. This close ties have continued under the reign of King Hussein's son, King Abdullah, who ascended to the throne after King Hussein died in 1999. "The unsung heroes in intelligence terms are the Jordanians," said terrorism expert Professor Magnus Ransdorp of St. Andrews University. "The Jordanian track record, given the size of the country, is mammoth, in terms of the contribution to our understanding of the al Qaeda network."[36]

The U.S. carefully coordinated its attacks on Afghanistan in 2001. Goaded by sentiment to punish the perpetrators who had attacked the World Trade Center Towers, the U.S. exercised diplomacy to court supports to form a coalition, performed intelligence gathering such as working with King Abdullah of Jordan[37] and consulting President Eduard Shevardnardze of Republic of Georgia who was involved in the Soviet-Afgan conflict of the 1980s, and worked with news media on psychology to avoid possible anti-war sentiment domestically. The Soviet-Afghan conflict of two decades earlier, on the other hand, was a stark different brawny arms-and-tank all-out attack. The December 1979 New Year eve Soviet invasion of Afghanistan was the most violent of power move by the Kremlin to ward off the developing urge of change and to alienate the generation of the 1980s.[38] As we know now, the conflict leads to an unpopular long drawn-out war that eventually resulted in the complete Soviet withdrawal on February 15, 1989.

9   What Is In Store?

Former CIA Director James Woolsey stated that with the end of the Cold War, the great Soviet dragon was slain. He wryly noted, however, that in its place the intelligence services of the United States are facing a "bewildering variety of poisonous snakes that have been let loose in a dark jungle; it may have been easier to watch the dragon".

The single greatest threat to world peace in the early part of this new century is the utilization of weapons of mass destruction — nuclear, chemical, biological and

---

[36] Mike Boettcher, "Jordanian intelligence helped thwart attacks, sources said", *CNN News*, November 19, 2001.
[37] Janine Zacharia, "King Abdullah promises Bush full support", *The Jerusalem Post*, September 30, 2001.
[38] Hedrick Smith, *The New Russians*, (Avon Books, New York, 1990).

digital — by fundamentalist terrorist organizations. These groups are already using the Internet to:

❑   recruit and communicate members with similar fundamentalist beliefs.
❑   coordinate terrorist activities with other aligned groups that share interests in a common outcome.
❑   raise money through computer based keyboard crimes.
❑   attack the national information infrastructures of hostile countries from thousands of miles away.

The CIA and other intelligence services must operate with shrinking budgets and manpower — the CIA will shrink 25 percent from its peak — but confront an array of new threats to national interests in different parts of the globe. To meet these challenges, all intelligence services will be forced to rely on digital solutions, massive computers and artificial intelligence in linked computer networks and databases to compensate for the reduction of people and resources.

The traditional world of spies such as James Bond exists now only in fiction. New intelligence services that most effectively identify, develop and implement the tools and techniques of the "cyber spy" will provide their citizens with an incalculable advantage in the new century.

Terrorism is an excellent example of how the focus of war has shifted toward civilian populations. For example, the September 11, 2001 horrendous attacks on the World Trade Center Towers. The aim of terrorism is not to destroy the enemy's armed might, but to undermine its will to fight. Terrorists seek to disrupt the daily life of their target nation by striking at the most vulnerable points in the society. Such vulnerable areas included transportation networks and public events, which insure good media coverage. By hitting the citizen just where the nation thinks is safest, the terrorists cause the greatest confusion and loss of morale.

Today, almost every aspect of our lives is dependent on information networks, terrorists have a whole new field of action. And while the technology to operate and protect these networks is quite costly, the means required to attack them are relatively cheap. In the simplest case, one needs only a computer, a modem, and a willing hacker. According to Alvin Toffler, "It's the great equalizer. You don't have to be big and rich to apply the kind of judo you need in information warfare, That's why poor countries are going to go for this faster than technologically advanced countries."[39]

According to Time Magazine,[40] the Defense Science Board at the Pentagon warned that annoying hackers trying to crack the Pentagon's computers were not the only things the defense strategists have to worry about. This threat arises from terrorist groups or nation-states, and is far more subtle and difficult to counter than the more unstructured but growing problem caused by hackers. A large, structured

---

[39] Alvin Toffler and Heidi Toffler, *War and Anti War, making sense of today's global chaos*, (Warner books 1993).
[40] Waller Douglas, "Onward Cyber Soldiers", *Time Magazine*, August 21, 1995 Volume 146, No. 8

attack with strategic intent against the U.S. could be prepared and exercised under the guise of unstructured 'hacker' activities. There is no nationally coordinated capability to counter or even detect a structured threat.

10   Ongoing And Future Business Practice

We have gone into length to talk about cyber warfare in military scenarios. Very similar scenarios can be played on business grounds and between competing companies.

10.1   Business Warfare

Since World War II, business has been customer-oriented, and King Customer has reigned supreme. In the plan of today and the future, a company has to be or will have to be competitor-oriented. The plan will carefully dissect each participant in the marketplace. There might even be a day when the plan will contain a dossier on each of the competitors' key people, their favorite tactics and style of operation. More and more, successful business campaigns will have to be planned like military campaigns. Companies will have to learn how to attack, to flank competition, to defend positions, and how and when to wage guerrilla warfare.

For these, we may learn from two great works on war[41]:

❑   Sun-tzu ping-fa or Sun Tzu the Art of War is one of those rare texts that transcends time. Though it was written in the 6th century B.C., it is arguably still one of the most important works on the subject of strategy today. Written by Sun Wu, Chinese general of the state of Wu, The Art of War was intended only for the military elite of his time period. However, this treatise would later be absorbed by others of influence, from the fearless samurai in feudal Japan to the shrewd business leaders of the 21st century.

❑   General Karl von Clausewitz (1780–1831) was one of the greatest writers on war. His *magnum opus*, On War, is carefully studied in military schools to this day, for its principles are as valid for nuclear as for conventional and guerrilla warfare. Weapons may have changed, but warfare itself is based on two immutable characteristics: strategy and tactics.

These great works have been applied to business: in trading,[42] in strategic planning,[43] in marketing,[44] and others.

---

[41] *The Book of War*, (Random House, 1999), paperback reprint of two books: Carl von Clausedwitz, *On War*, and Sun Tzu, *The Art of War*.

[42] Dean Lundell, *The Art of War for Traders and Investors*, (McGraw-Hill, 1996).

[43] B.H. Boar, *The Art of Strategic Planning for Information Technology*, (John Wiley & Sons, Inc., New York, 1993).

[44] Al Ries, and Jack Trout, *Marketing Warfare*, (McGraw-Hill, New York, 1986).

10.2   Business Cyber War

The business sector has become increasingly dependent on information for decision making and the Internet for dissemination. Internet users — business, consumers and home users inclusive — now use the Internet for many critical applications as well as online business transactions. A relatively short interruptions in service can cause significant economic loss and can jeopardize critical services.

It is not inconceivable that viruses, worms, Trojan horses, logic bombs, and back doors can be incorporated in software for use by clients so that a vendor can have a handle over the users. Hardware can also be designed with chipping feature for the same purposes. Indeed, in this new world of cut-throat business competition, it is not always economically feasible to develop all critical technologies and supporting software in-house. It is prudent to find a compromise between farming out projects and developing in-house. For example, in the biotechnology sector, it is very common for companies to license software from bioinformatics companies. In an effort for the licensee to avoid possible financial losses resulting from viruses, worms, Trojan horses, logic bombs and back doors, in the licensing agreement between licensor (bioinformatics company) and licensee (biotechnology company), a clause may be included. A sample is provided below:

> Viruses; Disabling Codes. Licensor represents and warrants that any Software and computer media furnished to Company pursuant to the License Agreement shall be free from computer viruses and any undocumented or unauthorized methods for terminating or disrupting the operation of, or gaining access to, software, computer systems or other computing resources or data, or other code or features which result in or cause damage, loss or disruption to all or any part of computer systems or other computing resources. Licensor shall not incorporate into Software any termination logic or any means to electronically repossess any Software licensed under the licensing agreement. "Termination logic" shall mean computer code that uses the internal clock of the computer to test for the date and/or time (e.g., Friday the 13th), use count, execution key, or any related techniques, as a trigger to render inoperable or otherwise disable the Software or any related computer system.

"Assassination" in the business sector can also be strategically exercised by hiring away key employees from chief competitors. This has happened. On January 24, 1997, Informix Software Inc., a Menlo Park database software firm, filed a lawsuit against its largest competitor, Oracle Corp. of Redwood Shores, claiming theft of trade secrets. The suit was filed in Oregon's Circuit Court for Multnomah County in Portland, Oregon, USA. The action stems from Oracle's recruitment and hiring of 11 employees from the Informix product-development laboratory in Portland. The suit charges Oracle and a former Informix employee with misappropriation of trade secrets and unfair competition. It was seeking injunctive relief and punitive damages.[45]

---

[45] "Informix sues Oracle over trade secrets", *Business Times*, January 24, 1997.

Just prior to the incident, Wall Street Journal had predicted Informix to be the database company of the new century. After the incident, and with a legerdemain trick from Oracle by turning news media coverage of the incident into a promotion campaign, Oracle came out the victor in the "assassination" and psychological campaign. During the heydays of dotcoms hype, Larry Ellison, the CEO of Oracle, was briefly the richest man on Earth. The stock of Informix plummeted after the incident and it never recovers to its luster.

"Assassinations" of employees have spurred a lucrative business. Head hunter and executive search agencies spring up to fill the gap. Head hunter agencies are for more routine and project level jobs. Executive search agencies are for management level positions. Essentially, these agencies move the work force from one sector, such as the academic sector, to another, such as the private sector. Or they move the work force from one company to another. At low unemployment, such as during the dotcom heydays, jobs and positions are bountiful and these agencies enjoy their best business. At a time of economic downturn, such as during the demise.com period of 2001, these agencies have a harder time to find vacancies to relocate those out of jobs.

Despite the faltering economy and a flurry of layoff across most industries in 2001, demand for high skilled workers with expertise in biotechnology remains strong. Most companies seem reluctant to relinquish any talent simply to shave costs, and the few employees that have been let go quickly find employment in competing companies.

"Friends and foes" can also be played in the business sector. This comes in the forms of strategic alliances, partnerships, and consortia to pool resources to achieve a goal that would otherwise be impossible or much harder to achieve alone. The International Human Genome Sequencing Consortium of the public sector and Celera Genomics of the private sector are excellent examples.

"Social engineering" is also common in the business sector. Workshops and conferences are fora for social engineering. End-of-the-meeting-day gatherings at pubs are the best places to tune into company secrets or new breakthroughs. It seems, after a few beers and drinks, experts, researchers and scientists talk more freely.

"Psychological wars" are usually waged by companies to their target audiences. For example, pharmaceutical companies use television advertisements to stretch their advertising dollars to the point of misleading viewers by instilling fear such as showing golden age inconveniences and ailments, or else by derogating viewers' ego such as promoting life-style drugs whose primary functions are to restore social faculties or attributes that tend to diminish with age.

After the September 11 blitz that turned civilian airliners into missiles, killing some 2,900 people, the United States must plan for new and different foes who will rely on surprise, deception and asymmetric weapons, or those meant to overcome the lopsided U.S. edge in conventional arms. "Asymmetric wars" may also be waged in the business sector by smaller, more savvy companies against monopolistic competitors. For example, in the 1980s, while others were losing money in the computer business, Digital Equipment Corporation was making a lot of profit by exploiting IBM's weakness in small computers.

It is thus not inconceivable that all the information warfare scenarios described above can be nefarious attempts by one company against the other to gain competitive advantage.

11   Ending Note

We hope this chapter will not make uncomfortable reading. Our intent is to bring to the readers increased and ongoing awareness and understanding of biowarfare and cyber-security issues, vulnerabilities, and threats to all stakeholders in physical and cyber spaces. We also hope to bring awareness that the business sector is also a war field.

We just cannot deny the deniability.

Chapter 14

TWO DECADES OF BIOTECHNOLOGY AND A DECADE OF BIOINFORMATICS

"By the time we enter the bio-economy, we will have accomplished the blending of genetics and computers... Turning inward, the bio-economy will see carbon-based organic substances that function like semiconductors... Turning outward, the bio-economy will also see silicon-based inorganic substances that have some brain-like functions..."
Stan Davis and Bill Davidson, 2020 Vision, Simon & Schuster, New York, 1991.

1   BT Meets IT

In the aftermath of the historic decision by the Supreme Court of the United States in June of 1980 to allow patenting of life forms,[1] biotechnology shed its pristine academic garb and plunged straight into the marketplace. Four months later, on October 14, 1980, Genentech offered over a million shares of stocks at $35 a share. By the end of the trading day, the company had raised $36 million and was valued at $532 million. Thus started the biotechnology industry.

In 1987, the neologism "bioinformatics" was coined.[2] In 1989, with the help of a graphics artist, the logo for a bioinformatics conference series — International Conference on Bioinformatics & Genome Research — was designed. This logo has since been accepted as the logo of the conference series, which is still being organized annually by Boston-based Cambridge Healthtech Institute. The logo consists of two double-stranded DNA fraying into computer chips to signify the confluence between biotechnology and computers, and the flow of bioinformation.

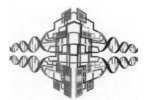

Figure 1. A replica of the logo designed for the very first international conference on bioinformatics in 1990. The black-and-white half is added to the original logo for use in lectures to test for color blindness. (Figure designed by David Poindexter and Hwa Lim, 1989).

In 1987, there was little overlap between computer businesses and institutions active in biotechnology. Computing was not a central part of biological research.

---

[1] Diamond, Sidney A., Commissioner of Patents and Trademarks, petitioner, v Chakrabarty, Ananda M., et al., 65 L ed 2d 144, June 16, 1980, pp. 148.
[2] http://www.d-trends.com/Bioinformatics/timeline.html

371

Computers were nice to have, but they were not critical to the process of discovery. That has changed dramatically.

In recent years, biotechnology has become very data-intensive and information-rich. Today, the frontiers of biology and computing are converging in ways that are revolutionizing biological research, genetics, drug discovery and medicine. In the new millennium of what many in the scientific and business communities are calling "The Biotech Century", two striking developments are taking shape. First, the genetic revolution and the computer revolution are just coming together to form a scientific, technological, and commercial phalanx, a powerful new reality that is going to have a profound impact on our personal and collective lives in the coming decades. Second, many of the scientific breakthroughs predicted more than a quarter of a century ago are now moving out of the laboratory and into widespread commercial use.[3]

Bioinformatics — a study of information content and information flow in biological systems — has also emerged from its humble beginning fifteen years ago and has changed from a special niche tool to an essential corporate technology. The scope has widened from a laboratory-based tool to an integrated corporate infrastructure.

This is why the logo, designed more than a decade ago, carries some historical significance.

2   Technology Convergence

The technological revolution, led by advances in information and communications technology, is changing the global economy by increasing the importance of knowledge as a factor of production. It is also changing the nature of markets, competition, and sources of comparative advantage. And it is providing solutions to the consequences of rapid population growth and resource depletion, informed consumers' demand for better quality of life and better healthcare. In providing solutions, technological revolution is offering hope for the sustainability of increased economic activity.[4,5]

In one sense, today's technological revolution is not new. Over the past 100 years, impressive advances in transportation, electrification, communications, and medicine have changed the way people live and work. What is different now is the convergence and interaction of many strands of technological change with social consequences far more profound, far more difficult to foresee.

---

[3] Jeremy Rifkin, *The Biotech Century*, (Jeremy P. Tarcher/Putnam, New York, 1998).
[4] Kristine Hallberg, and James Bond, "Revolutions in technology for development", World Bank, 1999. www.worldbank.org/html/fpd/technet/revol.htm
[5] Hwa A. Lim, and DaHsuan Feng, "A global economy without a global government", VERIZONTAL 7(1), 2000.

A cluster of innovations in telecommunications and informatics is feeding a revolution in information technology. Faster transmission speeds of optical fibers and new lightweight materials make construction faster and cheaper. Undersea telephone links now connect 51 countries with cellular telephone circuits, and 7.4 million miles of fiber optic cables were installed in Y2K alone. Information already is flowing faster, more generously, and less expensively throughout the planet, but this information technology revolution is still young, with full digitalization and intensive exploitation of bandwidth still years away.

The field of biotechnology is currently dominated by advances in pharmaceutical discoveries and health applications such as diagnostics and therapeutics. The field is quickly spreading to other areas. Marine biotechnology is allowing better disease prevention and control of reproduction in fish, leading to the creation of new possibilities in aquaculture. In the environmental field, bioremediation technology is improving waste site cleanup and forest restoration and providing new methods for waste and water treatment, environmental monitoring, and air quality management. In agriculture, biotechnology applications are allowing increased animal production and better survival in marginal areas, improved animal health, and greater use of biofertilizers and genetically manipulated microorganisms. Future developments are expected in biosensors, biomaterials, and bioelectronics.

3   BT-IT Convergence

The roots of biotechnology (BT) and information technology (IT) convergence (BT-IT convergence) lie in two seemingly unrelated events that took place six years apart about fifty years ago at the midpoint of the 20th century. The first was the discovery of the transistor in 1947 by three physicists at Bell Labs: John Bardeen, Walter Brattain and William Shockley. The second was the discovery in 1953 of the double-helix structure of DNA by James Dewey Watson and Francis Harry Compton Crick. These great advances in theoretical science — both of which were honored with the Nobel Prize — turned out to have significant practical applications as well. The invention of the transistor set the stage for the modern computing industry and information technology, while the pioneering work of Watson and Crick ultimately led to today's biotechnology industry.

The BT-IT convergence is driving breakthroughs that may have an even greater impact on our world. Today, it is increasingly difficult to separate the advances in biotechnology from advances in high performance computing. In fact, some leading scientists believe that high-end computing is the future of biology and medicine because biology is becoming an information science, and it will take increasingly more powerful computers and better software to gather, store, analyze, model and disseminate that information.[6] Traditionally, the most difficult task in biology has been to acquire data. Researchers were trained to design experiments that could extract the most amount of information with the least amount of effort. Today, in addition to the human genome data, public and private efforts to sequence genomes of different organisms are adding to the vast amounts of information. In addition, we also have protein data, protein-protein interaction data, and healthcare registry data.

4   Moore's Law Says More's Less

Moore's Law, an observation of the co-founder of Intel, Gordon Moore, states that the processing power of semiconductor devices doubles roughly every 18 months. This remarkable phenomenon has driven the information technology revolution. But information on gene sequences is growing at an even faster rate and in turn is driving a revolution in biology. It has even made indispensable bioinformatics in biotechnology. All this effort is really focused on three types of information:

❑   First, the linear information represented by genetic sequences.
❑   Second, the functions and the properties of the proteins associated with genes, and
❑   Third, how the whole complex system works.

The greatest challenge for biotechnology will be to turn that information into knowledge — knowledge that will help us understand how genes work and enable the development of new and more effective therapies and drugs.[7,8,9,10] Meeting that challenge will require increasingly powerful information technologies — from gene sequencers to genomics technologies, to proteomics technologies, and to high-performance computers.

The expanding role of computing in biotechnology today is very reminiscent of the changes that took place in business computing more than 30 years ago. That was when Digital Equipment Corporation introduced the minicomputer. Up to that point, computing was the domain of a select few who operated large, mainframe systems. But the minicomputer put more computing power into the hands of more

---

[6] Ben Rosen, "Compaq's commitment to bioinformatics", keynote lecture at BIO'99, May 18, 1999, Seattle, Washington, USA.

[7] Hwa A. Lim, "Biological and Biological-related information as a business commodity and the rise of bioinformatics" 1995. www.d-trends.com/webs/bio_business.html

[8] Hwa A. Lim, "Bioinformatics and cheminformatics in drug discovery cycle", In: *Lecture Notes in Computer Science, Bioinformatics*, R. Hofestaedt, T. Lengauer, M. Loffler and D. Schomburg (eds.), (Springer, Heidelberg, 1997), pp. 30–43.

[9] Hwa A. Lim and Tauseef R. Butt, "Bioinformatics takes charge", *Trends in Biotechnology*, March 1998, Vol. 16 No. 3 (170), pp. 104–107.

[10] T.V. Venkatesh, Benjamin Bowen, and Hwa. A. Lim, "Bioinformatics, pharma and farmers", *Trends in Biotechnology*, March 1999, Volume 17 No. 3 (182), pp. 85–88.

people than ever before. It made computing more affordable to more businesses. And it created new opportunities for software developers. The first successful minicomputer — the Digital PDP/8 — and its successor, the PDP/11, were mainstays of research laboratories around the world.

As recent as only a few years ago, one needed a supercomputer (such as a Cray YMP) or a massively parallel processing system (such as a Connection Machine) to perform the most compute intensive tasks in science. Only a few companies or research centers could afford that investment. But high-performance computing has become more and more affordable, even as systems have become more and more powerful. Microprocessor performance is traditionally measured in two ways:

❑   Integer performance, which measures the ability to manipulate strings of data or databases and to match patterns of data, and

❑   Floating-point performance, which is particularly important for modeling and simulation.

Integer operations are particularly suited for sequence comparisons, pattern recognition, and neural network. Floating-point operations are suited for protein fold simulation, molecular recognition and other in silico experiments. Because computers and information devices (infotronics) are becoming very affordable, biotechnology organizations are becoming very well equipped computationally.

In certain sense, biochips and similar information technologies in the biotechnology sector, when standardized, will be the equivalent of the personal computer of biotechnology.

5   The Internet — The Backbone of BT-IT Convergence

One thread that runs through all of these examples is the importance of the Internet to biotechnology research. The ability to access, compare and identify novel DNA and protein sequences is an integral part of the research process.[11] The Internet — together with specially designed, interactive search software like BLAST — makes it possible for scientists to access daily updates from genome projects and other sources. This brings an immediacy to science that did not exist before. In the past, scientists had to wait for papers to be published — a process that could take a year or more. Now information is available almost instantaneously. Research results can be shared and compared across a vast network of public and private institutions. In addition to Web sites that post data from genomic research, some specialized databases are now available only on the Web. This information is particularly useful for researchers and others involved in the diagnosis of human genetics, as well as for physicians and genetic counselors. In 1998, sixty millions people used online

---

[11] Hwa A. Lim, James W. Fickett, Charles R. Cantor and Robert J. Robbins (eds.), *Bioinformatics, Supercomputing and Complex Genome Analysis*, (World Scientific Publishing Co., New Jersey, 1993), 648 pages.

healthcare Web sites such as WebMD to understand their own health and to be better informed. And the National Library of Medicine recorded 120 million searches of its database in the same year — one-third of them by consumers. It is easy to believe that this voracious appetite for health information will extend to genetic information — once all the tools are available and the sites are friendlier for laypeople. One may agree or disagree with this vision, but the fact remains that any business — from a bookseller to a biotechnology company — that is not factoring the Internet into its plans will quickly find itself at a competitive disadvantage. The Internet is not just changing business — it is changing science and the ways that we disseminate scientific knowledge.[12]

6   Computers In Discovery Processes

Genome projects, the human and model organisms inclusive, give us what amounts to organism parts catalogs for different organisms. The most important thing is discovering what the parts do in each catalog — the roles they play individually and collectively in making each of the whole system run and the roles they play in making it break down.[13] These organism catalogs are not completely independent, for organisms descended from a common ancestry in the past, though some more proximal and others more distal. A more complete or understandable catalog can help understand the others. Thus these catalogs should be used side by side for cross-referencing. For most biologists and biotechnology companies, this is where the excitement really starts. It is also where biology and information technology become even more inseparable.

One key step will be to use sequence data and powerful graphics workstations to decipher the three-dimensional structure of proteins and to test the drug compounds that work on these proteins. This is one of the most challenging tasks in computational biology. The process for determining protein structure has traditionally been very complex and time-consuming. Computers have helped speed up the process. The data generated by genome projects is going to help speed it up even more. For example, once the genetic sequence for a protein is available, one will be able to search for other sequences that are similar. If the structures of those proteins have already been determined, one can more easily predict the structure of the first protein in question.

Computational techniques will make possible visualizations in three dimensions the interactions between a protein and drug compounds that may affect it in beneficial ways. This will go a long way toward making drug discovery more efficient and more productive. For example, instead of using random drugs to target

---

[12] DaHsuan Feng, and Hwa A. Lim, "Infotronics in a knowledge-base economy", *VERIZONTAL*, 8(1), 2000.
[13] See for example, functional reconstruction of disease pathways, http://www.genego.com

random molecules, one will be able to select the best targets and then use information systems to design exactly what is needed to attack them. It will create a much more systematic and predictable process for allocating scarce research and development dollars. This will increase the probability of success of projects.

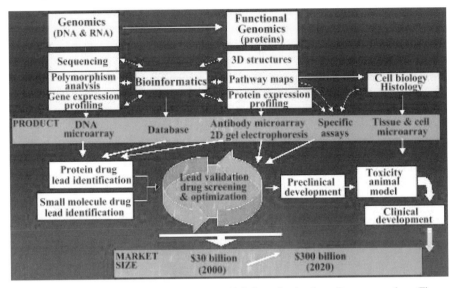

Figure 2. Genomics, functional genomics and bioinformatics in drug discovery cycle. (Figure: Courtesy of Dr. Jeff Wang, AbMetrix, Inc.).

The combination of genomic data, information technology and other advanced research tools will also give biologists the opportunity to think more broadly — to investigate not only the workings of a single gene but to study all of the elements of a complex biological system at the same time. That will be no small feat. There are 10 trillion cells in the human body, and each cell has 35,000 different genes.[14] Gene expression analysis, for instance, will shed light on which genes are active under the same conditions, which genes are important to major functions of the body, and which genes are active in a disease. This will narrow the focus from 35,000 genes to maybe 50 or 500 genes. Current microarray technology allows one to work with 10,000 genes at a time and determine which ones are turned on for a given function. Before long, one will be able to analyze all the genes simultaneously. The computational challenge will be to take this analysis and determine what has changed and how the changes are correlated. One will also be able to use one's knowledge of simpler biological organisms as a Rosetta Stone to decipher these complex systems. Lee Hood, William Gate III Professor at University of

---

[14] Walter Bodmer, and Robin McKie, *The Book of Man*, (Oxford University Press, Oxford, 1994), 259 pages.

Washington believes that systems analysis will be the driving force of biology in the 21st Century. It is also clear that the challenges will grow exponentially larger as we measure genetic variation in individuals (single nucleotide polymorphism or SNP) and gene expression. The best hope for truly understanding the human genetic code will be continuing advancement in high performance computing.

7   Challenges Of BT-IT Convergence

Now that biotechnology and information technology are converging, several challenges remain. None of these challenges, of course, is insurmountable. And they pale in comparison to the promise of biotechnology.

7.1   Bioinformaticist Dearth

First, there is a shortage of biological scientists with strong computing skills. There are a lot of smart biologists in universities, pharmaceutical companies and biotechnology companies and a lot of smart engineers in the computing industry, but there are not very many people who are knowledgeable in both fields.[15] Though the situation is now better than a few years ago, there is still a dearth. The new world of biology and biotechnology requires a new breed of scientist — one that is well-versed in information technology and biology, one that focuses on broad questions rather than on more narrow specialties. As the field of bioinformatics expands, becomes more mature and routinization sets in, more and more universities will build the kind of curricula that will prepare biologists to exploit the essential tools of information technology. When routinization has set in, what a laboratory will need is one or two bioinformaticists — the experts, and an army of bioinformaticians — the engineers or technicians who know how to operate the tools.

7.2   Bioinformatics Gorillas, Chimpanzees and Monkeys

The second challenge is the relative absence of a bioinformatics software industry. Most of the software available today comes from academic sources. The challenge is to find a viable business model for companies that are basically taking academic software and passing it through to the commercial world. One issue is keeping the commercial version current with the academic version.

Most biotechnology companies and pharmaceutical companies who are ultimate users of this software have set up bioinformatics components within the companies. However, more often than not, it is still economically more sensible for these companies to license some software. Several adventurous entrepreneurs have

---

[15] Hwa A. Lim, "Viva bioinformatics, but who survives?" 1999.
www.d-trends.com/webs/viva.html

tapped into the opportunities and set up bioinformatics software companies. To date, there is not a dominant player on the playing field. In the high-tech parlance of gorilla-chimpanzee-monkey terminology from Jeff Tarter, a software industry analyst and the editor of *SoftLetter*,[16] there are so far no gorillas, less than a handful of chimpanzees, and a number of monkeys. A few of the players, interestingly enough, attempted to surf the wave of the dotcom hype of Y1997–Y2000. These players are struggling as a consequence of dotcom demise of the late Y2000.

Rightfully, when their code is repackaged for sale, academic code developers can demand part of the action, such as compensation, deciding how the code is being used or upgrading the code. This might be where an open source code — much like the Linux model — is the best solution. An open environment will also help usher in routinization of bioinformatics.

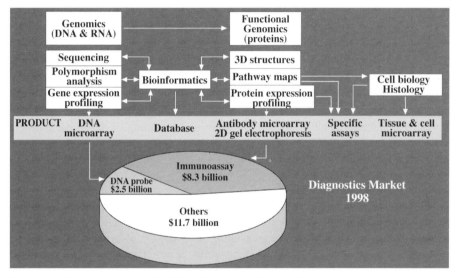

Figure 3. Diagnostics market for DNA microarray, antibody microarray, gel electrophoresis/mass spectroscopy and other technologies in 1998. (Figure: Courtesy of Dr. Jeff Wang, AbMetrix, Inc.).

7.3 Opening up a Framework

In response to the constant change in discovery process and the challenges mentioned above, many companies have opted for open systems as part of the informatics strategy. An ideal solution is to build a reusable components conforming to a plug-in architecture with universal compatibility. An example is the concept of framework. A framework is a skeletal structure upon which a more

---

[16] G. A. Moore, *Inside the Tornado: Marketing Strategies from Silicon Valley's Cutting Edge*, (Harper Perennial, New York, 1999).

complex program can be developed.[17] The use of a framework makes it relatively easy to introduce specific functionality in a domain without having to develop the applications from scratch. Thus within a framework, a well designed interface to software components and services should provide enough information about each component and service to facilitate communication, yet still has the flexibility to accommodate changes in the implementation without massive interface redesign.

An ideal framework will have the following salient features:

❏ The ability to integrate new tools, databases and data types into operation.
❏ The flexibility to support continual innovations.
❏ The ability to have enterprise-wide information management.
❏ The ability to move data across applications and is platform-transparent.

CORBA (Common Object Request Broker Architecture) provides the need for interoperability among hardware and software products by allowing applications to communicate with one another regardless of location or design. Java of Sun Microsystems is an object-oriented computing language developed for use on intranets and World Wide Web. It supports graphically rich programming that integrates well with web-server-based information delivery. Thus, CORBA and Java are two standouts among object-oriented software technologies that accomplish the above features.

As biotechnology companies come to depend more and more on computing, computer companies need to develop a deeper understanding of the biotechnology industry and the unique needs.

7.4   Creating a Language out of ATCG

To deliver promises of the genome project, biotechnology researchers need to be able to easily search all existing gene sequence information. But genome projects produce a massive amount of data, resulting in over 400 individual databases at companies and institutions around the world. Each contains sequences and information about the functions that genes and proteins perform, but each does so in its own programming language. Needless to say, this retards the ability of researchers to easily share their information, even in an open framework environment.

Without some kind of uniform code, researchers have to write storage rules that are different every time. Other attempts have been made at codifying the language, but there are still so many of them — about 14 at last count — that the only standard out there right now is chaos.

Rosetta Inpharmatics, which was recently purchased by Merck for a handsome $620 million, created the Genetic Expression Markup Language, or GEML. It

---

[17] Michael C. Dickson, and Daniel L. Hodnett, "Choosing a drug discovery framework: critical success factors", *NetGenics White Paper*, 2001.

received an endorsement from Nature magazine, which seemed to herald its widespread acceptance, but it never happened.

Recently, IBM has spearheaded an effort to write that common language using XML. Its Life Sciences has recruited Sun, Millennium Pharmaceuticals, Affymetrix, Accenture and 35 others in an effort called I3C, which promises to accelerate genetic research by helping researchers obtain data using a standard language.[18] So now IBM has surrounded itself with 40 key players in the hope of cutting the Gordian knot that Rosetta faced. IBM is banging on getting enough horsepower behind a consortium so that a set of standards becomes a standard by default.

Incyte Genomics, which is part of the consortium, has created a framework called the Genomic Knowledge Platform and will contribute what it can to a standard language. Members of the consortium expect to complete a language for expression array data in 15 months from mid Y2001.

So what is XML? Extensible Markup Language (XML) is the next generation HTML, the language currently being used to create Web pages. But XML reaches beyond the Web. It can be used to store any kind of structured information and to enclose or encapsulate information in order to pass it between different computing systems that would otherwise be unable to communicate. According to Charles Goldfarb, who coined the term "markup language" in 1970 and who invented Standard Generalized Markup Language (SGML), the mother tongue of HTML and XML, "HTML makes the Web the world's library. XML is making the Web the world's commercial and financial hub."[19]

Ron Schmelzer, founder of the Waltham, Massachusetts-based ZapThink, believes there are five principal forces driving XML:

❑ The increased need to integrate and share data outside the walls of a corporation.

❑ The desire to conduct ebusiness, thus reducing the cost of transactions and increasing the number of trading partners.

❑ The desires to archive, store, and accurately retrieve information that is overwhelming in volume.

❑ The presence of the Internet, which has increased organization's desire to interact and integrate with each other.

❑ People's growing familiarity with the Internet's protocols and standards.

However, though most industry experts applaud the impact, benefit and beauty of XML, not everyone agrees with the hype. Alex Karpovsky, founder of the Concord, Massachusetts-based Kanda Software, opines, "XML itself is nothing but a standard for defining 'dialects', sets of tags that concisely and precisely define the information they delimit. It is the first and easier step toward enabling diverse computer systems and applications to communicate between themselves with little

---

[18] Kristen Philipkoski, "A language out of ATCG", *Wired News*, June 27, 2001.
[19] Douglas Page, "XML: Creating a brave new World Wide Web", *High Technology Careers*, Vol. 18, No. 5, Oct/Nov/Dec 2001, pp. 7–10.

or no human involvement. The second and harder step of defining the actual universally understood industry or domain-specific dialects is just beginning to shape up and will take years to complete." Karpovsky also questions the maturity of XML's debugging tools, especially XSL, a companion that transforms XML into HTML or other formats.

8   All The Fuss, Where Is The Beef?

A recurring question in conversations, discussions and news interviews is "Bioinformatics is very popular. How come no bioinformatics company is known to make any profits?" To answer this question fully, we shall discuss information and knowledge in their broadest sense in the context of traditional economy, but shall quote examples from biotechnology.

Though usually used interchangeably, there are subtle differences between data, information and knowledge. Data is sensory or digitized input. For example, we are collecting data when we see or when we touch. When we key numbers into a computer, we are inputting digitized data. Information is the input, whether sensory or digitized, that has been placed in context. Thus data is stored fact that is inactive, while information is presented fact that enables doing. Knowledge arises from examining the relationship between pieces of information.[20] When information and knowledge are used as initial inputs, they play the role of data since then they enable the next level of doing. Information is, in some circumstances, quantifiable. Knowledge is not.

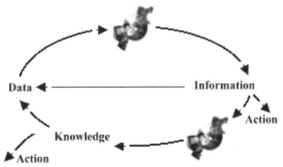

Figure 4. Data is input. When data is put in context by a human or a computer, it becomes information. The information can be put into a storage (disk or human brain) as data for further processing, or its relationship with other information can be examined to obtain knowledge. The knowledge can also be put into storage as data for further processing, or it can be used for decision and action. In general, when the dataset is large, a computer can process the data and information much better than a human.

---

[20] Daniel Kohanski, *Moths in the Machine: The power and perils of programming*, (St. Martin's Press, New York, 2000).

Nonetheless, we are inclined to confuse information with knowledge because until the advent of computers, the gathering of information was a uniquely human task — one that was not separated from knowledge for as the brain collects information, it also transforms it into knowledge. Recorded information, for example, in books, in scrolls and on cuneiform tablets, lies dormant until we come in contact with it.

The computer, however, is not a mere passive collector of data. It also possesses the capability to put the data in context, thus converting the data into information. Furthermore, it can manipulate the information in ways that approximate the human mind, acting as if it possesses knowledge as well. Stored information on computer disks lies dormant until the computer accesses it to process.

As an example, in a genome project, the DNA sequencer generates a lot data, the A, C, G, and T of an organism. When the data is put into context and analyzed in gene finding, we have information about the gene. The gene information can be stored in a computer disk as data for later comparison with genes of another organism, or the gene can be used in an action such as the design of a DNA microarray, or the gene function can be ascertained so that we have knowledge of the gene. The knowledge of the gene can be stored as data in a computer disk for later use to find the protein it encodes, or the knowledge can be used to effect a certain action such as finding the associated disease.

As another example, in proteomics, we have data from an antibody microarray experiment. The pattern of the microarray provides profiling information. From the profiling information, we know which proteins are expressed and which are not in the disease under study.

8.1 Information and Knowledge: Factors of Production

We see that information and knowledge are complex and multidimensional. Yet in models of economic growth — whether neoclassical or evolutionary — knowledge usually gets reduced to a single number: total factor productivity. Knowledge is an input, which combines with capital, labor, and other inputs to produce goods and services, and is thus a source of total factor productivity. But knowledge is also an output — the product of existing knowledge and investments in innovative activity. Knowledge has characteristics of a public good. As information spreads, the producer of knowledge cannot permanently and fully appropriate its value — though it is essential that the producer holds on to some of it, i.e., protect its intellectual property position.

In drug discovery, bioinformation and bioknowledge have helped make the discovery process more cost effective and less time consuming, have helped make drugs safer, and soon, will help make medication more personalized.

Fueled by research and the rapid generation of new knowledge, technological innovation has become the major factor behind increases in productivity. The "knowledge content" of goods and services from science, advanced design concepts, intelligent materials, automation, software, sensors, advanced services, new biotechnological concepts and new medical concepts has increased. And more and more of the goods and services in the world market are the result of complex production processes. The great speed of technological change and the rapid accumulation of new knowledge mean that firms failing to incorporate new knowledge lag behind in productivity and competitiveness.

For example, biochips are used in diagnostics. A DNA microarray chip can be designed to be disease-specific if the gene of the disease is known. In antibody chips, the chips can be designed to detect mutant variants. Thus the knowledge content and thus the analytical power of these diagnostic devices has increased from the good old days of a decade ago of "studying one gene at a time".

As the world economy becomes more knowledge-based, the value added by information propels the more developed countries. But the knowledge revolution also creates opportunities for developing countries to emerge from dependence on low-cost labor as a source of comparative advantage, increasing their productivity and incomes.

IT can empower and link people all over the world. It can sustain economic growth, enhance public welfare, and promote social cohesion.

8.2   *Technology, Markets, and Competition*

One effect of technological change is to alter the nature of goods, markets, and competition — nationally and internationally. This in turn changes business practices and the way that firms compete.

Technology is knocking down geographical boundaries, changing the structure of production and trade within and among countries. Previously non-traded goods (such as professional consulting services) are becoming internationally tradable through information technology. Previously immobile factors of production (labor) are becoming mobile as geographical barriers have less meaning. Some manufacturing production is becoming globalized, with different components produced simultaneously in different countries. For example, Motorola has a huge plant in Tianjin, China, and computer chips productions are farmed out to South East Asia countries. As the costs of communication fall, outsourcing and farming out services have become feasible, unbundling them from manufacturing activities to change the boundaries between firms. For example, financial institutions in the United States are farming out their soliciting jobs to cheaper labor countries like India, which speaks English. Software giants are farming out software development to India and Russia where the talent pool is high and labor cost is low.

Information technology and communications technology can make markets function more efficiently by reducing information asymmetries between buyers and sellers, eliminating the need for middlemen, and collapsing distance. Technology can also increase competition and market contestability by lowering barriers to entry, reducing the minimum efficient scale of production, and providing alternative production techniques. Industries previously thought to be natural monopolies, such as utilities, can become competitive. Often, this means that technology erodes the efficacy of regulatory frameworks, which become ineffective, inefficient, or incomplete.

We have thus entered an interesting time — a time of disequilibrium, a time of rapid change, a time of dynamics, a time of paradigm shift — however we prefer to call it.

9 The River Of Change

With the acceleration of information technology, telecommunications technology, biotechnology and transportation, we know only one thing for sure: change will leave nothing unscathed. If change is a river, we are sitting at the confluence of many of its tributaries and are made aware of all the factors funneling into the moment.[21]

History is nothing but the history of change: the history of a few communities of excellence that changed the world (for example, the civilization along the Ganges River, India in 5000 BC), the history of challenges and appropriate responses (for example, the Great Depression, 1929–1931), the history of turbulence (for example, the Second World War, 1941–1945), the history of a few companies that changed the way we live (for example, Visa Card, introduced in 1951), the history of a few revolutions that changed the world (the Eight Days that Change the World, Russia, August 19, 1991), the history of dangers and opportunities (for example, the dotcom hype, 1997–1999), and incidents that changed the way we view things (for example, the recent World Trade Center incident, September 11, 2001 that questioned the invulnerability of the United States, and changed the way we look at modern warfare).

Time and again, in each period of turbulent change, there are inherent risks and opportunities. No one has figured out a foolproof way to straighten out change yet. No one has resolved this humankind's dilemma yet. And fortunately for us, no one has used up the opportunities, and no one ever will. The mandate of history is clear: capitalize on the opportunities of change.

But in rapid change, we need radically different approaches. How-to methodologies and MBA programs are only effective when applied to stable and predictable environments.

---

[21] William M. Boast, *Master of Change*, (Executive Excellence Publishing, Provo, Utah, 1997).

9.1   Capability and Copability on the River of Change

Consider the fact that millions of years ago, Western Colorado was a swamp inhabited by, among many other things, dinosaurs and cockroaches. The dinosaurs had become tremendously successful. They were the largest creatures that ever roamed the earth. Not only were they great in size, they were also varied in sizes. They ranged from very large to very small, from the very swift to the very clumsy, and some even could fly. The dinosaurs had become very well adapted to living in the swamp — the epitome of success. Then one day, the dinosaurs became extinct. But the cockroaches are still here. There are many theories about how the dinosaurs disappeared, but that is a moot point. The cockroaches made it, indicative of their wider span of adaptability or copability.

Analogously, too many companies and too many individuals are dinosaurs with all the answers written into their genetic systems. The more workshops they attend and the more degrees they acquire, the more specialized they become and the more restricted their span of adaptability and range of flexibility. They become very capable, but they become less copable.

When gradual changes predominate, the job is to be sure that we have the appropriate specialists on staff to engineer and optimize the constant and predictable solutions. In gradual changes, we engineer. When rapid changes predominate, our job is to see that we are not the ones who inherit the dangers in any given crises but we are the ones to inherit the opportunities. In rapid changes, we create and innovate. That is why in gradual changes, we need capability. In rapid changes, we need copability.

In businesses, constant shifts in technological advances and in forces in the marketplace negate tried and true formulas, and once successful paradigms can suddenly sabotage any ability to anticipate market directions.

10   An Effective, Intelligent And Responsible Generalist With A Specialty

The twenty-first century will need a generalist with a specialty. And specialists need to know that the required specialty will change every 3–4 years. For example, in the biotechnology industry, during the early 1990s, we were talking about sequencing and we needed sequencing experts, A few years later, we were talking about gene hunting and were desperate for good programmers. Soon we were talking about functional genomics and we were searching for life scientists with programming skills. Now we are talking about proteomics and systems biology and we are seeking excellent bioinformaticists.

In the new world, we must constantly become a transformed specialist, standing solidly upon generalized knowledge that comes from a real education in its broadest sense.

To be truly successful, besides being a generalist with a specialty, one must be effective, intelligent, and responsible. Meeting only one or two of them will not be considered enough. Although sheer effectiveness is not enough, that does not diminish its importance. In fact, if one is not effective, one needs not concern with being intelligent or responsible. If one is in a biotechnology business to make money and one does not make a profit, it does not matter much if one is intelligent or responsible. Effectiveness is the pragmatic element of any endeavor and the first criterion that must be met. Doing things better is not the same as doing better things.

Effectiveness may pay off in the short term, but it will take intelligence to ensure that one is effective in the long term. Governments, companies, and individuals in the second half of the twentieth century have been quite satisfied with just one of the criteria — effectiveness. They have shown few signs of the other two. Governments are notorious for achieving effective foreign policies in the short term, but they often lack the intelligence to try to understand the culture and tradition of the other countries so that long-term success can be sustained. The stronger nations usually tend to export wholesale their ideologies. For example, the recalcitrant Middle East conflict. History has repeated itself time and again that every time the U.S. topples a regime and installs another in power, that same country, in a few years after they have gotten equipped with U.S. weapons, is the very country that the U.S. has to fight against. A company must be effective in selling products, but if, out of ignorance, the product immediately fails to perform as marketed, the company will gain a reputation for a poor design and lose its edge in the marketplace. An individual must complete jobs effectively, not in haphazard manners. A well thought out and well-planned project usually leads to an optimal solution.

Only responsibility can build the confidence and trust that are so essential for leadership and mastery. Governments are responsible for the effects their administrations have on the people who live in their countries and who depend upon their laws. For example, friendly laws and legal systems guiding cloning and stem cell research provide an environment conducive to breakthroughs. A company is responsible for the way its products affect customers. The company is responsible for the effect of the pricing of its products on the economies of the communities where the company does business. For example, tobacco companies are responsible for "smoking to death" half a million of their customers annually. Employers are responsible for their employees. Employees are responsible and have fiduciary duties for their job performances.

Companies in the twenty-first century must lead in intelligence and responsibility, as well as effectiveness. Success is the synergy of the whole, including leaders, project members, and those stakeholders who will benefit (or be harmed) from the projects. The same applies to governments as well as to businesses.

The Human Genome Project is an exemplary accomplishment. Through effective coordination, intelligent planning, and the hardwork of responsible consortium members, the project was completed ahead of schedule in Y2000. Other consortia: The SNP Consortium, The Human Proteome Project, and others are following the footstep.

The SNP Consortium Ltd. — Chairman and CEO, Arthur Holden — is a large non-profit organization. Sponsored by ten pharmaceutical company members, information gathered by the consortium is passed along to an informatics group of three sequencing laboratories.[22]

Currently, it has the following members: AP Biotech (Uppsala, Sweden), AstraZeneca Plc (Macclesfield, UK), Aventis Pharmacia AG (Frankfurt, Germany), Bayer AG (Leverkusen, Germany), Bristol-Myers-Squibb Co. (Princeton, New Jersey, USA), F. Hoffman-La Roche (Basel, Switzerland), Glaxo Wellcome Plc (Greenford, UK), IBM (Armonk, New York, USA), Motorola Inc. (Schaumburg, Illinois, USA), Novartis Pharma AG (Basel, Switzerland), Pfizer Inc. (New York, USA), Pharmacia Corp. Searle (Peapack, New Jersey, USA), SmithKline Beecham Plc (Brentford, UK), The Wellcome Trust (London, UK), Cold Spring Harbor Laboratory (Cold Spring Harbor, New York, USA), The Sanger Centre (Hinxton, UK), Stanford Human Genome Center (Palo Alto, California, USA), Washington University School of Medicine (St. Louis, Missouri, USA), Whitehead Institute for Biomedical Research (Cambridge, Massachusetts, USA).

The human genome has 3 billion bases, there is a mutation in every 1,300 base pairs, yielding about 2.5 million single nucleotide polymorphisms (SNPs) — a swap of a nucleotide for another one. By the end of Y2000, the Consortium has assembled about 1 million SNPs based on data from 24 people of different ethnic backgrounds. This is about one third of the estimated common SNPs, at an expense of about $50 million.

From the private sector, there are Celera Genomics (Rockville, Maryland, USA), and Incyte Genomics (Palo Alto, California, USA).

Most likely, these consortia will attain their respective goals.

10.1   The Gentlemen And The Technologist

There is a huge distinction between technology and manufactured goods. Many a myopic person would say that investment in technology has never generated a single penny in profit. Technology is not a package that can be bought off the shelf and become immediately productive: it is a cumulative process of learning. So, for developed and developing countries alike, the ability to realize knowledge-based productivity gains depends on the country's capacity to tap the global system of generation and transmission of knowledge, to generate indigenous knowledge, to

---

[22] Gail Karet, Julia Boguslavsky, Tim Studt, "Unraveling human diversity", *Drug Discovery & Development*, November/December, 2000, pp. S1–S14.

diffuse and transfer information, and to use that knowledge in productive activities. For firms, turning information into value depends on their ability to manage their knowledge assets.

Acquiring new technologies requires a system receptive to innovation, with incentives and mechanisms for translating knowledge into action. The process of diffusion and implementation is greatly strengthened if there is feedback from the users of technology to the generators of knowledge. To the end of achieving a receptive system, the social system must not be hostile, legal systems and policies must be technology-friendly, the information and telecommunications infrastructure must be well developed, and there must be human resource development.

For example, when the Human Genome Project first started in the early 1990s, the focus was on technology improvement. When the high throughput technology had been perfected to a point where the project could be completed within a reasonable time frame, about $1.9 billion had been sunk. The high throughput sequencing task was then divided among consortium members internationally, linked by the Internet technology. Daily generation of genetic data was accumulated at the National Center for Biotechnology Information (NCBI) for everyone to tap into. Every step was very goal-oriented, and the rest is history.

In retrospect and in contrast, nineteenth century England is a good precedent.[23] England started the Industrial Revolution. But by the 1850s England was losing its predominance and beginning to be overtaken as an industrial economy by the United States and then by Germany. It is generally accepted that neither economics nor technology was the major culprit. The main cause was social. Economically, and especially financially, England remained the great power until after the First World War. Technologically, it held its own throughout the nineteenth century. But the English did not accept the technologists socially. The technologist never became a gentleman. The English built first-rate engineering schools in India, but almost none at home. No other country so honored the "scientists". But the technologists remained a tradesman.

Nor did England develop the venture capitalist who has the means and the mentality to finance the unexpected and unproven. What might be needed to prevent the United States and other advanced countries from becoming the England of the twenty-first century are the social position of knowledge professionals and the social acceptance of their values. For them to remain traditional employees, be treated as such, or be regarded as not serious in the eyes of those hiding in ivory towers would be tantamount to England's treatment of technologists, and likely to have similar consequences.

Having said this, we have to stress that the England today is very different. The English have the most friendly laws guiding cloning and stem cell research, followed by Israel, Japan and Australia. The United States have imposed restrictions that many in the community suspect will stifle the research. With technologies tearing

---

[23] Peter F. Drucker, "Beyond the information revolution", *The Atlantic Monthly*, October 1999.

down geographical barriers, and consequent changes in the structure of research, it is likely that the United States stands to lose its best researchers to more receptive and accommodating countries.

10.2 *From Classroom to Boardroom and Vice Versa*

Business professionals and technologists work in a myriad of small, medium and large companies that rely for their financial health, if not their very survival, on the reactions of the market forces to their inventions. Traditionally, the academia competed in a very different environment. They were nurtured in research institutes and university laboratories, where in a tenured position, salaries were more or less assured and professional rewards and recognition were meted out through an elaborate system that included literature citations, research grants and prizes. The current generation of new academia are likely to be more entrepreneurial-minded, or to be reduced to making obsequious gestures toward those who hold the purse strings on their research. Notwithstanding, the academic sector can still be the nidus of creativity, provided the academic members have not already been restricted in their span of adaptability and range of flexibility by the workshops they have attended and the degrees they have acquired. The private sector, on the other hand, is always on the *qui vive* to look out for technologies or marketable innovations to transfer.

As of late, it has become a common place to have members of the academia joining the private sector, and the academia seeking experienced members from the private sector to direct research centers.

There are numerous reasons why academia members leave their academic positions for the private sector. Most popular responses are that they see working more with people or transferring an innovation through commercialization as the motivating force.[24]

Despite different backgrounds and experiences, former academicians find transferable skills advantageous in the business world: written and oral communication, being comfortable with numbers and critical thinking. Another portable quality is persistence and perseverance.

Culturally, there are also similarities and differences. In both the business world and the academic world, a hard work ethic and commitment are required to succeed. There is probably more accountability in the business world than in the academic world. Unless one is tenured at a university, the business world can be more flexible because risk aversion in the academic sector can be higher. In the business sector, gratification is more immediate. It is usually a direct function of revenues.

---

[24] Karen Young Kreeger, "From classroom to boardroom", *The Scientist*, 14(3), February 7, 2000, pp. 28–29.

The academia is a world of critique, not usually a world of action. So are many corporate managers who are no longer close to the frontline of the products or services their companies provide. They survive through their ability to critique, reorganize, and ensure that their stakeholders are satisfied. In time of gradual change, we can absorb a few critics, but in the dynamics — a time of rapid change or a paradigm shift — we need action-centered people. This is why corporations are getting thinner in the middle management, and we hear of people talking about a flat corporation. Most of the former academicians are hired into positions of R&D or into marketing where they can utilize their knowledge to sell products.

Academic leadership should begin to take clues from Olympics track coaches and stop relying on committees. After all, the job is to find one person who can jump seven feet high, not seven people who each can jump a foot. Too much time is wasted in committees where people talk, and not enough time is given to action. If organizations would simply devote more time to improving, and devote less time to talk, talk, goal, goal, talk, talk..., their results would be greatly improved. A committee is mostly for talk and occasional plans that might lead to action. Indeed, for a committee, meetings have become rests between coffee breaks. A team or a task force is for action. This is one of the reasons why entrepreneurial academic organizations are now hiring accomplished members of the private sector to run their research centers and institutes.

Intellectual property offers legal control over the creative productivity of the human brain, but the technical boundaries between what can be maintained under personal or professional control and what may be freely circulated for others to capture and manipulate are not yet clearly defined.[25] Consequently, the long-standing tradition of "each generation standing upon the shoulders of previous generations" is no longer tenable because of the tendency to patent, copyright or license everything, be it most insignificant! Intellectual property issue is taken more seriously in the private sector than in the academic sector. In Y2000 alone, IBM was awarded 2,922 patents in the United States — an average of more than 11 each working day, and 43 percent ahead of NEC of Japan, in second place. It was the eighth consecutive year in which IBM received the largest share of patents in the United States. The national total in Y2000 was 176,087.[26]

But in recent years, the academic sector is also getting into intellectual property protection of outcomes arising from research conducted by members of their organizations. For example, most research universities now have intellectual property or technology transfer offices. This is yet another reason why we are seeing more former private sector members at the helm of academic research centers and institutions.

---

[25] A.N. Branscomb, *Who owns information? From privacy to public access*, (Basic Books, New York, 1994).

[26] Barnaby J. Feder, "Eureka, IBM develops labs with profits", *New York Times*, September 9, 2001.

10.3 Smart Work, Hard Work and Hyper Work

It often seems that we have lost all common sense as human beings. Today the passions for productivity and cost effectiveness have blinded us to the human beings as the users of tools. These passions have led us to seek more and more "production capacity" in each employee and fit more and more into each job description, as though job descriptions were software with iterative loops and employees were computers upon which more and more iterative loops are executed. As a result, we have lost a clear distinction between smart work, hard work, and hyper work. We are willing to abandon smart work and institute hyper work under the illusion that it is harder work. We even contrive titles to make such hard work palatable. Have you ever wandered into a bank or financial institution and wondered how come all the employees are now personal bankers, managers, loan officers, assistant vice presidents, vice presidents and other exotic titles? They used to have titles like cashiers, tellers, and managers. Besides personal ego satisfaction for each employee, the banking institution makes each customer feel personal and good because an important officer is serving the customer.

In the hyper work of today's management, managers do not have time to step back and look at the multi-task focus. They hire consultant after consultant until the multitude of focuses at the top looks more like the compound eye of an insect, made up of at least four thousand lenses that break up the world into small pieces. As a result, the context is lost.

If the job to be accomplished is routine enough, then multi-tasking may be used. But in rapid change and ambiguous times, multi-tasking may be detrimental when trying to form a creative and appropriate response. Even for computers, multi-tasking is good only if the overhead in swapping tasks in and out does not more than offset the gain. For a human, the overhead in swapping is high. Besides the time wasted in going from one task to another, one also has to be in the right mindset to perform effectively.

While hard work can be productive, hyper work is quite the opposite. Fragmentation of focus is the nature of hyper work. In hyper work, one goes faster and faster on more and more. One is made into a computer judged on the teraHertz. Although quantitatively hyper work may turn out more production, the quality declines. Suddenly the human being is no longer a human being, but is now another part of the machinery instead of capitalizing on the machines.

It is time companies let automation machines and computers do all the hyper work and return human beings to smart work whenever possible, and hard work whenever necessary. Smart work begins in focusing on one thing at a time. By building a community of people who are effective, intelligent, responsible, and capable of the focus of their full capacities, they will create an environment for smart work. The smarter the people work, the more likely they will create a synchrony of intelligent design, effective process, and responsible results. They must learn to use human resources humanely and appropriately, just as they must learn to use technological tools effectively and appropriately.

Corporation must outgrow the megalithic mentality by becoming more humanist corporations and by making investments in human race. The belief in the potential of human beings to deal with their world effectively, intelligently, and responsibly, the belief that the world should exist to enrich the lives of human beings and that human beings should exist to enrich the world, is called humanism.

But employees must also help themselves. The Post-It notes yellow pad may be one of the greatest innovations for 3M,[27] but it serves many an employee a big disservice. It has become a means for procrastination. Ever notice that the number of yellow pads used to remind an employee tasks to be completed increases on the peripherals of computer terminals. The employee, each time when assigned a task, will just scribble on a yellow pad the task to be completed and paste the yellow pad to the terminal, and wait for "one of these days I will get to it". In actual fact, if the assignee would just think a little, be intelligent, and see if the task can be completed immediately. Some tasks require less effort, less thinking. By completing simple tasks, tasks that are not time-consuming nor require much brain juice, the assignee can purge the tasks off the brain. This way, the assignee will get more encouraged and be propelled further along, instead of getting drowned amidst tasks to be completed. By being intelligent, the assignee is more effective. And by being able to complete task in a timely manner, the assignee is being responsible.

The word "technology" is derived from the Greek word "techne", which means skill. Ancient Greeks believed that the goal of man was to free himself from labor so that he could have more time to think.

Though the Greeks were not a part of the project, the Human Genome Project was a scientific feat. Automated sequencing machines were working 24/7/365 non stop to churn out human DNA sequences, tended only by a few shift technicians. A project that would have taken a laboratory using the 1980's technology 300,000 years to complete was completed by an international consortium of 16 laboratories in less than ten years! The automated sequencing machines did the labor, freeing the humans to do other things.

11   Training And Education

The megalithic mentality emphasizes resources to achieve quantitative results, but the new emphasis must be on the human to achieve qualitative results. At a time of rapid technological breakthroughs, education — formal or informal — is the only way to develop the qualities that allow one to handle change and ambiguity effectively, intelligently, and responsibly.

---

[27] Damon Darlin, "Thank you, 3M", *Forbes*, September 25, 1995, pp. 86.

The convergence of information technology and biotechnology is presently focused on decoding the 35,000 or so genes in the human body. Scientists want to know the precise structure of these genes, what their functions are, and which genes are altered when diseases such as cancer spread through the body. Some people compare the efforts of categorization of genes to the formulation of the periodic table of elements in the nineteenth century, a scientific enterprise that spawned the chemical industry.

At the moment, the juncture of biology and computer is astoundingly complex, dominated by scientists with advanced degrees — roughly comparable to the Internet twenty years ago. These expert scientists are bioinformaticists, who are still quite difficult to come by. But as the technology pays off in new ways to diagnose diseases, and better ways to understand how the body functions, many jobs will be produced at many skill levels, and routinization sets in. As the industry grows, the skills involved become more and more routine. Laboratory technicians will be running the laboratory's highly automated equipment. Entry-level talents will then be sufficiently qualified to operate the automated processes.

This is where training and education differ. Training is the acquisition of skill and information that leads one to deal effectively with routinization or gradual changes. If one needs to learn to rivet a beam, measure the chemical content of a solution, wash windows of a skyscraper, fly a sub orbital jetliner, run a sequencing machine, or run a bioinformatics tool package, one needs training, not education. It is the greater part of applied science and engineering and of all specialization. It is the continual unfolding of the human being as homo faber (tool man).

Credentialed is not synonymous with qualified. Or specialists alone or generalist alone are not enough for the world we face. It would not be stretching the truth to refer to the present time as another sophistry. With more and more accelerated MBA and business degrees, and degrees of assorted specializations, such as bioinformatics and Microsoft Certified Engineers (MSCE) sprouting up at university extensions, we certainly are in the training age where programs promise to give you the skills to succeed. But they seldom give you the character, wisdom, perspective, or vision to use those skills intelligently and responsibly. These graduates are technicians. The bioinformatics graduates are bioinformaticians — bioinformatics technicians — not bioinformaticists.

Alfred North Whitehead, one of America's great philosophers, gave a succinct definition of education, "Education is the acquisition of the art of the utilization of knowledge". Notice that it is NOT "the acquisition of knowledge". Sir Winston Churchill put it very nicely, "The first duty of a university education is to teach wisdom, not a trade; character, not technicalities. We want a lot of engineers in the modern world, but we do not want a world of engineers." In gradualism, one works as an engineer. One develops formulas, principles, rules and methodologies and one seeks predictability and excellence in continuous recurrence. But in dynamic changes — ambiguity, inconsistency, and the action of the dynamic world — like the paradigm shift in biotechnology discovery we are experiencing now, one must continuously create new solutions for every new problem.

Education means, above all, mastery as a generalist. Both the specialist and the generalist — the narrow span and wide spans of tolerance, respectively — are valuable. Only humans can change from moment to moment, if they have not been limited already by their training or expectations. It is time to realize that no how-to nor training program is going to "do it" for us. The program is there only to support the individual, not the other way round. Somehow we have to equip individuals with the breadth of knowledge and the tools with which to think so that they can successfully maneuver through the whitewater of change. The object is not to take the white out of whitewater, but to put a master to steer the whitewater raft.

12 Above, At, Within the Laws

It has been the American way to change things by a process of case law, of case law made into new fashion, and ultimately, the new social order. It is in a way a form of gradualism. Change so gentle that we do not take alarm. For example, the way the censorship of literature in books or on stage was watered down until it has almost disappeared, that homosexuality has ceased to be a crime, that divorce has become a part of our new society, that mixed marriages have become blessed, that smoking in public places is no longer macho, and eventually drunkenness from roads and violence from screens. In a similar light, the issues of cloning and stem cell research are going through the equivalent of early stage literature censorship.

Decisions on issues with ethical and social implications are political tight ropes and judgment calls. But they should not be calls made solely in the court of laws. Judges do not think about what makes sense from the perspective of accelerating technological and economic progresses. Their concern is with how new areas of technology can be inserted into the legal framework with the least disruption to existing legal interpretations. Such lazy law writing practices in a technological paradigm shift environment do not make for good economics or sensible technology policies. The right approach would be to investigate the underlying economics of an industry to determine what division of incentives is necessary for its successful development.

A direct attack has not worked, is not going to work, and will not work. The goal is to make better things happen, not to stop bad things from happening. Horror stories do not work as motivational devices. Traffic schools show us videos of horrible automobile wrecks full of mingled bodies. They do nothing to make us better drivers. We simply block out the horror stories. We also get upset with the humiliation traffic schools put us through after having been cited and having to pay fines. We hardly become a better driver. Accidents do happen, but not to us. Similarly, public health official show us healthy lungs and cancerous lungs stiffened from smoking. There is no diminution of smoking. Cancer does happen, but in the far distant future to some yet unknown person.[28]

What should be done is something similar to the process that had actually brought smoking under control. Over a 25-year period, public education about the benefits accruing to non-smokers from living in a smoke-free environment radically changed social attitudes towards smoking. In the beginning, polite nonsmoker could not ask a smoker not to smoke. 25 years later a polite smoker could not smoke in the presence of nonsmokers without asking their permission. Over time, social pressure slowly shifted habits and definitions of what was acceptable.

This should be the policy to embrace the new biotechnology — by educating, not news media showing nightmarish scenarios, not outright banning.

13   History Of Science On Trial

A problem well known to historians is to define at what point the course of history is changed. It is a task that may be possible only in retrospect and in perspective.

The value of historical perspective is that it allows for the wisdom of hindsight, illuminating matters that were not at all obvious at the time.

To decide how important a world leader is, historians usually have to wait for a few years, or even decades. Most of these people are honored posthumously. During President Clinton's office, the United States enjoyed the longest prosperity in history. Personal affairs aside, in years to come, he may very well be regarded as one of the greatest presidents the United States has ever had.

To decide which event was epochal and which was not, historians may have to observe how subsequent events played out. For example, the Eight Days that Shook the World in the August Coup of 1991. Within months after the coup, on 21 December 1991, the Soviet Union ceased to exist. The great ideological experiment begun by Lenin's Bolshevik Revolution and constituted on 30 December 1922, disintegrated nine days short of its 70th year. The September 11, 2001 suicide attacks on the World Trade Center towers is no doubt an incident to remember. They could not have come at a worst time. The United States economy had been ailing for more than two quarters. The incident catalyzed the largest stock crash since the Great Depression, and more repercussions of the incident are yet to come.

If economic and political upheavals can be difficult to appreciate while they are unfolding, science is definitely harder to judge. Science can seem a world of its own, understood only by the elite few and specialists. The science involved can be so arcane that it is virtually impossible for anyone other than those directly involved to understand precisely where it is going. All too frequently, we are also confronted by the spectacle of dueling views. For example, the public effort and private sector

---

[28] Charles Handy, *The Age of Unreason*, (Harvard Business School Press, Boston, Massachusetts, 1989).

effort in human genome sequencing, or the current debate on human cloning and stem cell research.

Only when the dust finally settles can we see the road that took us to the astonishing present.

The Human Genome Project started in the mid 1980s, amidst strong opposition primarily from cottage industry researchers. At the rate of sequencing using the 1980 technology, it would have taken 300,000 years. The warriors marched on. Instruments were perfected. The project completed in 2000. To make the story spicier, there was the bitter duel between the private sector (Celera) and the public sector (International Human Genome Consortium). They both sprinted to the finish line, three years ahead of schedule, with no clear winner. The biggest winner of all is the humankind. Now the benefits of the project are undisputable.

The Human Proteome Project is beginning. Unlike the Human Genome Project, its beginning is rather different. It has a lot of support from the community and from the private sector. At our current rate of characterizing antibodies, we will need 1,000 years to complete the project. But the Human Genome Project has told us that with instrumentation improvement, the project is feasible. HPP will be completed.

Bioinformatics is now more than a decade old. Its humble beginning predates its popularity by almost a decade. The neologism "bioinformatics" was coined in 1987. Numerous international conferences on the subject had been organized by 1996, but still not too many people know about it. Then Science magazine published a series of articles on the subject. The area just took off. It is now very hot. In recent months, news media has hailed it as

"14 letters that spell the future" [referring to bioinformatics]
"In a bleak financing climate, bioinformatics are striking deals"
"It is one of the few technology sectors that is still winning favors from investors"
"Bioinformatics products is projected to reach $11.5 billion by 2004"
and others.

Bioinformatics

❑ Innovalues by creating bioinformatics products for the world market through innovation and adding value to existing products (database, information). It is an enabler, and in turn is enabled by the Human Genome Project, and will be enabled by the Human Proteome Project as well.

❑ Cloneverges — a combination of cloning and diverging — by taking an existing product and rapidly come up with a high-quality, lower cost alternative with some new wrinkles to the product. It is transdisciplinary.

In retrospect, bioinformatics would have been impossible had it not been the concurrent advances in computer technologies. During the humble beginning, it encountered no resistance from ethicists, nor hostility from Luddiths. In both innovaluing and cloneverging, BT-IT convergence is transparent. This eclectic subject is now ubiquitous.

Dolly the cloned sheep had her share of media coverage. But en route to Dolly, even people like Wilmut and Campbell, who were perhaps in the best positions to know what was happening, somehow missed the enormous significance of what they had done. Part of the problem with cloning Dolly was that so many scientists had convinced themselves that it was impossible to clone an adult. But the story is more nuanced than that. Most of the scientific community was looking the other way because sheep research was not popular then. As all scientists know, science is often times cliquish and trendy. Some fields of enquiry are popular, and every small advance in these fields are trumpeted by the media. Other fields are all but ignored by publicity.

In retrospect, Wilmut and Campbell were fortunate to be working in the backwater of science, and in a then unknown Roslin Institute. Not too many people paid much attention to their work. Had they been working in a mainstream area, or had they been working at an established university with an aggressive public relations staff, the press would probably have not left them alone. In that case, the history of cloning could have turned out to be very different. Pressure from ethicists might have side tracked their work, and Dolly might have not been cloned!

Agbiotechnology is getting a lot of media attention. Some liken genetically modified foods to frankenfoods. Most soldiers in the biotechnology revolution believe that the public will eventually accept genetically modified foods, thereby ending hostilities. However, science must first offer something of value, such as improved nutrition — functional food. Just making life easier for farmers with pest-resistant crops will not outweigh real or imagined risks to people.

Stem cell research and cloning are also under media chiaroscuro and are subjects of congressional debate. It is now up to those cloning and stem cell soldiers to prove that therapeutical cloning is in general good for humankind. Reproductive cloning may not be so bad if we look at the precedent case of *in vitro* fertilization.

An act as simple as educating the public can do wonders. Thus a sensible policy to embrace the new biotechnology is by educating, not news media showing nightmarish scenarios, not outright banning.

As for biowarfare and infowarfare, from the standpoint of international law, there is a big question to tackle before unleashing any kind of military response, whether it is clubs and spears, bullets and missiles, or bits and bytes. The question is whether a strike, including one in cyberspace, amounts to a "use of force" or an "armed attack" under international law. If it is, four distinct tests would have to be met before the use of cyber weapons or other arms would be considered lawful self-defense:[29]

❑   Discrimination — targeting combatants and not civilians.

---

[29] Jim Wolf, "U.S. prepares for cyberwar — The war next time", *Reuter News*, November 12, 2001.

❏ Necessity — using no more force than required to accomplish a mission nor using inhumane means such as chemical or biological weapons.
❏ Proportionality — balancing the military advantage against harm to civilians.
❏ Chivalry — the age-old principle of chivalry permits "ruses of war" to trick a foe but not "perfidy", which is defined as treacherous deceit about the legal status of the combatants. Tactical deception is okay but a legal deception is war crime.

All the above extend into cyberspace. Because cyber weapons are the newest weapons in the military arsenal, many of the questions surrounding their use are being confronted for the first time. They will have to be resolved on a case-by-case basis, much as new legal doctrines were developed for aircraft at the beginning of the last century.

14   Case Closed And The Future Is Rosy

There has never been a more exciting time to be part of the information technology industry or the biotechnology industry. Time magazine predicted that the 21st century will be the biotechnology century. But one can make an equally compelling case that it will be the information technology century. While such predictions are always risky, it is safe to say that the convergence of biotechnology and information technology will drive revolutionary changes in biology, in drug discovery, in medicine, and in humankind's own understanding of itself. One can only imagine the discoveries that await in the new century, discoveries that will begin to answer questions that humankind has been asking for hundreds of years, discoveries that will enable humankind to treat more people and conquer more diseases, discoveries that will tell humankind more than humankind ever thought possible about who humankind is and why, discoveries that will help resolve resource depletion, and others.

All these discoveries will come accelerated at the fertile delta of the convergence of biotechnology and information technology. But the most important duty lies in us, the global citizen and inhabitants on the delta. We have to act and conduct ourselves effectively, intelligently, and responsibly. Effectiveness ensures short-term pay off, but it is intelligence that sustains long-term effectiveness. Responsibility bonds us all together.

We should perhaps consider junking the Godzilla mentality of "bigger is better" and rely more on our intelligence by working smarter, and harder if necessary. Sheer size and brawn power are not friendly to dwindling resources. We should seriously rethink that in good old Westerns, Hollywood's projection of machoness is really having two sets of buildings as backgrounds — with one smaller to create the illusion so that actors appear taller in front of them, and another exactly the same but scaled bigger to create the illusion so that actresses appear more petite.

The day after the World Trade Center incident of September 11, 2001, I received an email from an acquaintance some 12,000 miles away from Ground Zero. She said that she saw TV footages of planes slamming onto the World Trade Center towers, just like in movies, and lamented that "where are the Tom Cruises, Arnold Shwartzneggers and Bruce Willieses when we really need them?" I know she was joking. But the fact of the matter is still we have to remember movies are movies, and we watch movies as a form of entertainment. We should not take everything we read or see literally.

In this highly competitive dog-eat-dog world, particularly in sports, too many are silly enough to want to win at all costs. These people turn to massive use of steroids, genetically engineered human growth hormones, and other well being enhancers, giving a creeping definition of what it means to be healthy.[30] Individuals should exercise discretion to decide if the drug is good. If the side effects outweigh the benefits, then the decision is clear.

It is interesting some of the features our medication (drugs and steroids) and technology (cloning) try to enhance are color of eyes, color of hair, and height and size. Size and height are particularly troubling. At a time of depleting resources, in order to save resources, we design more efficient machinery — cars that use less gas, car that use alternative fuels, and other examples. But we are extremely selfish. When it comes to ourselves, we never pause to look at ourselves, and think about making ourselves more efficient, such as designing ourselves smaller to consume less. We look outside of us to see how we can make others support our increasing appetite. Larger size is not necessary the same as effective, intelligent and responsible. Otherwise, dinosaurs would still roam the Earth. Otherwise, we *Homo sapiens* would not be at the top of the totem pole.

What is macho is our individual health. For this, as a global citizen, we should act responsibly by exercising preventive medicine, taking good care of our health, and perhaps taking the good old adage "prevention is better than cure" to heart.

It is time for us to listen carefully and interpret all the television commercials and advertisements. Many of them mislead us by instilling fear in us (such as golden age inconveniences and ailments), or else by derogating our ego (such as life-style drugs whose primary functions are to restore social faculties or attributes that tend to diminish with age). Despite these misleading advertisements and commercials, we are living in a better world. Whether we are talking about transportation, navigation, space travel, electricity, telecommunication, advances in technology have improved life expectancies, quality of life and prosperity. We are fortunate beneficiaries of advances and wonders of modern biotechnology and healthcare system.

Last but not least, biotechnology and healthcare system can only do so much if we do not help ourselves, if we do not conduct effectively, intelligently, and responsibly.

---

[30] Lyle Alzado, as told to Shelly Smith, "I'm sick and I'm afraid", *Sports Illustrated*, July 8, 1991.

INDEX

D

E

H

I

R